AF793278

ALLE · ZEIT · WACH
1842

Burkhard Lachmann (Ed.)

Surfactant Replacement Therapy

in Neonatal and Adult Respiratory Distress Syndrome

With 130 Figures

Springer-Verlag
Berlin Heidelberg New York
London Paris Tokyo

Burkhard Lachmann, MD, PhD
Department of Anesthesiology, Erasmus University
Postbox 1738, NL-3000 DR Rotterdam

Proceedings of the International Symposium on
Surfactant Replacement Therapy
Rotterdam/The Netherlands, November 14–15, 1987

ISBN-13:978-3-642-73307-9 e-ISBN-13:978-3-642-73305-5
DOI: 10.1007/978-3-642-73305-5

Library of Congress Cataloging-in-Publication Data
International Symposium on Surfactant Replacement Therapy (1987: Rotterdam, Netherlands) Surfactant replacement therapy in neonatal and adult respiratory distress syndrome / Burkhard Lachman (ed.).
p. cm.
"International Symposium on Surfactant Replacement Therapy, Rotterdam, the Netherlands, November 14–15, 1987"–T.p. verso.
ISBN-13:978-3-642-73307-9 (U.S.)
1. Respiratory distress syndrome, Adult–Chemotherapy–Congresses 2. Respiratory distress syndrome–Chemotherapy–Congresses. 3. Pulmonary surfactant–Therapeutic use–Congresses. I. Lachmann, Burkhard. II. Title.
[DNLM: 1. Pulmonary Surfactants–therapeutic use–congresses. 2. Respiratory Distress Syndrome–therapy–congresses. 3. Respiratory Distress Syndrome, Adult–therapy–congresses. WF 600 I6134s 1987]
RC776.R38I56 1987 618.92'24–dc19 DNLM/DLC 88–22439

Softcover reprint of the hardcover 1st edition 1988

2119/3140-543210 – Printed on acid-free paper

Preface

I am very happy and honoured to have so many famous contributors from around the world presenting their knowledge and experience in this book, including Dr. Rüfer who, as early as 1967, attempted to improve the pressure-volume curve of a child dying from RDS using dried surfactant from dogs; Drs. Enhorning and Robertson who were the first to treat rabbit fetuses with surfactant replacement and impressively demonstrated that surfactant improved lung mechanics; and Dr. Fujiwara who at the end of the 1970's had already successfully treated the first babies with surfactant replacement and published this work already in 1980. Thus, the idea of surfactant replacement therapy is now more or less 20 years old and almost 10 years have passed since the first publication by Dr. Fujiwara of his impressive clinical results.

Up to now (September, 1988) about 2,200 babies have been treated worldwide with tracheal instillation of surfactant, with most promising results. The type of surfactant used varies from pure natural extracts from bovine or pig lungs, to surfactant extracted from human amniotic fluid, to pure phospholipids. To date, however, there is still no surfactant commercially available (except in Japan), in spite of the fact that it is urgently awaited by neonatologists and intensive care doctors throughout the world.

This volume is based on the contributions presented at the international congress on Surfactant Replacement Therapy which was held in Rotterdam, The Netherlands, in November 1987. It represents the accumulated experience from researchers and clinicians who were involved in the development and first clinical trials of surfactant replacement therapy. Thus, this book is the first summary of practically all activity in the field of surfactant replacement therapy and provides a lot of information for everybody who wishes to treat the respiratory distress of patients by surfactant replacement.

I would like to thank all those collaborators who made this book possible. My thanks are due to the secretariat, Laraine

Visser and Trudy Drenth, and to my students Siegfried Armbruster, Geert-Jan van Daal, Diederik Gommers, Erwin de Ruyter, Wolfgang Schairer, Ralph So, Eric Stouten, Robert Tenbrinck, Michael Voets and Leon van Woerkens.

Rotterdam, September 1988 Burkhard Lachmann

Contents

List of Contributors

E. E. Aeberhard
Department of Pediatrics, UCLA School of Medicine
Los Angeles, CA 90024-1752, USA

J. L. Abraham
Upstate Medical Center, State University of New York
Syracuse, NY, USA

P. Arjomaa
Department of Obstetrics and Gynecology, University of Helsinki
00290 Helsinki, Finland

S. Armbruster
Department of Anesthesiology, Erasmus University
P.O. Box 1738, 3000 DR Rotterdam, The Netherlands

S. Bambang Oetomo
Academisch Ziekenhuis, Kindergeneeskunde
Oostersingel 59, 9713 EZ Groningen, The Netherlands

C. T. Barrett
Department of Pediatrics, UCLA School of Medicine
Los Angeles, CA 90024-1752, USA

P. Bartmann
Abteilung für Neonatologie, Universitäts-Kinderklinik
Prittwitzstraße 43, 7900 Ulm, FRG

A. Bartoletti
Department of Pediatrics, Albany Medical College
Albany, NY, USA

J. J. Batenburg
Laboratory of Veterinary Biochemistry, Utrecht University
P.O. Box 80177, 3508 TD Utrecht, The Netherlands

A. Bergin
Department of Pediatrics, Rotunda Hospital
Dublin 1, Ireland

P. Berggren
Department of Pathology, St. Gorans Hospital
112 81 Stockholm, Sweden

W. Bernhard
Medizinische Poliklinik, Philipps Universität
Baldingerstraße, 3550 Marburg, FRG

C. Berry
Department of Community Medicine, UCSD Medical Center
225 Dickinson Street, San Diego, CA 92103, USA

B. T. Bloom
HCA Wesley Medical Center
Division of Perinatal Medicine
550 N. Hillside, Wichita, KS 67214, USA

K. Bry
University of Helsinki Children's Hospital
00290 Helsinki, Finland

R. Burkhardt
Medizinische Klinik, Zentralkrankenhaus Bremen-Nord
Hammersbecker Straße 228, 2820 Bremen 70, FRG

L. A. Campbell
Department of Pediatrics, UCLA Medical Center
Los Angeles, CA 90024-1752, USA

S. Chida
Department of Pediatrics, Iwate Medical University
Uchimaru 19-1, Morioka, Iwate 020, Japan

T. Clarke
Department of Pediatrics, Rotunda Hospital
Dublin 1, Ireland

C. G. Cochrane
Section of Immunology, Scripps Medical Research Foundation
1066 N. Torrey Pines Road, La Jolla, CA 92037, USA

E. Cosmi
Universita deali studi di Perugia, Policlinico Monteluce
06100 Perugia, Italy

C. Cox
Division of Biostatistics, University of Rochester School of Medicine
601 Elmwood Avenue, Rochester, NY 14642, USA

T. Curstedt
Department of Clinical Chemistry, Karolinska Hospital
104 01 Stockholm, Sweden

J. B. D'Ambola
Department of Pediatrics, UCLA Medical Center
Los Angeles, CA 90024-1752, USA

J. M. Davis
Department of Pediatrics, University of Rochester School of Medicine
601 Elmwood Avenue, Rochester, NY 14642, USA

L. de Ley
Division of Neonatology, University Hospital
Oostersingel 59, 9713 EZ Groningen, The Netherlands

G. C. Di Renzo
Institute of Gynecology and Obstetrics, University of Perugia
06100 Perugia, Italy

B. G. Disse
Dr. Karl Thomae GmbH, Biologische Forschung Pharmakologie C
Postfach 1755, 7950 Biberach 1, FRG

H. S. Dweck
Department of Pediatrics, New York Medical College
Valhalla, NY, USA

H. Eberhardt
Dr. Karl Thomae GmbH, Biologische Forschung Pharmakologie C
Postfach 1755, 7950 Biberach 1, FRG

D. Edwards
Department of Radiology, UCSD Medical Center
225 Dickinson Street, San Diego, CA 92103, USA

H. R. Eistetter
Institut für Molekularbiologie (FB3), Byk Gulden Pharmazentika
Gottlieberstraße 25, 7550 Konstanz, FRG

J. Engel
Abteilung für Biophysikalische Chemie, Biozentrum der Universität Basel, 4000 Basel, Switzerland

G. Enhorning
Children's Hospital, Department of Obstetrics and Gynecology
219 Bryant Street, Buffalo, NY 14222, USA

B. J. Escudier
Laboratoire de Physiologie, Faculté de Médicine Kremlin-Bicêtre
94276 Le Kremlin Bicêtre Cedex, France

E. Evander
Department of Clinical Physiology, University of Lund
Lasarettet, 22185 Lund, Sweden

T. Fujiwara
Department of Pediatrics, Iwate Medical University
Uchimaru 19-1, Morioka, Iwate 020, Japan

M. Ganzuka
Department of Anesthesiology, Kanazawa University School of Medicine, 13-1 Takara-machi, Kanazawa 920, Japan

P. Gieles
Department of Physics, Eindhoven University of Technology
P.O. Box 513, 5600 MB Eindhoven, The Netherlands

N. Gilliard
Department of Pulmonary Medicine, UCSD Medical Center
225 Dickinson Street, San Diego, CA 92103-1990, USA

S. Gormally
Department of Pediatrics, Rotunda Hospital
Dublin 1, Ireland

L. Gortner
Abteilung für Neonatologie, Universitäts-Kinderklinik Ulm
Prittwitzstraße 43, 7900 Ulm, FRG

H.-L. Hahn
Medizinische Poliklinik der Universität Würzburg
Klinikstraße 8, 8700 Würzburg, FRG

H. L. Halliday
Neonatal Intensive Care Unit, Royal Maternity Hospital Belfast
Belfast BT 12 6BB, Northern Ireland

M. Hallman
University of Helsinki, Children's Hospital
Stenbäckinkatu 11, 00290 Helsinki, Finland

H. A. T. Hein
Department of Anesthesiology, Scott and White Clinic
2401 South 31st Street, Temple, TX 76508, USA

J. M. Hoeffel
Department of Medicine, University of California
San Francisco, USA

B. A. Holm
Medical Department of Biophysics, University of Rochester, School of Medicine
601 Elmwood Av., Rochester, NY 14642, USA

B. Hümmer
Medizinische Poliklinik der Universität Würzburg
Klinikstraße 8, 8700 Würzburg, FRG

M. Ikegami
Department of Pediatrics, Harbor-UCLA Medical Center
1000 West Carson Street, Torrance, CA 90509, USA

A. Jarvenpaa
University of Helsinki, Children's Hospital
Stenbäckinkatu 11, 00290 Helsinki, Finland

A. H. Jobe
Department of Pediatrics, Harbor-UCLA Medical Center
1000 West Carson Street, Torrance, CA 90509, USA

B. Jonson
Department of Clinical Physiology, University of Lund
Lasarettet 2, 22185 Lund, Sweden

H. Jörnvall
Department of Chemistry, Karolinska Institute
10401-Stockholm, Sweden

E. Kattner
Abteilung für Neonatologie, Universitäts-Kinderklinik
Heubnerweg 6, 1000 Berlin 19, FRG

J. W. Kendig
Division of Neonatology, University of Rochester
Medical Center
601 Elmwood Avenue, Rochester, NY 14642, USA

T. Key
Department of Physics, Eindhoven University of Technology
P.O. Box 513, 5600 MB Eindhoven, The Netherlands

M. King
Department of Pediatrics, Rotunda Hospital
Dublin 1, Ireland

T. Kobayashi
Department of Anesthesiology, School of Medicine
Kanazawa University, 13-1 Takara-machi, Kanazawa 920, Japan

M. Konishi
Department of Pediatrics, Iwate Medical University
Uchimaru 19-1, Morioka, Iwate 020, Japan

K. Kosack
Abteilung für Neonatologie, Universitäts-Kinderklinik
Heubnerweg 6, 1000 Berlin 19, FRG

B. Lachmann
Department of Anesthesiology, Erasmus University
P.O. Box 1738, 3000 DR Rotterdam, The Netherlands

I. Lopez de Heredia
Department Pediatrics, Hospital de Cruces
Cruces-Baracaldo 48930 Vizcaya, Spain

L. Lützen
Dr. Karl Thomae GmbH, Biologische Forschung
Pharmakologie C, Postfach 1755, 7950 Biberach 1, FRG

H. Maeta
Department of Pediatrics, Iwate Medical University
Uchimaru 19-1, Morioka, Iwate 020, Japan

S. Matalon
Department of Physiology, State University of New York
at Buffalo, Buffalo, NY 14214, USA

T. G. Matthews
Department of Pediatrics, Rotunda Hospital
Dublin 1, Ireland

G. McClure
Department of Neonatology, Royal Maternity Hospital
Belfast BT12 6BB, Northern Ireland

B. McCord
Neonatal Intensive Care Unit, Royal Maternity Hospital
Belfast BT12 6BB, Northern Ireland

T. A. Merritt
Department of Pediatrics, UCSD Medical Center
225 Dickinson Street, San Diego, CA 921030, USA

C. Morley
Department of Paediatrics E8, Addenbrookes Hospital
Hills Road, Cambridge CB2 200, Great Britain

B. Müller
Medizinische Poliklinik, Philipps Universität
Baldingerstraße, 3550 Marburg, FRG

S. Murakami
Department of Anesthesiology
School of Medicine, Kanazawa University
13-1 Takara-machi, Kanazawa 920, Japan

J. F. Murray
Department of Medicine, University of California
San Francisco, CA, USA

R. H. Notter
Department of Pediatrics, Box 777
University of Rochester School of Medicine
601 Elmwood Avenue, Rochester, NY 14642, USA

M. Obladen
Kinderklinik und Poliklinik (WE 16)
FUB Universitätsklinikum Charlottenburg
Heubnerweg 6, 1000 Berlin 19, FRG

A. Okken
Department of Pediatrics, University Hospital
Oostersingel 59, 9713 EZ Groningen, The Netherlands

A. Pettenazzo
Department of Pediatrics (Building A17)
Harbor-UCLA Medical Center
1000 W. Carson Street, Torrance CA 90509, USA

O. Pitkänen
Department of Obstetrics and Gynecology, University of Helsinki
00290 Helsinki, Finland

M. Pohjavuori
Children's Hospital, University of Helsinki
00290 Helsinki, Finland

F. Pohlandt
Abteilung für Neonatologie, Universitäts-Kinderklinik Ulm
Prittwitzstraße 43, 7900 Ulm, FRG

I. Purnama
Medizinische Poliklinik, Universität Würzburg
Klinikstraße 8, 8700 Würzburg, FRG

H. Redl
Ludwig Boltzmann Institut für Experimentelle Traumatologie
Donaueschingerstraße 13, 1200 Wien, Austria

MMcC. Reid
Department of Neonatology, Royal Maternity Hospital
Belfast BT12 6BB, Northern Ireland

S. D. Revak
Scripps Medical Research Foundation, Section of Immunology
1066 N. Torrey Pines Road, La Jolla, CA 92037, USA

P. Richman
Department of Pulmonary Medicine, UCSD Medical Center
University of California, San Diego, CA 92103-1990, USA

H. M. Risemberg
Department of Pediatrics, Albany Medical College
Albany, NY, USA

B. Robertson
Department of Pathology, St. Gorans Hospital
Box 12500, 11281 Stockholm, Sweden

R. Rüfer
Institut für Pharmakologie und Toxikologie
Medizinische Fakultät Mannheim
Maybachstraße 14-16, 6800 Mannheim, FRG

M. Samuels
Cardiothoracic Institute, Brompton Hospital
Fulham Road, London SW3 6HP, Great Britain

P. Sanjurjo
Department of Pediatrics, Hospital de Cruces
Cruces-Baracaldo 48930 Vizcaya, Spain

K. P. Schäfer
Abteilung für Molekularbiologie, Byk Gulden Pharmazeutika
Postfach 6500, 7750 Konstanz, FRG

W. Schairer
Department of Anesthesiology, Erasmus University
P.O. Box 1738, 3000 DR Rotterdam, The Netherlands

G. Schlag
Ludwig Boltzmann Institut
für Experimentelle Traumatologie
Donaueschingerstraße 13, 1200 Wien, Austria

C. Schoots
Department of Clinical Immunology, University Hospital
Oostersingel 49, 9713 EZ Groningen, The Netherlands

S. Seidner
Department of Pediatrics, Harbor-UCLA Medical Center
1000 West Carson Street, Torrance, CA 90509, USA

D. L. Shapiro
Department of Pediatrics
University of Rochester School of Medicine
601 Elmwood Avenue, Rochester, NY 14642, USA

M. P. Sherman
UCLA Medical Center, Department of Pediatrics
Los Angeles, CA 90024-1752, USA

J. M. Smit
Department of Anesthesiology, Erasmus University
P.O. Box 1738, 3000 DR Rotterdam, The Netherlands

D. P. Southall
Cardiothoracic Institute, Brompton Hospital
Fulham Road, London SW3 6HP, Great Britain

R. G. Spragg
Division of Pulmonary and Critical Care Medicine
UCSD Medical Center, University of California
225 Dickinson Street, San Diego, CA 92103-1990, USA

A. Steeghs
Department of Physics, Eindhoven University of Technology
P.O. Box 513, 5600 MB Eindhoven, The Netherlands

P. Stevens
Abteilung für Neonatologie, Universitätskinderklinik
Heubnerweg 6, 1000 Berlin 19, FRG

D.S. Strayer
Department of Pathology, University of Texas, Medical School
P.O. Box 20708, Houston, TX 77225, USA

W. Strohmaier
Ludwig Boltzmann Institut für Experimentelle Traumatologie
Donaueschingenstraße 13, 1200 Wien, Austria

D. G. Talbert
Cardiothoracic Institute, Brompton Hospital
Fulham Road, London SW3 6HP, Great Britain

J. Taniguchi
Department of Anesthesiology, School of Medicine,
Kanazawa University, 13 Takara-machi, Kanazawa 920, Japan

T. Ueda
Department of Anesthesiology, School of Medicine,
Kanazawa University, 13 Takara-machi, Kanazawa 920, Japan

A. Valls-Soler
Department of Pediatrics, Hospital de Cruces
Cruces-Baracaldo 48930 Vizcaya, Spain

L. M. G. van Golde
Laboratory of Veterinary Biochemistry, University of Utrecht
P.O. Box 80177, 3508 TD Utrecht, The Netherlands

J. van Santvoort
Department of Physics, Eindhoven University of Technology
P.O. Box 513, 5600 MB Eindhoven, The Netherlands

Y. E. Vaucher
Department of Pediatrics, UCSD Medical Center
225 Dickinson Street, San Diego, CA 92103, USA

T. Voss
Abteilung für Molekularbiologie, Byk Gulden Pharmazeutika
Gottlieberstraße 25, 7750 Konstanz, FRG

F. J. Walther
Department of Pediatrics, University of Southern California,
School of Medicine
1240 North Mission Road, Los Angeles, CA 90033, USA

E. Weller
Dr. Karl Thomae GmbH
Biologische Forschung Pharmakologie C
Postfach 1755, 7950 Biberach 1, FRG

P. von Wichert
Medizinische Poliklinik, Philipps Universität
Baldingerstraße 3550 Marburg, FRG

B. Widjaja
Institut für Pharmakologie und Toxikologie
Medizinische Fakultät Mannheim
Maybachstraße 14–16, 6800 Mannheim, FRG

P. Wollmer
Department of Clinical Physiology, University of Lund
22185 Lund, Sweden

J. Wuthe
Institut für Pharmakologie und Toxikologie
Medizinische Fakultät Mannheim
Maybachstraße 14–16, 6800 Mannheim, FRG

M. Zelter
Laboratoire de Physiologie, Faculte de Medecine
94276 Kremlin Bicêtre Cedex, France

H. Ziegler
Dr. Karl Thomae GmbH
Biologische Forschung Pharmakologie C
Postfach 1755, 7950 Biberach 1, FRG

Pulmonary Surfactant: Evolution of Functional Concepts

G. Enhorning

The first and clear indication that surface tension offers resistance to the initial lung aeration after birth came with von Neergaard's study in 1929 [26]. He found that lungs that were collapsed would open up more readily if the effect of surface tension was completely nullified by using liquid rather than air as the expanding medium. Following this basic study, it took considerable time until Pattle in 1955 [19] could declare that he assumed there was an agent present in the lung that had the ability to depress surface tension to extremely low values. He extruded bubbles from the sectioned lung and found that when they were surrounded by saline, so that they could be studied under a microscope, they would quickly shrink 30 % from their initial size, when their diameter was around 40 μ, but then the bubbles would persist for long periods of time (Fig. 1). Pattle reasoned that the high surface tension of water, 72 mN/m, ought to give an enormously high pressure in the tiny bubble, so that the gas inside the bubble should have been absorbed quickly by the surrounding saline solution. When this did not happen, but the bubble persisted, Pattle concluded that there must be a lining layer exerting surface pressure, almost completely counteracting the collapsing effect of water surface tension. The net surface tension according to Pattle must be very close to 0.

The Wilhelmy balance gave an exact account of how surface tension varies with surface area. It was noted that surface tension was very much reduced as the surface area was compressed to 20 % of its original size. The surface tension-area loop showed considerable hysteresis, which was felt to be a main reason for the one observed when the lungs' pressure-volume loop was studied. Clements pointed out that the change in surface tension with area would offer the lung stability [4]. The law of Laplace makes it clear that if there is no change in the value of surface tension, an alveolus would have to be surrounded by a great negative pressure when it reaches minimal size at end of expiration. With the Wilhelmy balance, Clements had demonstrated, however, that surface tension diminishes with area, a phenomenon that would protect the small alveoli from collapsing during expiration [4]. The concept that surfactant protects the small alveoli from becoming smaller or collapsing gained general acceptance. With this principle it was possible to explain that with a surfactant deficiency and the Respiratory Distress Syndrome (RDS) has developed, there are large areas of atelectasis and the alveoli remaining open tend to become overexpanded. It is also conceivable, however, that the areas of atelectasis developed already when the lungs initially became aerated. From the law of Laplace ($\Delta P = 2 \gamma/R$) it is clear that the minisci forming in the

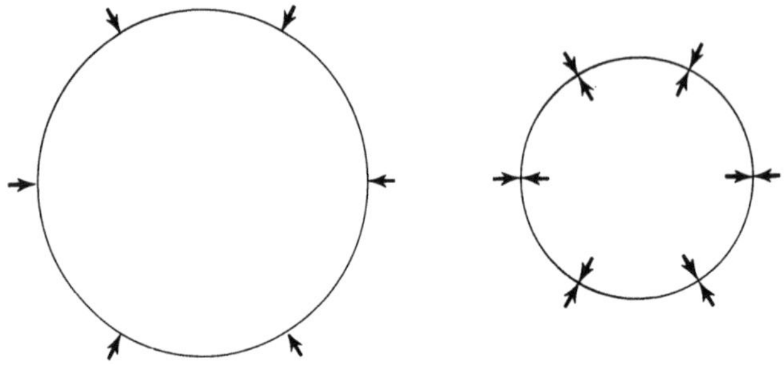

Fig. 1. Pattle (1955) noted that bubbles, extruded from the cut surface of a lung, would quickly shrink from their original size *(left)* to a smaller size *(right)* that they would maintain for a long time. He concluded that surface tension of the surrounding water would raise pressure in the bubble, but an inner layer of surfactant would offer high surface pressure when compressed to a small surface area and would counteract the shrinking effect on the surrounding water *(right)*

airways during the crucial first breath will move in the direction of the alveoli, only if the pressure difference (Δ P) is more than 2 γ/R. The premature infant with a surfactant deficiency will have a high value of γ and the value of R is likely to be small, since the airways are underdeveloped. For aeration to occur, the pressure difference created by the neonate itself or the attending neonatologist, needs to be extensive and it is probable that minisci in narrow airways will be halted and, instead, expanded alveoli and airways will become overexpanded. With such an excessive dilatation, there will be an increased leakage of serum proteins which will inhibit the surfactant and exacerbate the deficiency. Eventually, fibrinogen among the invading proteins will be converted to fibrin and form the hyaline membranes. The first notion that RDS was due to surfactant deficiency came with a publication by Avery and Mead, 1959 [2]. A report by Adams et al. offered further support [1]. They found that the lung lavage of infants succumbing from RDS had less of the phospholipids that are essential components of pulmonary surfactant. It was easy to accept that RDS could be caused by a surfactant deficiency, and since it is well known that the main component of pulmonary surfactant is dipalmitoylphosphatidylcholine (DPPC) it was felt that it might perhaps be possible to alleviate the infants' problems by supplying only this main surfactant component, DPPC. Important reasons for utilizing DPPC only were the commercial availability of the phospholipid and the fact that when its physical properties were studied with the Wilhelmy balance, they were found to be very similar to those of the whole pulmonary surfactant [3]. Nevertheless, early attempts to cure neonatal RDS by instilling DPPC into the trachea with the hope that this would replace the missing surfactant were very disappointing [3, 22, 24].

When DPPC was studied with the bubble surfactometer [7], it did not seem the least surprising that the expected beneficial effect had been missing during the clinical trials. The surface properties of the phospholipid alone were very different from those of natural surfactant. The dominating difference was that DPPC, when suspended alone, required a very long time to form a surface

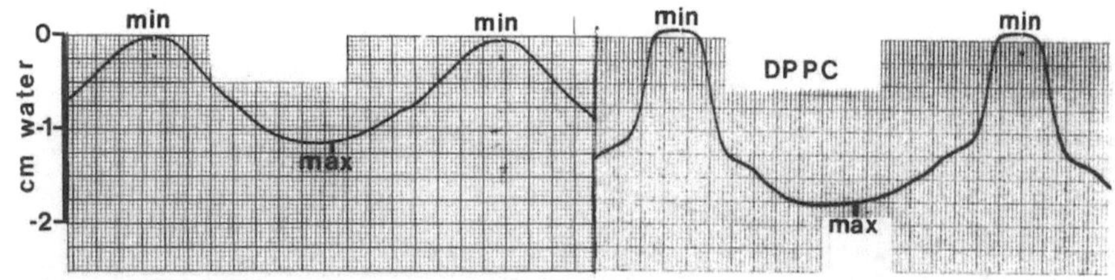

Fig. 2. Natural surfactant, or in this case, calf lung surfactant extract *(CLSE)*, has a tracing characterized by a Δ P of zero at minimal bubble size *(min)* and of ~ 1 cm H_2O at maximal size *(max)*. For DPPC, the Δ P at maximal bubble size is much greater, i.e., surface tension is correspondingly grater

film, i.e., adsorption rate was extremely low. Another difference was the greater change in the value of Δ P, noted with DPPC (Fig. 2). Once a film had formed, the value of Δ P, at minimal bubble size was 0, just as it would be with natural surfactant, but at maximal bubble size, the value of Δ P was much greater when the film consisted of DPPC only than when it also contained other components of natural surfactant. Perhaps those other components move into the film as it expands. When studied with a pulsating bubble surfactometer the properties of natural surfactant, obtained from lung lavage of adult rabbits, proved to be such that one could say: If they were present when the neonate takes its first breath, the aeration should be facilitated, the expansion should become even and remain stable. That was the reason Robertson and I studied the effect of instilling natural surfactant into the airways of premature rabbit neonates [8]. Later, we learned that Rüfer [23] had been working successfully on the same concept: supply the missing surfactant directly into the airways. The results obtained with our study were very promising [8]. The lung expansion was greater and improved. This was seen with pressure-volume loops of the lungs, with radiology, and with histology [8, 9]. In later studies we found that gas exchange improved, as did survival rate [10, 27]. We could not apply the principle clinically since the surfactant we used was very crude and inadequate for human use. While we were trying to develop an adequate preparation, Fujiwara et al. in Japan beat us to it by publishing how they had successfully treated ten infants with severe RDS by instilling surfactant into the airways [12]. Eventually, a surfactant preparation was prepared from the lung lavage of calves. The lipids were extracted and resuspended in 100 mM NaCl and 0.5 mM $CaCl_2$. This yielded an active preparation that tolerated autoclaving without loss of surface activity [17]. This preparation, called CLSE (i.e. calf lung surfactant extract) has been used successfully (Fig. 3) to prevent the development of RDS in preterm infants [10, 15, 25]. In several institutions, it is now used prophylactically for all infants born at a gestational age of less than 33 weeks. Since this routine was introduced, the neonatal mortality at Children's Hospital of Buffalo has been dramatically reduced (Fig. 4).

The principle of surfactant supplementation to prevent or treat neonatal RDS is becoming generally accepted [11, 13–16, 18, 21, 25]. However, several unanswered questions remain: Which preparation is optimal? Is a very

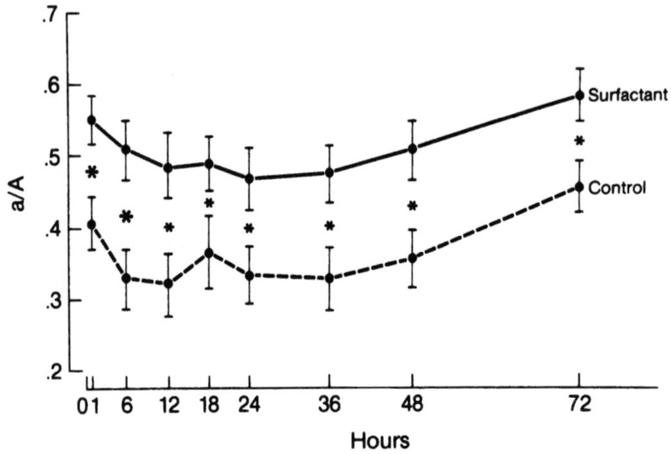

Fig. 3. In a randomized clinical trial [11] it was shown that calf lung surfactant extract (CLSE) can be used successfully to prevent neonatal RDS. The ratio of $_2$ in artery and alveolus is significantly higher among the infants treated with surfactant (CLSE) prior to the first breath. *Large asterisk* indicates $P < 0.005$; *small asterisk* indicates $P < 0.05$. (From Enhorning et al. [11])

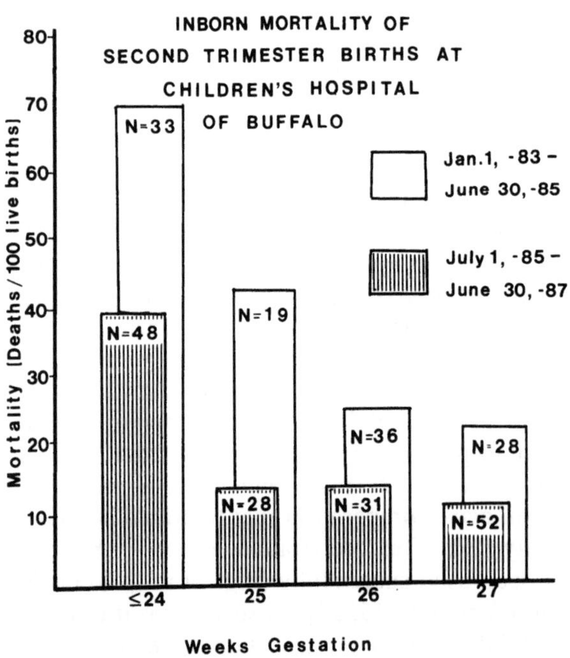

Fig. 4. Prophylactic surfactant treatment was introduced in Buffalo in 1985. In extremely prematurely born infants, mortality has been dramatically reduced. (Data kindly supplied by E. A. Egan)

hydrophobic, low molecular weight protein (5,000–18,000 daltons) the only one needed? Which phospholipids are needed and which neutral lipids have to be present? Should the surfactant of necessity be able to form myelin figures? Should the surfactant be given to prevent RDS, or to treat the manifest condition? Are there any toxic side effects? If used to prevent RDS, should it be given prior to the first breath and can it be given into the pharynx? Some of these questions will be discussed in this book.

Neonatal RDS is a fairly simple condition of surfactant deficiency that well lends itself to treatment by supplementation. If the infant is born before alveolar cells type II have synthesized and released an adequate amount of surfactant, then the principle is to give the infant that which is missing by instilling surfactant into the upper airways, prior to the first breath. The infant's lung maturity is thereby upgraded instantaneously and the first breath can be taken without the difficulty characterizing a surfactant deficiency, and the secondary changes, that of injury to the lungs, can be avoided. Lung injury may allow proteins to leak into the airways and those proteins may further inhibit the surfactant, which was already inadequate. A vicious circle may thus be started, which could be prevented if the surfactant is supplied at an early stage, preferably prior to the first breath. The question then arises: Are there pathological conditions other than neonatal RDS that might be due to a surfactant deficiency, albeit not as pure as in the neonatal period?

Adult RDS

Adult RDS might be such a condition when, for 'various reasons' proteins have invaded the airways and inhibit the surfactant action. Alveoli may then collapse so that the exchange of blood gases is disrupted, allowing hypoxia and acidosis to develop. A vicious circle has started that often will end with death. If surfactant treatment is to be considered, it probably would have to be preceded by a cleansing lavage of the airways with saline, perhaps in one lobe at a time, followed by instillation of a massive quantity of surfactant (see chapter on ARDS).

Pulmonary Edema

Pattle was the first to suggest that pulmonary surfactant is needed to prevent the development of pulmonary edema [20] and Clements published an article on this subject [5]. The line of thought was that if surface tension was excessive, a greater negative pressure would be required in the space surrounding the air-liquid interface of the alveolus and liquid could thereby be sucked into that space. There is probably a surfactant deficiency in most cases of pulmonary edema, but it would be difficult to prove that the deficiency caused the edema and not vice versa.

Edema develops when the hydrostatic pressure in the capillaries exceeds the transcapillary colloidal osmotic pressure. Normally, the hydrostatic capillary pressure is far from being in excess and the direction of flow is from alveoli to capillaries. Primarily a change in surface tension will not have an effect on this direction of flow. If surface tension increases, so that a greater negative pressure is required on the convex side of the interface, inspiratory muscles would have to supply that pressure change. A fall in the total intrathoracic pressure will have no influence on the difference in pressure between capillary and alveolar hypophase but it might result in a dilatation of pulmonary capillaries. This, in turn, could lead to a less obstructed passage of proteins and alter the transcapillary colloid osmotic pressure, allowing edema to develop. The proteins invading the alveolar space might also inhibit the surface activity of phospholipids at the air-liquid interface.

Pneumonia

Pneumonia, caused by virus or bacteria, might also be a condition of surfactant deficiency. Not because the total amount of phospholipids is subnormal, but because phospholipases have interfered with their action. For instance, an excess of lysophosphatidylcholine, that acts as a detergent, may have formed. Furthermore, there may be an excess of inhibiting proteins. Therapy might include an attempt to correct the surfactant deficiency as for Adult RDS.

Asthma

When discussing the effect surfactant might have on the etiology of asthma I am truly out on thin ice. One might start with a generally accepted concept. When surfactant is deficient so that surface tension remains at a fairly high value during compression of the surface area, i.e., at end of expiration, small alveoli are considered to be at risk of collapsing and emptying their air into larger alveoli (Fig. 5). I am not denying the existence of that risk, but judged from a review of the lungs' structure and dimensions, the risk of collapse would seem greater in small cylindrical airways. The law of Laplace is $\Delta P = 2\gamma/R$ and $\Delta P = \gamma/R$ for spherical and cylindrical surfaces, respectively. That means that if surface tension (γ) is the same at the two surfaces the pressure difference (ΔP) required to counteract collapse would have to be greater in the cylinder than in the sphere if the radius of the cylinder (R_c) is less than half that in the sphere (R_s). Very often the radius of the cylinder is less than half that in the sphere and, furthermore, it is quite conceivable that during expiration the surface film will be more compressed, reaching a lower value in the alveolus than in the small cylindrical airway. Collapse of the latter, resulting in air trapping, is therefore threatening and has been observed with photography when there was a deficiency of surfactant, and it disappeared when the missing surfactant had been replaced [6].

If we assume that the patient with asthma has a surfactant deficiency leading to air trapping, how can that cause an airflow obstruction which is dominating during expiration? During normal breathing, the airflow is due to

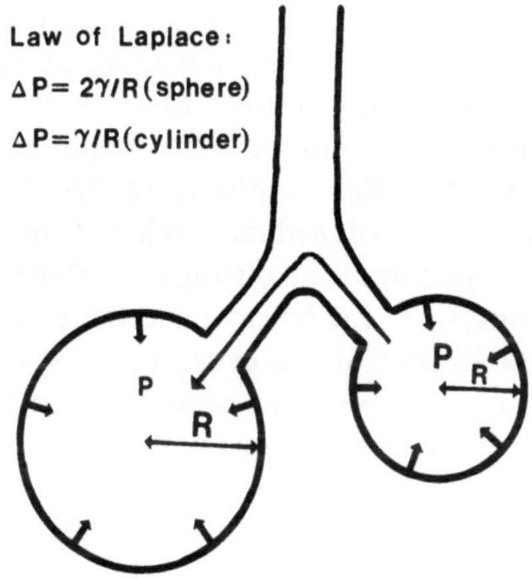

Fig. 5. According to a widely accepted concept, one risk of surfactant deficiency is that the small alveolus will build up a larger pressure than the large alveolus and empty into the latter. This will be prevented, though, if there is a surfactant film building up a high surface pressure which counteracts the surface tension of water. However, according to the law of Laplace, the small cylindrical airway is at greater risk of collapsing if its radius is less than half the size of the radius in the spherical alveolus

the difference in pressure between alveoli and atmosphere. Resistance to that flow is mainly in the larger airways that have cartilage rings preventing them from changing their width when transmural pressure is altered. Since the main resistance is in the larger airways, only a moderate pressure difference will build up across the walls of alveoli and small airways, even when pleural pressure undergoes considerable changes such as during vigorous breathing. If, however, there is collapse of the smallest cylindrical airways due to a surfactant deficiency with air trapping in the alveoli, then the main resistance will shift from airways with cartilage rings to the smallest cylindrical airways, now filled

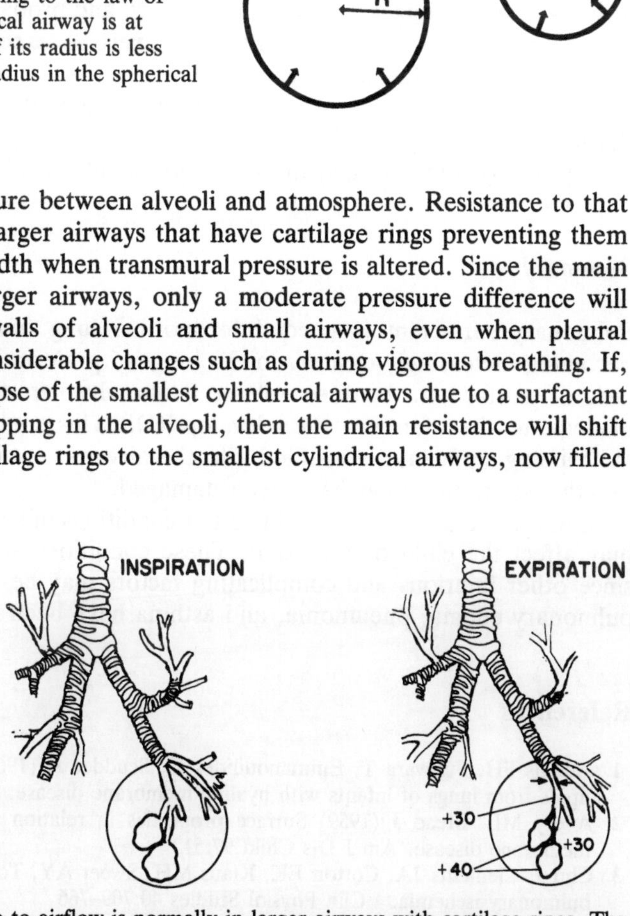

Fig. 6. The main resistance to airflow is normally in larger airways with cartilage rings. The more distant airways are thereby protected from developing a transmural pressure that is dilating during inspiration and collapsing during expiration. Normally, during expiration, pressure is higher in the alveoli than in the surrounding parenchyma due to alveolar recoil. During vigorous expiration the pressures may be 40 and 30 cm water, respectively. Pressure in airways, too small to have cartilage rings, is also higher than in the surrounding parenchyma. With a surfactant deficiency, causing collapse of small cylindrical airways, the main resistance has shifted to those airways from the more centrally located. Pressure in the larger airways, without cartilage rings, may then fall to that of surrounding parenchyma and the airway will be compressed

with fluid. Under these circumstances, a greater pressure difference will develop across the walls of larger airways, but since they do not have cartilage rings that will prevent them from changing their width, they will dilate during inspiration and become compressed during expiration (Fig. 6). They will thereby offer considerable resistance to expiration and perhaps cause the wheezing of asthma. When salbutamol, terbutaline, or other β-adrenergic agonists are used during an asthma attack, these drugs are said to relax smooth muscles of the bronchi and thereby dilate them. It is well known, however, that the drugs will also cause a release of surfactant into the alveolar space. Perhaps such a release will prevent the previous collapsing of the smallest cylindrical airways. Air trapping would no longer occur, and probably there would be not further need for the asthma patient to exert so excessive an effort during expiration. If a surfactant deficiency is indeed causing an asthma attack in the way described above, the surfactant deficiency should obviously be prevented. Perhaps this could be achieved by continuously supplying surfactant to the airways or by stimulating synthesis of surfactant, or inhibiting its catabolism.

Summary

Pulmonary surfactant is needed for normal lung function, and its role is becoming increasingly clearer. In the lungs of preterm infants there may be a lack of surfactant, because release from alveolar cells type II was still inadequate; the infant may then develop RDS. This can be avoided, however, if the missing surfactant is instilled into the airways, optimally prior to the first breath, before the lungs have been damaged.

Later in life there are probably other conditions of surfactant deficiency that may affect the child or the adult. These conditions are less straightforward, since other injurious and complicating factors may be involved. Adult RDS, pulmonary edema, pneumonia, and asthma have been discussed.

References

1. Adams FH, Fujiwara T, Emmanouilides G, Scudder A (1965) Surface properties and lipids from lungs of infants with hyaline membrane disease. J Pediatr 66:357–364
2. Avery ME, Mead J (1959) Surface properties in relation to atelectasis and hyaline membrane disease. Am J Dis Child 97:517
3. Chu J, Clements JA, Cotton EK, Klaus MH, Sweet AY, Tooley WH (1967) Neonatal pulmonary ischemia. I Clin Physiol Studies 40:709–766
4. Clements JA (1957) Surface tension of lung extracts. Proc Soc Exp Biol Med 95:170–172
5. Clements JA (1961) Pulmonary edema and permeability of alveolar membranes. Arch Environ Health 2:280
6. Enhorning G (1977) Photography of peripheral pulmonary airway expansion affected by surfactant. J Appl Physiol 42:976–979
7. Enhorning G (1977) Pulsating bubble technique for evaluating pulmonary surfactant. J Appl Physiol 43:198–203
8. Enhorning G, Robertson B, Milne E, Wagner R (1972) Lung expansion in the premature rabbit fetus after tracheal deposition of surfactant. Pediatrics 50:58–66

9. Enhorning G, Robertson B, Milne E, Wagner R (1975) Radiologic evaluation of the premature newborn rabbit after pharyngeal deposition of surfactant. Am J Obstet Gynecol 121:475–480
10. Enhorning G, Hill F, Sherwood G, Cutz E, Robertson B, Bryan C (1978) Improved ventilation of prematurely delivered primates following tracheal deposition of surfactant. Am J Obstet Gynecol 132:529–536
11. Enhorning G, Shennan A, Possmayer F, Dunn M, Chen CP, Milligan J (1985) Prevention of neonatal respiratory distress syndrome by tracheal instillation of surfactant: a randomized clinical trial. Pediatrics 76:145–153
12. Fujiwara T, Maeta H, Chida S, Morita T, Watabe Y, Abe T (1980) Artificial surfactant therapy in hyaline membrane disease. Lancet 1:55–59
13. Gitlin JD, Soll RF, Parad RN, Horbar JD, Feldman HA, Lucey JF, Taeusch HW (1987) Randomized controlled trial of exogenous surfactant for the treatment of hyaline membrane disease. Pediatrics 79:31–37
14. Hallman M, Merritt A, Jarvenpaa AL, Boynton B, Mannino F, Gluck L, Moore T, Edwards D (1985) Exogenous human surfactant for treatment of severe respiratory distress syndrome: a randomized prospective clinical trial. J Pediatr 106:963–969
15. Kwong MS, Egan EA, Notter RH, Shapiro DL (1985) Double-blind clinical trial of calf lung surfactant extract for the prevention of hyaline membrane disease in extremely premature infants. Pediatrics 76:585–592
16. Merritt TA, Hallman M, Bloom BT, Berry C, Benirschke K, Sahn D, Key T, Edwards D, Jarvenpaa AL, Phojavouri M, Kankaanpaa K, Kunnas M, Paatero H, Rapola J, Jaaskelainen J (1986) Prophylactic treatment of very premature infants with human surfactant. N Engl J Med 315:785–790
17. Metcalfe IL, Pototschnik R, Burgoyne R, Enhorning G (1982) Lung expansion and survival in rabbit neonates treated with surfactant extract. J Appl Physiol 53:838–843
18. Morley CJ, Bangham AD, Miller N, Davis JA (1981) Dry artificial lung surfactant and its effect on very premature babies. Lancet 1:64–68
19. Pattle RE (1955) Properties, function and origin of the alveolar lining layer. Nature 175:1125–1126
20. Pattle RE (1958) Properties, function, and origin of the alveolar lining layer. Proc R Soc Lond [Biol] 148:217–240
21. Raju TNK, Bhat R, McCullough KM, Maeta H, Vidasagar D, Sobel D, Anderson M, Levy PS, Furner S (1987) Double-blind controlled trial of single-dose treatment with bovine surfactant in severe hyaline membrane disease. Lancet 1:651–656
22. Robillard E, Alarie Y, Dagenais-Perusse P, Baril E, Guilbeault A (1964) Microaerosol administration of synthetic β-γ-dipalmitoyl – L-α-lecithin in the respiratory distress syndrome: a preliminary report. Can Med Assoc J 90:55–57
23. Rüfer R (1967) Der Einfluß oberflächenaktiver Substanzen auf Entfaltung und Retraktion isolierter Lungen. Pflügers Arch 298:170–184
24. Shannon DC, Kazemi H, Merril EW, Smith KA, Wong PSL (1969) Restoration of volume-pressure curves with a lecithin fog. J Appl Physiol 28:470–473
25. Shapiro DL, Notter RH, Morin FC, Deluga KS, Golub LM, Sinkin RA, Weiss KI, Cox C (1985) Double-blind, randomized trial of a calf lung surfactant extract administered at birth to very premature infants for prevention of respiratory distress syndrome. Pediatrics 76:593–599
26. von Neergaard K (1929) Neue Auffassungen über einen Grundbegriff der Atemmechanik. Z Ges Exp Med 66:373–394
27. Wallin A, Burgoyne R, Enhorning G (1977) Oxygen consumption of the newborn rabbit treated with pulmonary surfactant. Biol Neonate 31:245–251

Part I
Experimental Evaluation of Surfactant for Replacement Therapy

Evaluation of Pulmonary Surfactant

G. Enhorning

It is established that pulmonary surfactant lining the lungs has several essential properties (Table 1), and any preparation to be used as a replacement must have those properties and, furthermore, should probably simulate those of the natural product as far as possible.

If the liquid in a newborn infant's lungs has an adequate amount of surfactant with the properties listed in Table 1, the lungs will probably expand with ease during the first breath. The expansion will be even, i. e., few areas of atelectasis will remain, and there will be no alveolar overexpansion. During expiration, alveolar sacs and cylindrical airways will remain open so that secondary atelectasis and air trapping can be avoided. When the physical properties result in the morphology just mentioned, the mechanics of breathing should be facilitated. The initial expansion will meet minimal resistance, compliance will increase, and the lungs will retain air at the end of expiration without the occurrence of air trapping.

When the physical properties result in this beneficial type of breathing, blood gas exchange and survival chances should be optimal.

Physical Properties

For several years, the physical properties of surfactant have been evaluated with the modified Wilhelmy balance. The lipids examined with this method are usually dissolved in hexane when spread on the surface of the water in the trough. Since the size of the surface area is exactly known, the area occupied by each molecule can be calculated with accuracy. This is of great value when studying the intermolecular forces of a film consisting of one or few molecular components of surfactant. This advantage cannot be made use of, however, when studying the whole pulmonary surfactant. Another advantage with the Wilhelmy balance is the ease with which an agent can be added to the hypophase, i. e., the liquid under the surface film. The Wilhelmy balance is the instrument with which the most important data on the physics of pulmonary surfactant has been obtained. Clements used it to demonstrate how surface tension varies with surface area and, from this data, he concluded that the lung is stabilized, due to pulmonary surfactant [3]. It was a fairly simple, homemade Wilhelmy balance that Avery and Mead [1] used when they found that infants dying from the Respiratory Distress Syndrome (RDS) had a surfactant deficiency.

Table 1. Necessary physical properties of pulmonary surfactant

Ability to lower surface tension quickly to an equivalent value of 25 mN/m, i. e., adsorption rate is fast
The film formed at the air-liquid interface should lower surface tension to close to 0 when compressed to half the area. When the surface area is enlarged, surface tension should increase, but to no more than 35 mN/m
Minimal hysteresis

Another instrument for evaluating the physical properties of pulmonary surfactant is the pulsating bubble surfactometer [5]. With this instrument a bubble is expanded in the liquid to be examined. The bubble communicates with ambient air and can be compared to a model of an alveolus. Pressure around the bubble, corresponding to intrathoracic pressure, is measured continuously. Since the bubble communicates with ambient air, the absolute pressure measured is equal to ΔP, the pressure difference across the bubble surface. The volume of liquid surrounding the bubble is changing sinusoidally, causing the bubble to pulsate in a regular fashion so that its radius (R) varies from a maximum of 0.55 mm to a minimum of 0.4 mm. This makes it possible to calculate surface tension, from the law of Laplace ($\Delta P = 2\,\gamma/R$). Previously, it has been quite tedious to calculate γ at other than the extreme sizes of the radius. The surfactometer is now being modified, however, so that the recorder gives a tracing of P as before and, with a microprocessor, a continuous tracing is also obtained of γ, or if one so wishes of surface pressure π. Instead of these analog tracings of γ, or π surface tension or surface pressure can be given as digital printouts.

The bubble surfactometer offers several advantages. The examination is fast and only a small sample, around 30 µl, is needed. The speed with which a surface tension lowering film is formed, i. e., the rate of adsorption, can be studied. The bubble is then created instantaneously and is not pulsating for 15 seconds (Fig. 1). From the pressure recorded around the bubble, not pulsating, surface tension can be calculated and is the equivalent value (normally ~ 25mN/m). As soon as pulsation has started surface tension at minimal bubble size should approach zero if the sample is natural pulmonary surfactant, or an acceptable surrogate, at a concentration of at least 3 mg/ml. When the bubble comes to maximal size, surface tension should increase, but to no more than 35 mN/m.

The effect inhaled solvent vapors may have on a surfactant film can be observed when they are allowed to enter the pulsating bubble [8]. They are found to be profoundly damaging when they originate from solvents that are able to dissolve disaturated phospholipids. However, as soon as the solvent has evaporated and disappeared from the lumen of the bubble, the surfactant film returns to its previous properties. Provided the vapors have a similar effect on the surface film of the avleoli, an inhalation of the vapors, even when of short duration, may seriously disrupt the lung's structure. The solvent vapors assault the surfactant film from the airphase, but the film can also be attacked from the

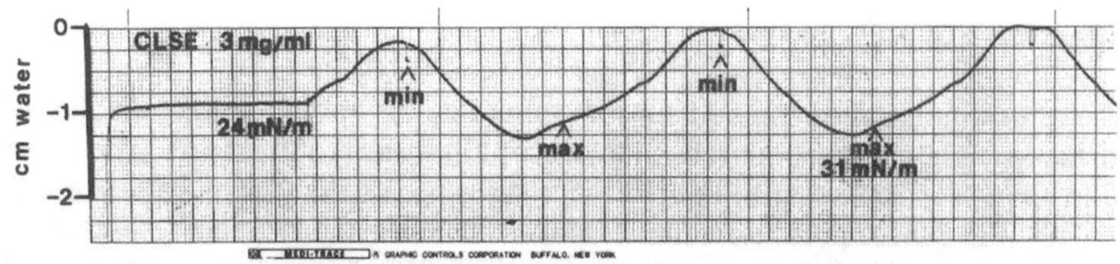

Fig. 1. Calf lung surfactant extract *(CLSE)* evaluated with the pulsating bubble surfactometer. The tracing shows pressure around the bubble as it was created and initially not pulsating. The pressure corresponds to a surface tension of 24 mN/m. When pulsation was started, surface tension at minimal bubble size was soon zero and at maximal size 31 mN/m

other side, from the hypophase, and that can also be studied with the bubble surfactometer [9]. This requires a special arrangement (Fig. 2) by which the liquid around the bubble is interchanged without altering the bubble size. This arrangement for hypophase exchange opens up the possibility of studying the buildup of a surfactant film, and a whole range of potential inhibitors can be evaluated. Firstly, proteins are of interest since they tend to invade the airways under various pathological conditions and have a suspected surfactant inhibiting effect. Phospholipases should also be evaluated since, when occurring in excessive amounts as during pneumonia, they may be destructive to the surfactant film. The possibility of evaluating the inhibition caused by protein may be decisively important. A synthetic surfactant preparation may seem to have the right physical properties when studied alone, but may be very sensitive to the assaulting effect of protein. There will always be potentially damaging proteins present in the airways. A synthetic surfactant preparation, with less ability to withstand the inhibition of these proteins than natural pulmonary surfactant, cannot be accepted as a replacement and can be discarded during the evaluation of physical properties.

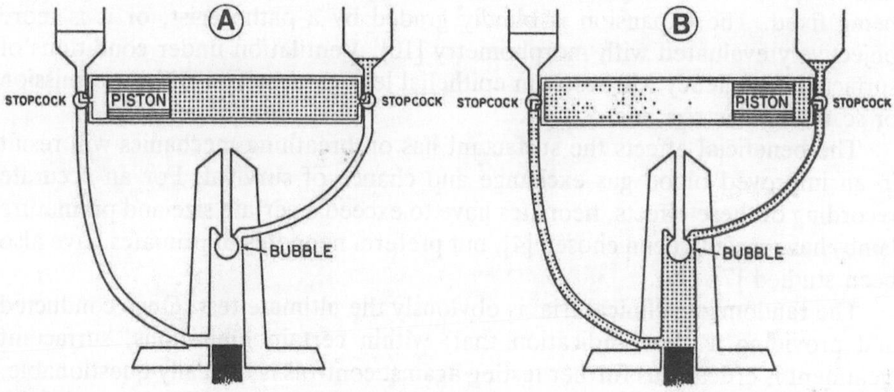

Fig. 2A, B. Arrangement for replacing the hypophase around the bubble without changing the bubble size. **A** Right section of glass cylinder has been filled from reservoir to the right. **B** Stopcocks have been turned so that, when piston moves to the right, liquid in right part of glass cylinder is moving into bubble chamber, as the same volume of liquid is moving out

Effect on Lung Expansion

The animal lungs to be studied should contain no air and a minimal amount of surfactant. The lungs from preterm fetuses fulfill these requirements. The effect of instilling surfactant into the upper airways can be studied with radiography. Usually the lungs' pressure-volume loops are studied, and the fetuses of a litter are examined individually for evaluation of various qualities and quantities of surfactants. To save time and minimize the risk that postmortem effects may vary between fetuses, a system has been developed for simultaneous evaluation of all fetuses of one rabbit litter [6]. They are submerged in a temperature-controlled water bath. A tube is introduced into the trachea of each fetus and is connected to a horizontal glass tube containing 2 ml of air. The other end of the glass tube contains red stained water and communicates with an open jar, also containing the dyed water. Step by step, pressure is successively raised and lowered by moving the jar and when the lungs of a fetus become expanded, the volume of air entering the lungs from the glass tube is replaced with red stained water, forming horizontal "bars" in a graph. The glass tubes with the "bars" are photographed 15 seconds after the jar has been placed at a new level. In this way pressure-volume loops are obtained simultaneously from all fetuses of a litter.

Lungs with no air and minimal amount of surfactant can also be obtained from adult animals, provided there has been a thorough lavage [2]. The pressure-volume loops will show how readily the lungs were expanded and how they were able to retain air. Poor expansion and minimal air retention at a deflation pressure of 0 or 10 cm H_20 has usually been interpreted as indicating a surfactant deficiency.

The effect surfactant has on lung mechanics can also be studied in the living animal by evaluating the dynamic lung thorax compliance, which can be correlated with blood gas exchange. The effect surfactant has on lung mechanics of dead or living animals can also be correlated with the histology. The lungs are then subjected to a certain expanding pressure while they are being fixed. The expansion is blindly graded by a pathologist, or it is more objectively evaluated with morphometry [10]. Ventilation under conditions of surfactant deficiency will result in epithelial lesions observed with transmission or scanning electron microscopy.

The beneficial effects the surfactant has on breathing mechanics will result in an improved blood gas exchange and chance of survival. For an accurate recording of these effects, neonates have to exceed a certain size and premature lambs have usually been chosen [4], but preterm neonates of primates have also been studied [7].

The randomized clinical trial is obviously the ultimate test. Once conducted and providing a clear indication that, within certain limitations, surfactant treatment is efficacious further testing against controls is ethically questionable. If the surfactant is in short supply, a continued randomization against a placebo is justified, but not if there is enough surfactant to treat every infant that would benefit from such treatment. Unanswered questions, however, still remain. For example, we do not know if surfactant treatment is justified at very low gestational age when not only the lungs are affected by serious immaturity. Nor do we know which surfactant preparation is the most active and has the least side effects when used for treatment of premature neonates. It is also unclear whether prophylactic surfactant treatment can be recommended for infants born at a gestational age of 30 weeks or more. To find answers to these questions randomized clinical trials may be necessary and they may also be needed for an evaluation of the efficiency of surfactant treatment for conditions other than those affecting the neonate, for example, adult RDS.

Summary and Conclusions

Due to the physical properties of pulmonary surfactant the mechanics of breathing and the lung histology will be normalized. This improves lung function, as judged by blood gas exchange and survival rate. An evaluation of any preparation considered as a replacement for pulmonary surfactant should certainly start with an assessment of the physical properties. If they are acceptable, even when the surfactant has been exposed to a moderate dose of inhibiting of proteins, in vivo studies are essential before the surfactant can be considered acceptable for clinical use.

References

1. Avery ME, Mead J (1959) Surface tension properties in relation to atelectasis and hyaline membrane disease. Am J Dis Child 97:517
2. Bermel MS, McBride JT, Notter RH (1964) Lavaged excised rat lungs as a model of surfactant deficiency. Lung 162:99–113
3. Clements JA (1957) Surface tension of lung extracts. Proc Soc Exp Biol Med 95:170–172
4. Egan EA, Notter RH, Kwong MS, Shapiro DL (1983) Natural and artificial lung surfactant replacement therapy in premature lambs. J Appl Physiol 55:875–883
5. Enhorning G (1977) Pulsating bubble technique for evaluating pulmonary surfactant. J Appl Physiol 43:198–203
6. Enhorning G, Robertson B (1972) Lung expansion in the premature rabbit fetus after tracheal deposition of surfactant. Pediatrics 50:58–66
7. Enhorning G, Hill D, Sherwood G, Cutz E, Robertson B, Bryan C (1978) Improved ventilation of prematurely delivered primates following tracheal deposition of surfactant. Am J Obstet Gynecol 132:529–536
8. Enhorning G, Pototschnik R, Possmayer F, Burgoyne R (1986) Pulmonary surfactant films affected by solvent vapors. Anesth Analg 65:1275–1280
9. Hardy P, Lysikiewicz A, Enhorning G (1984) Hypophase effect on a DPPC monolayer. Prog Respir Res 18:10–17
10. Robertson B (1980) Surfactant substitution. Experimental models and clinical applications. Lung 158:57–68

Some Remarks on the Use of Physico-Chemical Methods in Lung Surfactant Research*

P. Gieles, T. Key, J. Van Santvoort, and A. Steeghs

Introduction

Since the introduction of a modified Wilhelmy balance by Clements [1] physico-chemical methods have been of major importance in the development of the concepts concerning lung surfactant (LS) function and the occurrence of RDS [2]. The Langmuir – Wilhelmy method (LWM) has proven a strong means in determining surface properties of single and multicomponent synthetic monolayers [3, 4], as well as of natural surfactants [5, 6]. The LWM showed its value in experiments revealing the influence of many factors – including, temperature, ions, pH, subphase composition, the surface tension (σ) – area (A) relation [7, 8]. However, objections were also raised, both against the LWM and against what might be called the "classical" model of LS function. The former objections mainly concerned leakage of monolayer material and contact-angle problems at the Wilhelmy plate (WP). The objections to the model concerned the existence of a continuous liquid lining layer, the geometry of the alveoli and the relevance of surface tension [9, 10]. In the mid-sixties the oscillating bubble method (OBM) was introduced [11]. This method has certain advantages over the LWM: its small dimensions allow for a better control of temperature, and the required (surfactant) volume is far less than for trough measurements. In addition, no measurement device disturbs the interface. Finally, an often emphasized advantage is the shape similarity between alveolus and bubble [12–14].

Yet, the OBM is not without problems, and results that were different from those obtained with the LWM invoked more questions than answers [15, 16]. We therefore studied problems in both LWM and OBM and attempted to devise set-ups so that comparative measurements could be performed [17].

Contact-Angle Problems At Wilhelmy Plate

Users of the WP method generally consider the contact angle as an artifact although it should be regarded as a fundamental property of solid-liquid interactions. Many factors affect these interactions: chemical inhomogeneity of

*Parts of this paper are described in more detail in Gieles [17].

the liquid or solid, surface roughness of the solid, evaporation/adsorption effects, pH, ions present in the liquid, etc. In equilibrium situations the height of the contact line depends on surface tension (σ) and contact angle (Θ). By consequence, in a monolayer compression-expansion experiment the contact line travels up and down the plate, and monolayer molecules may be deposited on the WP. Consequently, the meniscus meets a chemically different surface which may drastically influence the wetting behavior. It is difficult to understand that monolayer deposition should be absent in the WP method, as this deposition is the basis of the Langmuir-Blodgett technique [18]. Often the surface of a WP is treated mechanically or chemically in order to minimize contact angle phenomena.

From the meniscus behavior it can be understood that one can only try to minimize contact-angle effects on the WP, but the measures taken cannot guarantee a LWM to be free of contact-angle phenomena. To this end much can be done by measurements additional to the LWM. We shall refer to these additional measurements as capillarography [19].

In capillarography a WP is moved up and down through an air-liquid interface with or without monolayer. The recording of the resulting force change versus penetration depth is called a capillarogram (Fig. 1). If the contact angle is and remains zero during immersion and retraction, the capillarogram will be a straight line coinciding with the buoyancy line, and the area of the capillarogram (A_c) will be zero. However, in general, on lowering the WP the (advancing) contact angle will deviate from zero and cause a lower force. On retraction of the plate the meniscus is pulled upward and in many cases the force line then coincides with the buoyancy line implying a (receding) contact angle $\Theta \simeq 0$. So contact angle hysteresis is then found through a nonzero A_c value. A_c will depend on – a. o. – penetration speed and surface tension. We have assumed that measuring capillarograms in dependence of penetration speed, surface tension, pH and evaporation/adsorption time gives a good impression of the wetting properties of the combination of WP and surfactant during a σ – A measurement.

Of five materials – microscope cover glass, grounded glass, sandblasted platinum, filter paper and glass filter – capillarograms were measured under different circumstances including variations in wetting speed (0-0.915 mm/s), evaporation time (0–40 min), pH (4–9), and surface tension in the presence of a

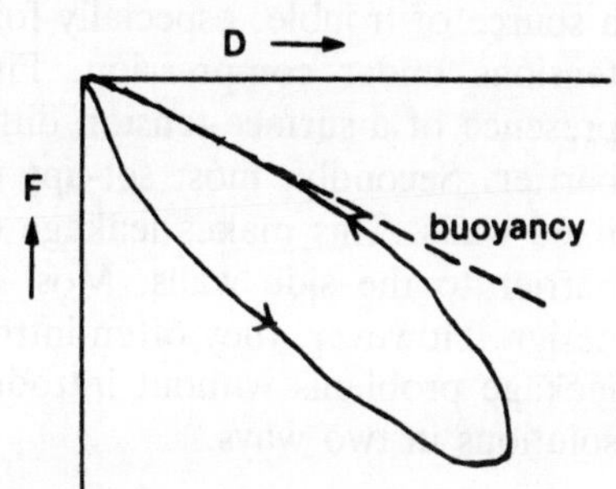

Fig. 1. A capillarogram. *F*, force; *D*, penetration depth of the WP into the liquid (see text)

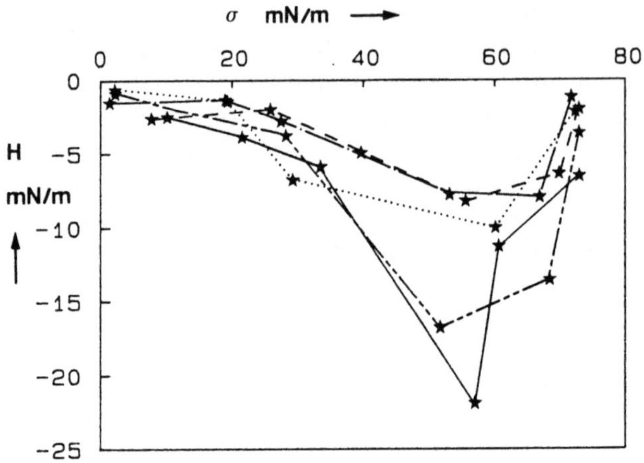

Fig. 2. H as a function of σ for the glass plate in the presence of DPPC. Wetting speeds: 0.061 mm/s (– –), 0.244 mm/s (– . –), 0.488 (– - - –), 0.915 mm/s (—),. *Dotted line* indicates static measurement.

DPPC monolayer (0–72 mN/m). The methods, results and discussion are extensively presented elsewhere [17]. Figure 2 illustrates the effect of wetting speed and surface tension on the wettability of the microscope cover glass. Use is made of the quantity $H = A_c/p.D$, in which p is the plate's perimeter (30.1 mm) and D the maximum penetration depth (4.0 mm).

The main conclusions from this study were [17]:

1. Capillarograms are a suitable means to study the wetting properties of WPs.
2. Filter paper and glass filter showed the least H values under all circumstances ($H > -2$, cf. values Fig. 2). This implies average contact angle effects of less than 13 degrees, i. e. less than 3 % error in a σ – A measurement. Since filter paper is easy to handle it seems the most suitable material to be used for WP.

Leakage Problems in Langmuir-Wilhelmy Method

Leakage of molecules flowing across the moving barrier has continuously been a source of trouble, especially for surfactants that can attain very low surface tensions under compression. First and principal cause for leakage is the presence of a surface-tension difference between the two sides of the moving barrier. Secondly, most set-ups make use of a moving barrier sliding along fixed walls. This makes leakage depend on the tightness of the fitting of the barrier to the side walls. Most solutions concentrate on a different barrier design. However, they often introduce new complications. In order to remove leakage problems without introducing new complications we have sought for solutions in two ways.

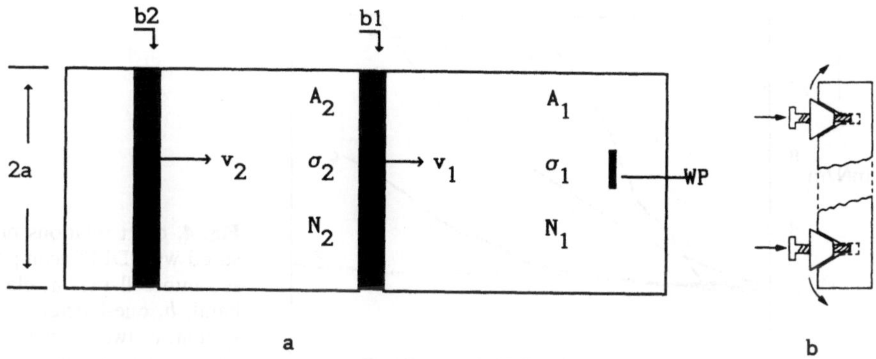

Fig. 3.a The second-barrier method. *2a*, trough width; *b*, barrier; *v*, barrier speed; *A*, surface area; *n*, number of monolayer molecules. **b** Detailed view of the second barrier with wedges. If the wedges are screwed into the barrier, the ends of the barrier will be displaced towards the side walls of the trough

Second-Barrier Method. We start from the schematic trough division and nomenclature as shown in Fig. 3. Initially we have two surfaces (areas A_{10}, A_{20}) covered with identical monolayer molecules (numbers N_1 and N_2 respectively). The quantities spread on each surface are chosen to yield equal surface tensions in each compartment ($\sigma_{10} = \sigma_{20}$): $N_1/A_{10} = N_2/A_{20}$. We assume linear barrier speeds v_1 and v_2 respectively. In order to minimize leakage effects we try to realize equal surface tensions throughout a compression-expansion cycle by an appropriate choice of v_2. We assume identical circumstances for both monolayers which implies that the surface tensions are determined by the specific surface area ($A_s = A/N$) only. The requirement of equal surface tensions then leads to $v_2 = v_1.(1 + A_{20}/A_{10})$.

Elastic Band Method. The use of a closed elastic band might be another way to reduce the leakage problem. We therefore studied the properties of natural latex and the possibility for its use in monolayer studies.

The closed bands (17 mm wide) were cut from veterinary gloves supplied by Rubber, Holland. The gloves were stretched around a perspex cylinder and taped. The bands were then carefully cut from the gloves during rotation of the cylinder.

After removing surface active impurities in a Soxhlet apparatus filled with ethanol a band with appropriate properties results [17]. A disadvantage may be the possible deposition of monolayer material [20] which hampers the accurate measurement of σ – A curves. It should be noted that the band may well be used to determine the lowest possible surface tension under compression in measurements where excess quantities of surfactant are applied, or in measurements where σ instead of A_s is the independent quantity (e. g. elasticity vs. surface tension).

Test Experiments. A reliable leakage test is the measurement of the surface tension of DPPC as a function of time – at constant area – after it has been

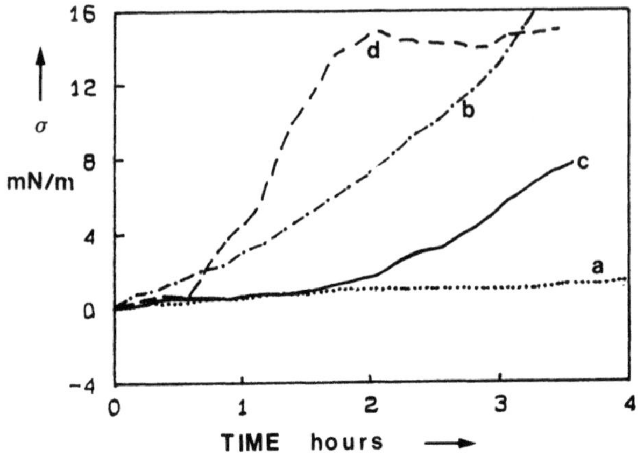

Fig. 4. σ – t relations measured with DPPC compressed into collapse. *a*, elastic band; *b*, one-barrier system; *c*, two-barrier system, first compartment; *d*, two-barrier system, second compartment

compressed into collapse. Such "relaxation" measurements were carried out on DPPC (chemically pure) spread from chloroform solution on a subphase of triple distilled water at T = 20.0 C. Three different barrier systems were used. In the two barrier system use is made of a specially constructed second barrier (Fig. 3b). The results (Fig. 4) show that the elastic band is indeed closed since DPPC shows hardly any relaxation. This implies that the relaxation in both one and two barrier systems is mainly due to leakage. The application of the second barrier is a distinct improvement: the leakage problem is now transferred to the second barrier.

Oscillating Bubble Method

With the purpose of comparative measurements between LWM and OBM in mind (see Introduction), we came to the following requirements for the experimental set-up:

1. A continuously determined σ – A loop should result during compression and expansion of the monolayer.
2. This requires on-line measurement of bubble radius and Laplace pressure, and (on-line) calculation of σ and A.
3. The compression ratio should have a sufficiently large range (up to at least 50 %).
4. The period of the oscillations should be continuously variable from 1 to at least 20 s.

The criteria have led to a set-up which is schematically shown in Fig. 5.

The bubble chamber is a cylindrically shaped perspex chamber (inner diameter 5.0 mm), with a glass capillary (radius 250 μm). Its volume amounts to 200 μl. A temperature bath – made of perspex – surrounds the bubble chamber. Temperature can be kept constant within 0.1° C in the range 20–50° C by means of a proportional temperature control. In normal use the water surrounds the bubble chamber, thereby undoing the optical distortion

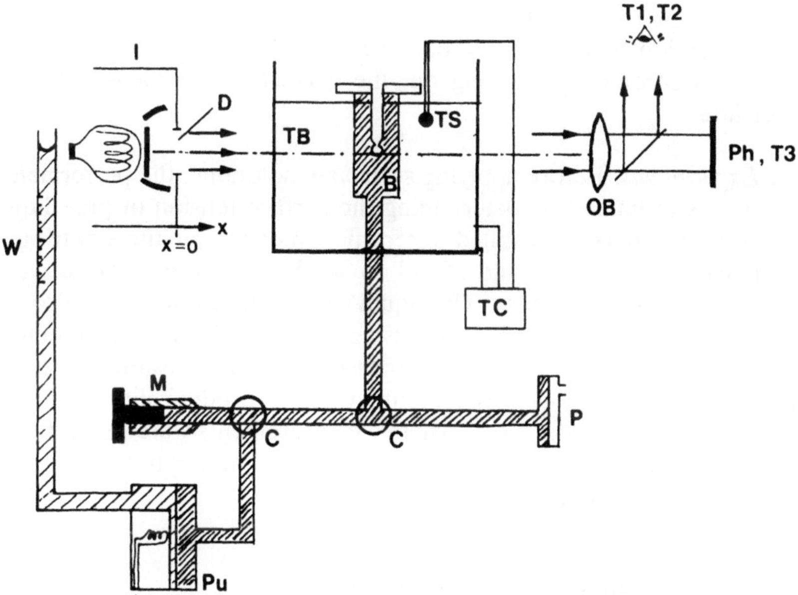

Fig. 5. Schematic presentation of the set-up of the OBM. *I*, illumination system of the microscope; *D*, diaphragm; *dashed line*, optical axis; *TB*, temperature bath; *B*, bubble chamber; *TS*, temperature sensor; *TC*, temperature control; *OB*, objective of microscope; *M*, microsyringe; *C*, cock; *Pu*, pulsator; *P*, pressure transducer. The bubble is viewed through two tubes **(T1, T2)** of the stereomicroscope, while its size is measured using a photocell *(Ph)* located in the third tube *(T3)*. *W*, water column for calibration

caused by the bubble chamber. The temperature bath is provided with an overflow to keep the water at a constant level. The bubble chamber and temperature bath are mounted on an x-z table which enables correct positioning of the bubble with respect to the optical system.

A microsyringe (Hamilton, total capacity 50 μl) is used to create a bubble in the bubble chamber. The pulsator unit is a differential pressure transducer (Validyne, MP45) which is used the other way round. The displacement- and frequency ranges of this pulsator (0–6 μl, and 0–40 Hz respectively) meet with the demands mentioned above. The pressure sensing device is of the silicon type (Honeywell microswitch 142 PC-01-D), and has a compliance of approx. 5.10^{-8} μl/Pa. This leads to negligible volume changes.

The bubble is viewed through a stereomicroscope (Olympus X-TR), which has a working distance of 85 mm and is provided with a third tube. In this extra tube a photovoltaic cell (United Detector Technology S100, area 100 $mm^{2)}$ is placed for the measurement of the bubble size.

The set-up has been automated: calibrations, measurements, calculations, drawing σ – A curves, background storage etc. are performed by means of a M68000 computer connected to the host (PDP11/23).

The OBM also contains several sources of error [17]: (a) the bubble is an incomplete sphere and the missing bubble cap varies during cycling, (b) the variation of the center of the bubble during cycling causes variations in the hydrostatic pressure contribution, and (c) deformation of the bubble occurs if surface tension and buoyancy become competitive. The first two errors are taken into account by carrying out the corrections by means of a computer procedure.

Test Experiments. Before applying surfactant solutions, the performance of the system was examined by determining the surface tension of pure liquids. The liquids used were triple distilled water, methanol and mixtures of triple distilled water and methanol (25 and 50 vol % methanol). Temperature was 20.0 ± 0.5° C for all measurements. The liquid's surface tensions were also measured by means of a Wilhelmy balance using FP as WP. All measurements were performed in duplo. In all cases the measured surface tensions agreed very well with the literature values. The accuracy and reproducibility of the measurements were quite satisfactory. Dynamic measurements were performed in order to check for the absence of unwished phase shifts between σ and A. In none of the measurements phase difference was found.

Comparative Experiments on Monolayers

In a pilot study comparative measurements in both LWM and OBM were performed on monolayers of sodium-dodecylphosphate (SDS) and on sheep surfactant recovered by lung lavage of healthy adult sheep at a slaughterhouse. The extraction and purification procedure of the sheep surfactant are described elsewhere [17].

The measurements were carried out at T = 20.0 ± 0.5° C, on a subphase of triple distilled water. In the LWM filter paper was used as Wilhelmy plate. The cycle speeds used were 6 cycl/min (sheep surfactant), 1.5 and 15 cycl/min (SDS). In all cases, except for sheep surfactant, in the LWM monolayer formation occurred through adsorption. The compression ratio was 25 %. In the

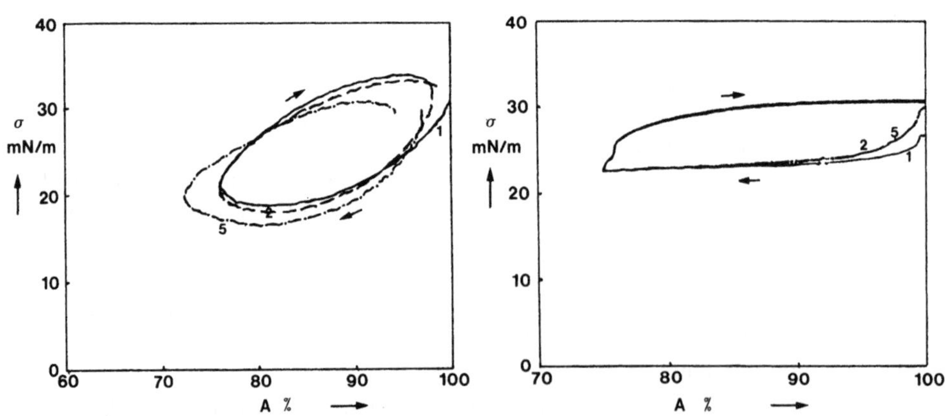

Fig. 6. First, second and fifth σ – A loop of SDS at T = 20° C measured in OBM *(left)* and LWM *(right)* at 1.5 cycl/min

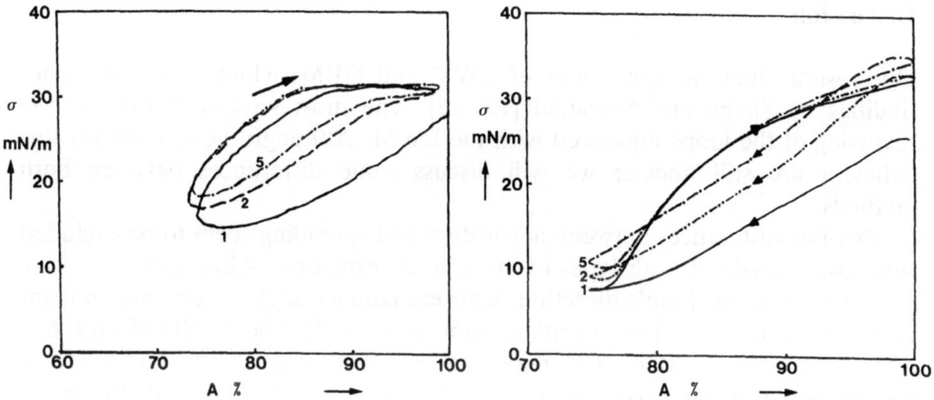

Fig. 7. First, second and fifth σ – A loop of SDS at T = 20° C measured in OBM *(left)* and LWM *(right)* at 15 cycl/min

LWM initial areas were 10.0 × 16.6 cm^2 (sheep surfactant) and 10.0 × 20.0 cm^2 (SDS). In the OBM initial areas were 10.6 mm^2 (sheep surfactant) and 8.38 and 6.79 mm^2 (SDS, 15 and 1.5 cycl/min respectively). The measurements were performed in duplo.

The difference between the results of OBM and LWM are striking (Figs. 6–8). Not only do the minimal surface tensions at 25 % compression differ, also the clockwise direction of the loops in the OBM changes to anti-clockwise direction in the LWM (Figs. 7, 8).

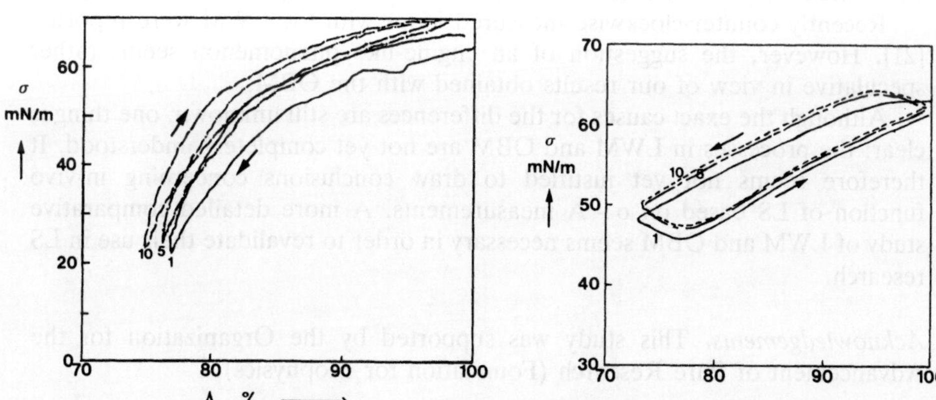

Fig. 8. First, fifth and tenth σ – A loop of sheep surfactant at T = 20° C in OBM *(left)* and LWM *(right)* at 6 cycl/min

Discussion

Our results show no agreement of LWM and OBM, which supports earlier findings by Slama and Schoedel [15, 16]. The most striking feature is the reversing of the loops measured with the LWM. Although the reasons for this behavior are still unclear we will discuss some differences between both methods.

Possible differences between adsorption and spreading seem to be excluded since use is made of both types of monolayer formation. Also, leakage effects cannot reverse the loop's direction, whereas contact angle effects are thought to be absent in view of the capillarogram results [17]. The results of Figs. 6–8 suggest a frequency effect. This may be due to two things. First, the waveform was sinusoidal in the OBM whereas it was triangular in the LWM; the latter thus also contains higher harmonics. Second, since surface deformation is principally nonuniform in the LWM due to one-sided compression, this may give rise to inhomogeneities in which the effect increases with frequency. After all, in the LWM it is assumed that the variation of the total area equals the local area variation at the site of the WP. Since in the OBM the spherically shaped bubble mostly deforms uniformly no surface tension gradients are invoked. It is also noted that in the LWM monolayer molecules move in the plane of the interface whereas they move perpendicular to it in the OBM. This might lead to different monolayer formations.

Since the reversing of the loops is developing during cycling (SDS) the frequency dependence cannot be the only cause. Other differences between the methods are the absolute area, the absolute compression speed, the applied quantities, geometry (of importance in diffusion processes). It is still unclear to what extent they can be held responsible for the strikingly different $\sigma - A$ loops.

Recently counter-clockwise measured loops with the LWM were reported [21]. However, the suggestion of an engine-like phenomenon seems rather speculative in view of our results obtained with the OBM.

Although the exact causes for the differences are still unknown, one thing is clear: the processes in LWM and OBM are not yet completely understood. It therefore seems not yet justified to draw conclusions concerning in-vivo function of LS based on $\sigma - A$ measurements. A more detailed comparative study of LWM and OBM seems necessary in order to revalidate their use in LS research.

Acknowledgements. This study was supported by the Organization for the Advancement of Pure Research (Foundation for Biophysics).

References

1. Clements JA (1957) Surface tension of lung extracts. Proc Soc Exp Biol Med 95:170–172
2. Avery ME, Mead J (1959) Surface properties in relation to atelectasis and hyaline membrane disease. Am J Dis Child 97:517–523

3. Phillips MC, Chapman D (1968) Monolayer characteristics of saturated 1,2 diacyl phosphatidylcholines (lecithins) and phosphatidylethanolamines at the air-water interface. Biochim Biophys Acta 163:301–313
4. Villalonga F (1968) Surface chemistry of L-α-dipalmitoyl lecithin at the air-water interface. Biochim Biophys Acta 163:290–300
5. Burnell JM, Kyriakides EC, Edmonds RH, Balint JA (1978) The relationship of fatty acid composition and surface activity of lung extracts. Respir Physiol 32:195–206
6. Bienkowski R, Skolnick M (1972) Dynamic behavior of surfactant films. J Colloid Interface Sci 39(2):323–330
7. Colacicco G, Basu M, Scarpelli EM (1976) pH, temperature, humidity and the dynamic force-area curve of dipalmitoyl lecithin. Respir Physiol 27:169–186
8. Kobayashi T, Robertson B (1983) Surface adsorption of pulmonary surfactant in relation to bulkphase concentration and presence of CaC12 Respiration 44(1):63–70
9. Hills BA (1982) Water repellency induced by pulmonary surfactants. J Physiol 325:175–186
10. Reifenrath R (1975) The significance of alveolar geometry and surface tension in the respiratory mechanics of the lung. Respir Physiol 24:115–137
11. Adams FH, Enhorning G (1966) Surface properties of lung extracts. I A dynamic alveolar model. Acta Physiol Scand 68:23–27
12. Adams FH, Enhorning G (1966) Surface properties of lung extracts. II Comparison of fetal and adult rabits. Acta Physiol Scand 68:28–36
13. Adams FH, Enhorning G (1966) Surface properties of lung extracts. III Changes appearing during first ten minutes after surface formation. Acta Physiol Scand 68:37–42
14. Schoedel W, Slama H, Hansen E (1971) Zeitabhängiges Verhalten von Filmen von oberflächenaktivem Material aus Lungalveolen. Pflügers Arch 322:336–346
15. Slama H, Schoedel W, Hansen E (1971) Bestimmung der Oberflächeneigenschaften von Stoffen aus den Lungenalveolen mit einer Blasenmethode. Pflügers Arch 322:255–263
16. Slama H, Schoedel W, Hansen E (1973) Lung surfactant: film kinetics of the surface of an air bubble during prolonged oscillation of its volume. Respir Physiol 19:233–243
17. Gieles P (1987) Methods of measurement for the evaluation of monolayer properties. Development and applications. Thesis, University of Technology, Eindhoven, The Netherlands
18. Blodgett KB (1935) Films built by depositing successive unimolecular layers on a solid surface. J Am Chem Soc 57:1007–1022
19. Bayramli E, van de Ven TGM, Mason SG (1981) Tensiometric studies on wetting IV. Contact angle and surface pressure relaxation. Colloid and Surfaces 3:279–293
20. van Liempd J, Boonman A, Demel R, Gieles P, Gorree T (1987) Nonselective squeeze-out of DOPC and DOPG from binary mixed monolayers with DPPC. Biochim Biophys Acta 897:495–501
21. Hills BA, Barrow RE (1984) An "engine" phenomenon displayed by monolayers of a pulmonary surfactant cycled to steady state. Phys Med Biol 29(11):1399–1408

In Vivo Tests for Evaluation of Pulmonary Surfactant

B. Lachmann

Introduction

Before using surfactant preparations in animal experiments or in clinical trials, it is necessary that they fulfill specific physical properties [5, 9]. Some of these essential physical requirements for an effective lung surfactant have been reviewed by many workers [1, 4, 7, 27]. However, different clinical results concerning arterial oxygen tension have been reported. A marked improvement in arterial oxygen tension after surfactant replacement has been reported by Fujiwara et al. [6], whereas no effect on arterial oxygen tension was observed by Milner et al. [22], Morley et al. [23] and Wilkinson et al. [32]. All these workers used a surfactant which was highly effective in in vitro studies. This poses the question as to whether the physical characteristics of surfactant, obtained through in vitro studies, can be also be considered as a method of testing a preparation which has to be physiologically active in the lung.

In earlier investigations [12] no prognosis could be made from surface tension characteristics (Langmuir balance) of an exogenous surfactant on its effects on the improvement of thorax-lung compliance in immature rabbit fetuses. When testing pure artificial surfactant giving effective in vitro results, Obladen et al. [28] could not guarantee its in vivo effects.

In order to exclude the possibility that the lack of correlation between effects in vitro and in vivo are not specific to the premature rabbit fetus model, we made similar investigations in the lung lavage model and the viral pneumonia model [17]. These studies also showed that the shape of the surface area diagram of an exogenous surfactant, or any other parameter from the surface tension area diagram, did not correlate with the improvement of arterial oxygen tension in the lung lavage model or the improvement of thorax-lung compliance in mice infected with viral pneumonia. Our results are in agreement with other authors and lead us to conclude that the efficacy of various preparations of exogenous surfactants should mainly be evaluated in living animals with surfactant-deficient lungs [8, 17, 26, 30].

In recent years such experimental models of surfactant deficiency have been developed [20, 29], but conclusive criteria for the effectiveness of a surfactant have not yet definitely been defined. For this reason some standard levels for improvement of lung function, or threshold levels, must be established in standardized animal models with surfactant deficiency.

Immature Rabbit Fetuses (Day 27)

Using immature rabbit fetuses we developed a method which allows study of lung mechanics during spontaneous [15] as well as artificial ventilation in up to ten rabbit fetuses simultaneously [16]. In this model, especially under standardized conditions, compliance is a very sensitive parameter, and the difference between controls and animals treated with high-quality exogenous surfactant varied from 10- to 15-fold (Fig. 1).

Other criteria which can easily be analyzed in this model include the effect of instilled surfactant on compliance in relation to time and the decrease of the critical opening pressure under dynamic conditions. A very effective surfactant is, for example, characterized by maintaining the compliance constant for up to 1 h during pressure-controlled ventilation and decreasing the critical opening pressure to 15–20 cm H_2O (Fig. 1; Table 1).

To prevent different results concerning the improvement of lung compliance when testing surfactant, the lungs of all rabbit fetuses should have the same degree of stiffness, or immaturity, before surfactant replacement. This avoids high standard deviations in thorax-lung compliance, especially when using a non-optimal surfactant (Fig. 2). Other parameters which should be kept constant when testing exogenous surfactant are the instilled volume and the concentration of total phospholipids. With a constant instilled volume, the lungs of animals treated with a surfactant of higher concentration were more

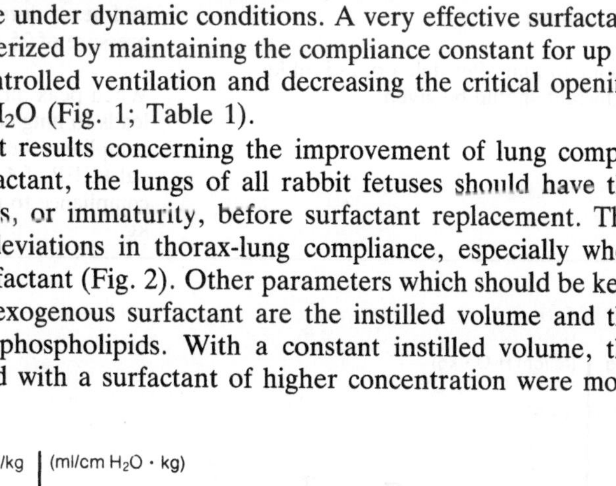

Fig. 1. Experimental protocol for testing exogenous surfactant in immature rabbit fetuses from gestational day 27. Course of the compliance shows the effects of a rather optimal exogenous surfactant. Note that tracheal instillation of surfactant leads, both in already treated animals and in controls, to an improvement in lung compliance after 55 min. *Paw*, peak airway pressure; *PEEP*, positive end-expiratory pressure. Mean values of eight treated animals and six controls

Table 1. Experimental conditions for testing exogenous surfactant and functional characteristics of effective lung surfactant in premature newborn rabbits (day 27)

Experimental conditions	Functional improvements
Ventilator settings: pressure-controlled ventilation peak airway pressure 25 cm H_2O; 50 %–60 % inspiratory time; frequency 30–40/min	At 15 min after surfactant instillation compliance should be higher than 1.0 ml/cm H_2O kg
	Compliance should be stable or
Before surfactant instillation compliance has to be less than 0.1 ml/cm H_2O kg in each animal	Increase over an observation period of 1 h and never decreases when peak airway pressure of 25 cm H_2O is used
About 10 min after delivery 2–5 ml/kg surfactant should be given	When peak airway pressure of 15 cm H_2O leads to lung collapse and decrease in compliance below 0.2 ml/H_2O kg additional PEEP of 4–5 cm H_2O should improve compliance to more than 0.5 ml/cm H_2O kg

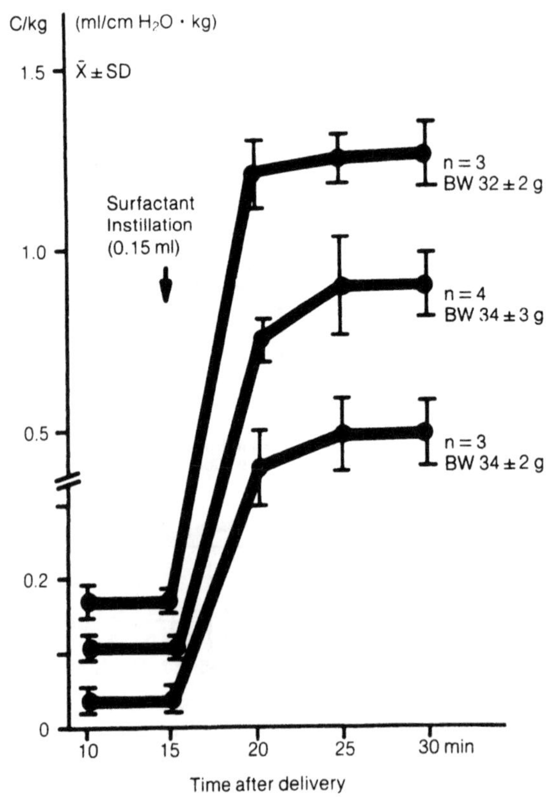

Fig. 2. Lung-thorax compliance in surfactant-treated immature newborn rabbits with different compliance immediately after delivery (gestational age 27 days) at various intervals after onset of pressure-controlled ventilation with standardized peak airway pressure of 25 cm H_2O. Note that the animals with the highest compliance after delivery show the greatest improvements

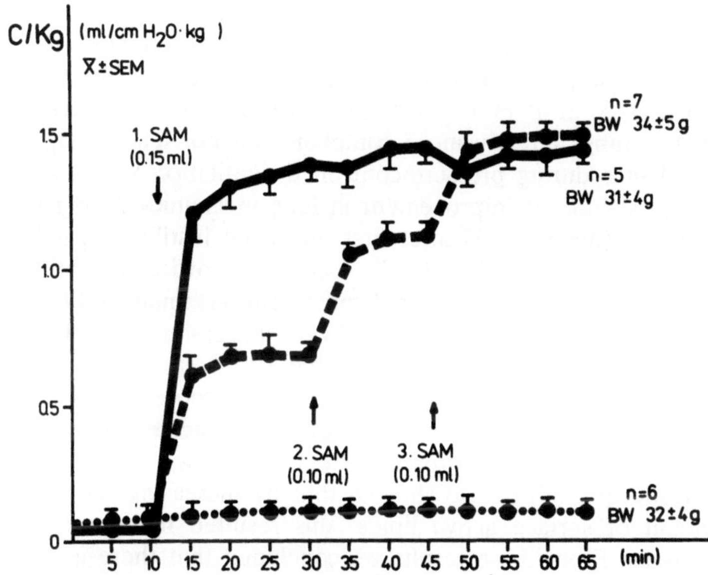

Fig. 3. Lung-thorax compliance in immature rabbit fetuses (gestational age 27 days) at various intervals after onset of artificial ventilation (peak airway pressure, 25 cm H_2O). Animals were treated with one preparation of exogenous surfactant, 80 mg/ml total phospholipids *(solid line)* and 30 mg/ml *(dashed line)*. Surfactant administration *(SAM)* is indicated with *arrows*. Note that compliance reached the same level when animals treated with less concentrated surfactant received the same amount of phospholipids as animals treated with the more concentrated surfactant

Table 2. Dependence of thorax-lung compliance on the volume instilled of a rather optimal (but not excellent) exogenous surfactant (total phospholipid concentration 60 mg/ml)

	Min after delivery					
	10	20	30	40	50	60
Total volume of surfactant	–	50	100	150	200	250
Group A compliance ml/cm H_2O kg Mean ± SD	0.06 ± 0.02	0.45 ±0.12	0.72 ±0.28	1.11 ±0.26	1.45 ±0.36	1.61 ±0.34
Total volume of surfactant μl	–	200	–	–	–	–
Group B compliance ml/cm H_2O kg Mean ± SD	0.07 ±0.02	1.43 ±0.26	1.51 ±0.28	1.55 ±0.27	1.55 ±0.29	1.54 ±0.28

Group A (six animals, BW 38 ± 7 g) received five doses of surfactant (50 μl) at 10-min intervals. Group B (six animals, BW 33 ± 8 g) received one dose of 200 μl surfactant.

compliant than those treated with a lower concentration (Fig. 3, at 25 min). However, two additional surfactant instillations (lower concentration) finally led to the same degree of lung compliance compared with the higher concentration (Fig. 3, at 65 min). Additional instillations of surfactant (high concentration) in lungs already more compliant did not lead to significant changes in compliance during pressure-controlled ventilation with 25 cm water, but did lead to significant improvement in lung mechanics during ventilation with 20 cm water (mean ± SD after first surfactant instillation, 0.78 ± 0.19; second, 1.22 ± 0.24; third, 1.42 ± 0.17). Also, the stepwise instillation of surfactant (50 μl with a constant phospholipid concentration) finally led to the same improvement in lung mechanics compared with thorax-lung compliance when an initial large volume (200 μl) was used (Table 2).

Our observations are in agreement with those of other workers [2, 8, 24] who found that higher concentrations of surfactant improved the clinical and functional status to a greater extent than lower concentrations. Moreover, Metcalfe et al. [21] also showed that by increasing the instilled volume and amount of surface active lipids, this resulted in improved pressure-volume diagrams. From these results we concluded that there is no danger from the instilled volume in a range from 2–10 ml/kg, when experimental animals are artificially ventilated. Therefore, we believe that for reaching a threshold level of improvement in functional and clinical parameters, an excess of surface active material should be given. For a good intrapulmonal distribution the instilled volume must be larger than anatomical dead space.

In Vivo Lung Lavage

We have used guinea pigs, rabbits, and dogs to develop a model of adult respiratory distress syndrome (ARDS) in which alveolar surfactant phospholipids are selectively removed by in vivo lung lavage [11, 13, 14]. Severe respiratory insufficiency was defined as a fall in PaO_2 below 60 mmHg during pressure-controlled ventilation with pure oxygen, positive end-expiratory pressure (PEEP) of 6–8 cm H_2O, peak airway pressure of 26–30 cm H_2O, and inspiratory time 33 %–50 % (Table 3).

We found the lung lavage model (particularly in small animals) useful for a variety of experimental purposes, especially for the testing of alternative surfactant preparations [10, 18, 19], as lung mechanics and arterial blood gases can be measured in up to eight guinea pigs simultaneously.

Viral Pneumonia in Mice

Viral pneumonia results in a clinical situation equivalent to ARDS. Important functional changes in lungs infected by influenza virus are induced by surfactant deficiency secondary to destruction of type II cells [31]. Therefore, we used viral pneumonia as an additional surfactant-deficiency model for testing exogenous surfactant [20]. Mice infected with influenza virus, accord-

Table 3. Experimental conditions for testing exogenous surfactant and functional characteristics of effective lung surfactant in lung lavage model

Experimental conditions	Functional improvements
Ventilator settings: pressure-controlled ventilation peak airway pressure 27–29 cm H_2O; PEEP 6–8 cm H_2O; inspiratory time 50 %; frequency 30/min; inspiratory O_2 concentration 100 %	PaO_2 should rise to more than 200 mmHg within 15 min after first surfactant instillation
At 10 min after lung lavage PaO_2 should be below 60 mmHg	After second surfactant instillation PaO_2 should be stable or increase within the following 30 min
At 10 and 40 min after lung lavage 3–4 ml/kg surfactant should be given	

ing to Noack et al. [25], were used on day 6 after infection, when the compliance had diminished to 20 %–30 % of the initial value [3]. Compliance measurements were made according to Lachmann et al. [16] at a peak airway pressure of 25 cm H_2O during pressure-controlled ventilation (Table 4).

Conclusions

Although, to date, some workers have achieved excellent clinical results with their surfactant preparations (in contrast to others), we believe that before use of other exogenous surfactant preparations in clinical trials, some of the criteria discussed in this paper should be studied in different surfactant-deficiency models [30].

Our experience has shown that the respiratory distress syndrome (RDS) model of premature rabbit fetuses is the most sensitive model, while the model of viral pneumonia is the least sensitive when testing exogenous surfactant. With a less effective exogenous surfactant, a small improvement in lung function in the rabbit fetuses model can be observed (Fig. 4) while in the viral pneumonia model almost no improvement occurred.

Table 4. Experimental conditions and functional characteristics of exogenous surfactant in virus pneumonia model

Experimental conditions	Functional improvements
Ventilator settings: pressure-controlled ventilation peak airway pressure 25 cm H_2O; inspiratory time 50 %; frequency 40/min	Compliance should improve more than 2.5-fold within 5 min after tracheal surfactant instillation

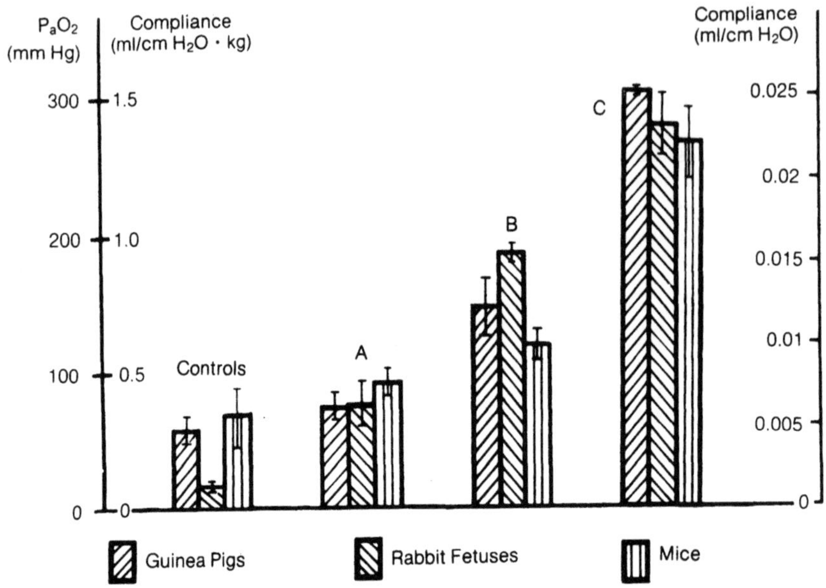

Fig. 4. Changes in PaO_2 and compliance in guinea pigs with ARDS, immature rabbit fetuses (gestational age 27 days), and mice with viral pneumonia after tracheal instillation of different effective exogenous surfactants *(A, B, C)*. Compliance is standardized by body weight in rabbit fetuses but not in mice. Number of animals in each group varied between six and eight

A suboptimal surfactant led to slight improvements in the three animal models and only a functionally effective surfactant fulfilled all our criteria for functional improvements in lung function.

References

1. Bangham AD, Morley CJ, Phillips MC (1979) The properties of an effective lung surfactant. Biochim. Biophys Acta 573:552–556
2. Berggren P, Curstedt T, Grossman G, Nilsson R, Robertson B (1985) Physiological activity of pulmonary surfactant with low protein content; effect of enrichment with synthetic phospholipids. Exp Lung Res 8:29–51
3. Bergmann K CH, Lachmann B, Noack K (1984) Lung mechanics in orally immunized mice after aerolized exposure to influenza virus. Respiration 46:218–221
4. Clements JA (1977) Functions of the alveolar lining. Am Rev Respir Dis 115:67–71
5. Fujiwara T, Tanaka Y, Takei Y (1979) Surface properties of artificial surfactant in comparison with natural and synthetic surfactant lipids. IRCS Med Sci 7:311
6. Fujiwara T, Chida D, Watabe Y, Maeta H, Morita T, Abe T (1980) Artificial surfactant therapy in hyaline-membrane disease. Lancet 1:55–59
7. Goerke J (1974) Lung surfactant. Biochim Biophys Acta 344:241–261
8. Ikegami M, Adams FH, Towers B, Osher AB (1980) The quantity of natural surfactant necessary to prevent the respiratory distress syndrome in premature lambs. Pediatr Res 14:1082–1085
9. King RJ and Clements JA (1972) Surface active materials from dog lung. II. Composition and physiological correlations. Am J Physiol 223:715–726

10. Lachmann B, Fujiwara T, Chida S, Morita T, Konishi M, Nakamura K, Maeta H (1981) Improved gas exchange after tracheal instillation of surfactant in the experimental adult respiratory distress syndrome (abstract) Crit Care Med 9:158
11. Lachmann B, Robertson B, Vogel J (1980) In vivo lung lavage as an experimental model of the respiratory distress syndrome. Acta Anaesth Scand 24:231–236
12. Lachmann B (1982) Animal models for the in vivo testing of exogenous surfactant (abstract). Mackinac workshop on exogenous surfactant, October 6–9, 1982
13. Lachmann B, Jonson B, Lindroth M, Robertson B (1982) Modes of artificial ventilation in severe respiratory distress syndrome. Lung function and morphology in rabbits after wash-out of alveolar surfactant. Crit Care Med 10:724–732
14. Lachmann B, Danzmann E, Haendly B, Jonson B (1982) Ventilator settings and gas exchange in respiratory distress syndrome. In: Prakash O (ed) Applied physiology in clinical respiratory care. Nijhoff, The Hague, pp 141–176
15. Lachmann B, Grossman G, Nilsson R, Robertson B (1981) Effect of supplementary surfactant on in vivo lung mechanics in the premature rabbit neonate. Eur J Pediatr 136:173–179
16. Lachmann B, Grossman G, Freyse J, Robertson B (1981) Lung-thorax compliance in the artificially ventilated premature rabbit neonate in relation to variations in inspiration:expiration ratio. Pediatr Res 15:833–838
17. Lachmann B (1986) New aspects in pathophysiology and therapy of respiratory distress syndrome: criteria for characterisation of exogenous surfactant in surfactant deficient animal models. In: Cosmi EV, Di Renzo GC (eds) Selected topics in perinatal medicine. Rome, CIC, pp 177–198
18. Lachmann B, Fujiwara T, Chida S, Morita T, Konishi M, Nakamura K, Maeta H (1983) Surfactant replacement therapy in the experimental adult respiratory distress syndrome (ARDS). In: Cosmi, Scarpelli (eds) Pulmonary surfactant system. Elsevier, Amsterdam, pp 231–235
19. Lachmann B, Berggren P, Curstedt T, Grossman G, Robertson B (1984) Surfactant replacement in experimental respiratory distress syndrome induced by lung lavage. Prog Resp Res 28:251–256 In: Robertson B, Van Golde LMG, Batenburg JJ (eds) Pulmonary surfactant.
20. Lachmann B, Danzmann E (1984) Adult respiratory distress syndrome. Elsevier, Amsterdam, pp 505–548
21. Metcalfe IL, Burgoyne R, Enhorning G (1982) Surfactant supplementation in the preterm rabbit: effects of applied volume on compliance and survival. Pediatr Res 16:834–839
22. Milner AD; Vyas H, Hopkin IE (1983) Effects of artificial surfactant on lung function and blood gases in idiopathic respiratory syndrome. Arch Dis Child 58:458–460
23. Morley CJ, Miller N, Bangham AD, Davis JA (1981) Dry artificial lung surfactant and its effect on very premature babies. Lancet 1:64–68
24. Nilsson R, Grossman G, Berggren P, Robertson B (1981) Surfactant treatment in experimental hyaline membrane disease. Eur J Respir Dis 62:441–449
25. Noack K, Bergmann K CH, Eckert H, Luther P, Tischner H, Pohl WD (1982) Tierexperimentelles Modell einer aerogenen Influenza-Infektion. Z Erkrank Atm-Org 159:155–165
26. Nohara K, Berggren P, Curstedt T, Grossman G, Nilsson R, Robertson B (1986) Correlation between physical and physiological properties of various preparations of lung surfactant. Eur J Respir Dis 69:321–335
27. Notter RH, Taubold R, Mavis RD (1982) Hysteresis in saturated phospholipid films and its potential relevance for lung surfactant function in vivo. Exp Lung Res 3:109–127
28. Obladen M, Kampmann W, Zimmerman I, Lachmann B (1984) Artificial surfactant in preterm rabbits with and without RDS: difference of in vitro and in vivo activity. Eur J Pediatr 144:195–199
29. Robertson B (1980) Surfactant substitution; experimental models and clinical applications. Lung 158:57–68
30. Robertson B, Lachmann B (1988) Experimental evaluation of surfactants for replacement therapy. Exp Lung Res 14:279–310

31. Stinson SF, Ryan DP, Hertweck MS, Hardy JD, Hwang-Kow SY, Loosli CG (1976) Epithelial and surfactant changes in influenzal pulmonary lesions. Arch Pathol Lab Med 100:147–153
32. Wilkinson AR, Jenkins PA, Jeffrey JA (1985) Two controlled trials of dry artificial surfactant: Early effects and later outcome in babies with surfactant deficiency. Lancet 2:287–291

Efficacy and Standardisation of SF-RI 1: A Preparation from Bovine Lung Surfactant

B. Disse, L. Gortner, E. Weller, H. Eberhardt, and H. Ziegler

Preparations from natural sources of pulmonary surfactant are subject to biological variability. Therefore Tanaka and colleagues [1] standardized their reconstituted lung surfactant by adjusting the lipid composition. The product, Surfactant TA, revealed a low variability of in vitro surface properties [2]. However, to date a convincing correlation between in vitro surface properties and physiological surfactant activity, determined as dynamic or static compliance of premature ventilated animal fetuses, has not been shown [3–5]. Nevertheless, in vitro surface properties have extensively been used for the development of artificial surfactants [6–10].

SF-RI 1, a purified surfactant from bovine lung, is prepared in a process which has been optimized by testing the product in premature rabbit and lamb fetuses. A method for the standardisation of testing surfactant activity in premature rabbit fetuses is presented.

Methods

Physiological surfactant activity was tested in premature rabbit fetuses of 27.0 days ± 2 h gestational age in a modification of the procedure of Nilsson [10] described in more detail elsewhere in this book [11]. The premature fetuses received one intratracheal instillation of 3.33 ml/kg surfactant suspension before the first breath. The concentration of phospholipids ranged from 8–32 mmol/l and was determined as described by Eibl and Lands [12].

Dry surfactant lipid from one large batch was used as a reference standard (# 872005). This batch was divided into approximately 50 mg portions and stored under nitrogen in sealed vials at –70° C. The day before the experiment the material was suspended in distilled water by sonication in a standardized manner.

Surfactant activity was further determined in premature lamb fetuses of 125 days gestational age (Merino Landschaf, normal duration of gestation 145–150 days). The lambs were ventilated with a time-cycled pressure-controlled ventilator (Sechrist, Small Infant Ventilator) with 100 % O_2, a frequency of 60 min^{-1} and 40 % inspiratory time. The inspiratory pressure was regulated in order to adjust $paO_2 > 40$ mm Hg and $paCO_2 < 50$ mm Hg. PEEP was 2 cm H_2O.

Control lambs received one intratracheal instillation of 1 ml/kg 75 mmol/l NaCl solution, treated lambs 1 ml/kg surfactant suspension (46 µmol/kg pholipids, batch # 160325). Blood pressure, inspiratory airway pressure, respiratory flow and blood gases were determined. The animals were ventilated for more than 5 hours.

Results and Discussion

Fetuses from chinchilla rabbits of this age are so premature that an untreated control animal cannot be adequately ventilated. The tidal volume of these animals at 25 cm H_2O inspiratory pressure after 2 min of ventilation is 0.71 ± 0.29 ml/kg (mean ± SD, n = 16, dynamic compliance: 0.028 ± 0.012 ml/kg/cm H_2O). Replacement therapy with SF-RI 1 shows a dose-dependent effect.

None, or nearly no effect was seen after 2 min of ventilation, if the administered dose was equal to or less than 27 µmol/kg (20 mg/kg, lot # ZR1515, Fig. 1). A maximal effect was already obtained with 80 µmol/kg (60 mg/kg). Several preparations of the bovine surfactant SF-RI 1 resulted in similar dose-response curves with only a factor of 3–4 between the minimal and maximal effective dose after 2 min (see also data in [11]).

A determination of surfactant activity is, of course, only possible if dosages in the steep part of the dose-response curve are applied, bringing about a tidal volume between 2 and 6 ml/kg at an inspiratory pressure of 25 cm H_2O. We determined the dosage necessary to effect a tidal volume of 4.5 ml/kg ($ED_{4.5\ ml}$). This volume represents the mean half-maximal tidal volume from several dose-response curves.

Repeated determination of the reference standard # 872005 showed an $ED_{4.5\ ml}$ ranging from 43–99 µmol/kg (33–75 mg/kg) after 2 min of ventilation

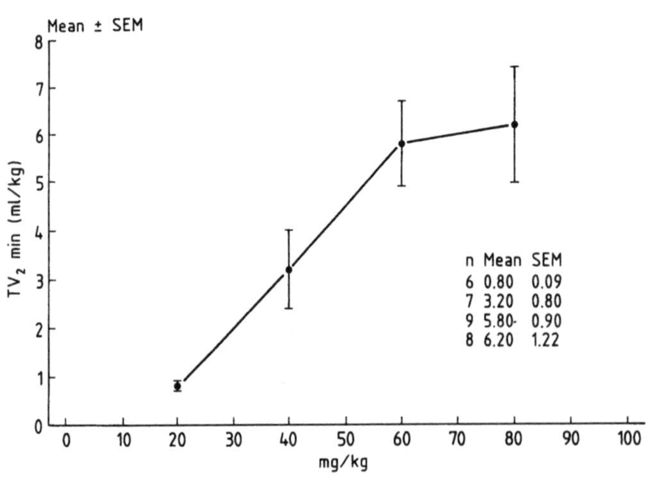

Fig. 1. Dose-response curve for surfactant replacement therapy. Tidal volume of premature rabbit fetuses at 25 cm H_2O inspiratory pressure 2 min after start of ventilation. The animals received one dose of 3.33 ml/kg of surfactant suspension lot no. ZR 1515 of different concentrations before the first breath

(Table 1). In this example the rabbit fetus model demonstrated a variability by a factor of 2.3 although the following parameters were well controlled:

1. The gestational age was defined as ± 2 h.
2. We have never found mature fetuses at 27.0 days in this strain of rabbit.
3. The lungs were ventilated under vision in order to detect and exclude pneumothoraces.
4. Ventilation before surfactant replacement was avoided.

Therefore we determine physiological surfactant activity by comparing the ED's of an unknown sample and of freshly prepared reference standard. The fetuses of the litters in one experiment become equally distributed between standard and sample concentrations. The relative activity is then calculated according to the formula:

Rel.activity [%] = $ED_{4.5\ ml}$ (standard) × 100 / $ED_{4.5\ ml}$ (unknown)

The relative activity of lot # ZR1515 was calculated as 148 % based on 2 min of ventilation or as 134 % based on maximal ventilation.

The procedure described here is a new attempt to more accurately define physiological surfactant activity by excluding the variability of the rabbit fetus model. To our knowledge, to date only Lachmann [4] has tried to reduce the variability of this model by measuring the compliance before surfactant replacement and excluded the more mature fetuses. Although his method is simpler, it requires however, that the fetuses be ventilated before surfactant replacement, and does not offer a control of the remaining variability.

In a study with premature lambs the necessary inspiratory pressure after 120 min of ventilation was significantly more favourable in the SF-RI 1-treated animals (32,8 ± 6,1, n = 12) than in controls (38,5 ± 4,7 cm H_2O, n = 10). The arterial oxygen tension was significantly improved in comparison to the control (Fig. 2).

Table 1. Dose of surfactant necessary to effect a tidal volume of 4.5 ml ($ED_{4.5\ ml}$) in premature rabbit fetuses after 2 min of ventilation or after 10–30 min (maximal volume in this time interval). The $ED_{4.5\ ml}$ was determined graphically from the linear part of a dose-response curve

Batch no.	Study no.	$ED_{4.5\ ml}$2 min (µmol/kg)	$ED_{4.5\ ml}$max (µmol/kg)
872005	1	43	37
872005	2	82	59
872005	3	73	54
872005	4	99	63
ZR1515	4	67	47
Activity compared to reference standard (%)			
872005		148	134

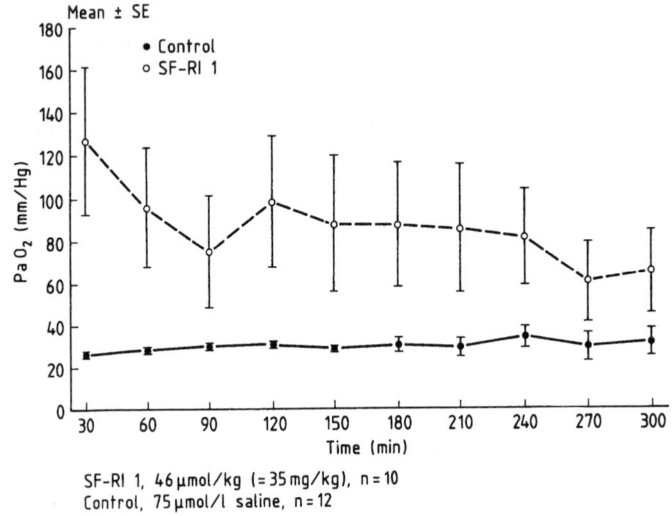

Fig. 2. Arterial oxygen tension of premature ventilated lamb fetuses

The decline of the difference in paO_2 between controls and treated lamb fetuses and the high variability of the latter group suggests that the administered dose of 46 µmol/kg (35 mg/kg) was borderline low.

Conclusions

Repeated intratracheal instillations of exogenous surfactant into premature rabbits and rats have shown that a high cumulative dose may mean a potential risk of phospholipid and macrophage accumulation [11]. Therefore the physiological activity of surfactant for replacement therapy per unit phospholipid should be as high as possible. A reliable and reproducible method of determination is important. The direct comparison to a reference standard in premature fetuses is recommended.

Acknowledgements. We would like to thank Ms. A. Hoffmann and Ms. R. Rupp for excellent technical assistance.

References

1. Tanaka Y, Takei T, Kanazawa Y, Kiuchi A, Fujiwara T (1982) Reconstitution of lung surfactant by adjusting chemical composition. J Jpn Med Soc Biol Interface 13:43–50
2. Taeusch HW, Keough KMW, Williams M, Slavin R, Steele E, Lee AS, Phelps D, Kariel N, Floros J, Avery ME (1986) Characterization of bovine surfactant for infants with respiratory distress syndrome. Pediatrics 77:572–581
3. Nohara K, Berggren P, Curstedt T, Grossmann G, Nilsson R, Robertson B (1986) Correlations between physical and physiological properties of various preparations of lung surfactant. Eur J Respir Dis 69:321–335

4. Lachmann B (1986) New aspects of pathophysiology and therapy of respiratory distress syndrome. In: Cosmi EV, Di Renzo GC (eds) Selected topics in perinatal medicine. CIC, Rome, pp 177–198
5. Ikegami M, Agata Y, Elkady T, Hallman M, Berry D, Jobe A (1987) Comparison of four surfactants: in vitro surface properties and responses of preterm lambs to treatment at birth. Pediatrics 79:38–46
6. Reifenrath R, Zimmermann I (1976) Dynamic surface tension properties of mixed lecithin-cholesterol films related to the respiratory mechanics. Respiration 33:303–314
7. Bangham AD, Morley CJ, Phillips MC (1979) The physical properties of an effective lung surfactant. Biochim Biophys Acta 573:552–556
8. Keough KMW (1984) Physical chemical properties of some mixtures of lipids and their potential for use as exogenous pulmonary surfactants. Prog Respir Res 18:257–262
9. Tanaka Y, Takei T, Aiba T, Masuda K, Kiuchi A, Fujiwara T (1986) Development of synthetic lung surfactants. J Lipid Res 27:475–485
10. Nilsson R (1982) The artificially ventilated preterm rabbit neonate as experimental model of hyaline membrane disease. Acta Anaesthesiol Scand 26:89–103
11. Disse B, Weller E, Lützen L, Ziegler H, Eberhardt H (1987) Comparison between natural and artificial surfactant preparations in premature rabbit fetuses. (This book)
12. Eibl H, Lands WEM (1969) A new, sensitive determination of phosphate. Anal Biochem 30:51–57

Comparison Between Natural and Artificial Surfactant Preparations in Premature Rabbit Fetuses

B. Disse, E. Weller, L. Lützen, H. Ziegler, and H. Eberhardt

The effectiveness of several surfactant preparations for the treatment of hyaline membrane disease in premature newborns has been shown in clinical studies [1–7]. Some of these studies allow for more than one application or show a decline in the beneficial effect of replacement therapy after several hours, suggesting that at least some of the very premature babies require more than one dose of surfactant. The potential risk of substitution with exogenous surfactant may depend on the number of applications and the dose applied. Thus an optimal surfactant for replacement therapy should have high activity at a low phospholipid concentration. Artificial (chemical synthesis, cell culture synthesis) and natural (from biological sources) surfactant preparations have been extensively optimized and qualified by physical methods, such as surface tension/surface area-measurements (e. g. [8–10]). However, a convincing correlation between biochemical or physical parameters and biological surfactant activity could not be established [11, 12].

Therefore we tested the influence of several artificial and natural surfactant preparations on the ventilation of premature rabbit fetuses, which is believed to be a dynamic model of high relevance to the clinical situation of premature infants with hyaline membrane disease [13]. The potential toxicity of exogenous surfactant was tested in a histological study after multiple intratracheal administrations of extracted surfactant lipids to adult rabbits and rats.

Methods

The biological surfactant activity was tested in premature rabbit fetuses by a modification of the method of Nilsson [13]. Briefly, rabbit fetuses at 27.0 days ± 2 h gestational age were delivered by hysterotomy from chinchilla rabbits (term 31 d ± 2 h). The fetuses were relaxed by pancuronium and received one dose of 100 μl surfactant suspension. The animals were ventilated in a constant pressure, temperature-controlled, humidified body-plethysmograph with an inspiratory pressure of 35 cm H_2O in the first min, thereafter with 25 cm H_2O. The tidal volume was measured with a Fleisch tube 0000 2 min and 5–30 min after administration of surfactant. The maximum tidal volume is defined as the maximum value in this time interval.

The toxicity of surfactant was tested by daily application of 35, or 70 mg/kg surfactant to Himalayan rabbits for seven days through an implanted intratracheal catheter. The toxicity was further tested in rats, receiving 40 mg/kg daily for 4 days through an endotracheal tube under brief anesthesia.

Results and Discussion

As our version of the rabbit fetus model uses very well defined, very premature fetuses (27.0 days ± 2 h) only very active surfactants can improve the ventilation in these animals. The dynamic compliance of untreated controls is only 0.028 ml/kg/cm H_2O. A tidal volume of 3 ml/kg or more in treated rabbit fetuses indicates sufficient biological surfactant activity, though lower values may already be statistically significant in comparison to control. Sufficient activity and rapid distribution is indicated by the tidal volume after 2 min ($TV_{2\,min}$). Borderline activity, or less favourable distribution of the surfactant, or the requirement for endogenous factors to contribute may be detected by an isolated improvement in the maximal tidal volume (TV_{max}) without an increase in the 2 min value.

Table 1. Tidal volume (mean ± SD) of artificially ventilated premature rabbit fetuses (gestational age 27 days)

	Dose (mg/kg)	*n*	$TV_{2\,min}$ (ml/kg)	TV_{max} (ml/kg)
Control	0	16	0.71 ± 0.29	1.07 ± 0.52
SF-RI 1	34	9	1.3 ± 1.2	3.9 ± 1.8*
Lot # 160715-7/9	46	7	2.3 ± 2.3*	4.5 ± 2.5*
	58	7	5.3 ± 5.5*	10.0 ± 7.7*
	77	8	7.6 ± 4.2*	11.2 ± 6.2*
	150	10	12.9 ± 3.9*	20.3 ± 3.4*
	250	9	9.0 ± 4.2*	11.0 ± 3.4*
SF-mbl	100	11	0.5 ± 0.4	1.6 ± 1.6
	180	7	1.9 ± 1.6*	3.9 ± 2.0*
SF-LH 20	150	8	1.02 ± 0.36*	1.55 ± 0.46*
SF-LH 20 +p	120	6	2.20 ± 0.66*	3.87 ± 1.87*
DPPC-PG	120	6	0.88 ± 0.27	1.26 ± 0.34
DPPC-PG-p	120	6	2.87 ± 1.47*	4.45 ± 2.65*
DPPC-Chol	1000	9	0.6 ± 0.4	1.0 ± 0.6
DPPC-Chol-DPPG-DPPS	60	10	0.70 ± 0.68	1.06 ± 0.51
DPPC-Chol-DPPG-DPPS	120	21	1.27 ± 1.12	2.19 ± 2.05*
SF-Tylox	75	10	1.08 ± 0.74	2.42 ± 1.42*
SF-Tylox	150	6	0.32 ± 0.26	1.11 ± 1.14
SF-Tylox-Ca	75	5	0.30 ± 0.21	1.41 ± 0.88

The tidal volume (TV) was measured at an inspiratory pressure of 25 cm H_2O 2 min and between 5 and 30 min after administration of surfactant. TV_{max} is defined as the maximum value in this time interval.

* $P < 0.05$ or better in comparison to control.

Abbreviations: DPPC, synthetic dipalmitoylphosphatidylcholine; DPPG, synthetic dipalmitoylphosphatidylglycerol; PG, phosphatidylglycerol from egg yolk, Sigma; DPPE, synthetic dipalmitoylphosphatidylethanolamin; DPPS, synthetic dipalmitoylphosphatidylserin; Chol, cholesterol, Serva.

A surfactant preparation of bovine lung lavage was found to be very active (SF-RI 1, lot # 160715-7/9, Dr. K. Thomae, FRG, Table 1). The lowest dose, 34 mg/kg, had a beneficial influence only on the maximal tidal volume. 46 mg/kg was the threshold dosage for immediate improvement of ventilation ($TV_{2\,min}$). In the dose range of 58–150 mg/kg there was a dose dependent improvement in immediate and maximal ventilation. 250 mg/kg induced a submaximal effect. This shows that for assessment of surfactant quality and determination of biological surfactant activity examination of more than one dose is required. In the following text, the threshold dose, which induces a $TV_{2\,min}$ and/or TV_{max} of ≥ 3 ml/kg, is used for comparison of activities.

Extracted surfactant from minced bovine lungs (SF-mbl), prepared as described by Nilsson et al. [14], required a high dose of 180 mg/kg to gain activity. Extracted bovine surfactant, from which the associated proteins had been removed by Sephadex LH-20 chromatography (SF-LH 20) retained significant, but very low activity at a high dose of 150 mg/kg. When reconstituted with the protein the approximate threshold dose was 120 mg/kg.

An aqueous suspension of the artificial mixture DPPC-PG (3 : 1, weight/weight, Table 1), which was used in similar proportions but in a dry crystalline state by Bangham and Morley [7], was not active in the highest concentration tested here (120 mg/kg). However, after addition of surfactant associated proteins from bovine lungs, the preparation (DPPC-PG-p) revealed considerable biological surfactant activity. The approximate threshold dose was 120 mg/kg. The latter findings confirm the importance of the surfactant associated proteins and are in agreement with recent literature [15, 16].

The artificial binary mixture DPPC-Chol (4 : 1), described by Reifenrath and Zimmerman [8] was not active as surfactant substitute, though tested in a very high dose of 1000 mg/kg. The four component mixture DPPC-Chol-DPPG-DPPS (9 : 1 : 1 : 1) revealed a low but significant activity. The activity at a dose of 120 mg/kg was still below the limit.

An artificial surfactant, composed of DPPC, hexadecanol and the detergent tyloxapol (13.5 : 1.5 : 1, SF-Tylox), which has been reported to be nearly as active as natural surfactant in premature lambs [17], showed in our model only low but significant activity with delayed onset. If applied with 2 mmol/l calcium (SF-Tylox-Ca) or in a higher dose, no activity was found.

The daily intratracheal instillation of 35 or 70 mg/kg extracted bovine surfactant for 7 days into adult rabbits was acutely well tolerated, only rhonchi could be heard for 20 min. Histopathologic examination of the animals revealed intrabronchial and intrapulmonary focal accumulation of lipid laden macrophages, in part with consecutive atelectatic areas. A slight increase in lung weight was observed. There were no signs of inflammatory reactions.

When rats received a daily intratracheal instillation of 40 mg/kg for 4 days, the treatment induced a moderate increase in the number of lipid laden macrophages.

Conclusions

These findings demonstrate a potential risk of surfactant replacement therapy, namely the accumulation of lipids in the lungs. This risk may be negligible if few low dosages of a very active preparation are administered, but may be relevant in case a considerable lipid load is necessary with less active surfactants. Dose-response studies of the ventilation of premature rabbit fetuses are proposed as a means of assessment of biological surfactant activity, which reflects the quality of a preparation.

References

1. Enhorning G, Shennan A, Possmayer F, Dunn M, Chen CP, Milligan J (1985) Prevention of neonatal respiratory distress syndrome by tracheal instillation of surfactant: randomized clinical trial. Pediatrics 76:145–153
2. Kwong MS, Egan EA, Notter RH, Shapiro DL (1985) Double-blind clinical trial of calf lung surfactant extract for the prevention of hyaline membrane disease in extremely premature infants. Pediatrics 76:585–592
3. Shapiro DL, Notter RH, Morin III FC, Deluga KS, Golub LM, Sinkin RA, Weiss KI, Cox C (1985) Double-blind, randomized trial of a calf lung surfactant extract administered at birth to very premature infants for prevention of respiratory distress syndrome. Pediatrics 76:593–599
4. Merrit TA, Hallmann M, Bloom BT et al. (1986) Prophylactic treatment of very premature infants with human surfactant. N Engl J Med 315:785–790
5. Raju TNK, Vidyasagar D, Bhat R, Sobel D, McCulloch KM, Anderson M, Maeta H, Levy PS (1987) Double-blind controlled trial of single-dose treatment with bovine surfactant in severe hyaline membrane disease. Lancet 1:651–656
6. Gitlin JD, Soll RF, Parad RB, Horbar JD, Feldman HA, Lucey JF, Taeusch HW (1987) Randomized controlled trial of exogenous surfactant for the treatment of hyaline membrane disease. Pediatrics 79:31–37
7. Ten centre study group (1987) Ten centre trial of artificial surfactant (artificial lung expanding compound) in very premature babies. Br Med J [Clin Res] 294:991–996
8. Reifenrath R, Zimmermann I (1976) Dynamic surface tension properties of mixed lecithin-cholesterol films related to the respiratory mechanics. Respiration 33:303–314
9. Bangham AD, Morley CJ, Phillips MC (1979) The physical properties of an effective lung surfactant. Biochim Biophys Acta 573:552–556
10. Keough KMW (1984) Physical chemical properties of some mixtures of lipids and their potential for use as exogenous pulmonary surfactants. Prog Respir Res 18:257–262
11. Nohara K, Berggren P, Curstedt T, Grossmann G, Nilsson R, Robertson B (1986) Correlations between physical and physiological properties of various preparations of lung surfactant. Eur J Respir Dis 69:321–335
12. Lachmann B (1986) New aspects of pathophysiology and therapy of respiratory distress syndrome. In: Cosmi EV, Di Renzo GC (eds) Selected topics in perinatal medicine. CIC, Rome, pp 177–198
13. Nilsson R (1982) The artificially ventilated preterm rabbit neonate as experimental model of hyaline membrane disease. Acta Anaesthesiol Scand 26:89–103
14. Nilsson R, Berggren P, Curstedt T, Grossmann G, Renheim G, Robertson B (1985) Surfactant treatment and ventilation by high frequency oscillation in premature newborn rabbits; effect on survival, lung aeration and bronchiolar epithelial lesions. Pediatr. Res 19:143–147
15. Revak SD, Merrit AT, Hallman M, Cochrane CG (1986) Reconstitution of surfactant activity using purified human apoprotein and phospholipids measured in vitro and in vivo. Am Rev Respir Dis 134:1258–1265

16. Suzuki Y, Curstedt T, Grossmann G, Kobayashi T, Nilsson R, Nohara K, Robertson B (1986) The role of the low-molecular weight (< 15000 daltons) apoproteins of pulmonary surfactant. Eur J Respir Dis 69:336–345
17. Durand DJ, Clyman RI, Heymann MA, Clements JA, Mauray F, Kittermann J, Ballard P (1985) Effects of a protein-free, synthetic surfactant on survival and pulmonary function in preterm lambs. J Pediatr 107:775–780

Part II
Diagnostics of Surfactant Deficiency

Respiratory Distress Syndrome: Requirements of Perinatal Diagnosis, Prevention and Treatment

E. V. Cosmi and G. C. Di Renzo

In the last ten years the strategy for the prevention and causal treatment of the respiratory distress syndrome (RDS) of the newborn has been directed towards the acceleration of fetal lung maturity *in utero* by means of drug administered to the mother and to the development of surfactant substitutes for the treatment of surfactant deficiency at birth or following the development of RDS. Data derived from several sources demonstrate that babies at special risk for RDS include: preterm babies, males more than females, caucasian more than blacks, babies of diabetic (not euglycaemic, poorly controlled) mothers, babies with birth asphyxia and babies born by cesarean section.

Prenatal Diagnosis

An important prerequisite for the management of high risk pregnancies is the accurate prediction of fetal lung maturity. Assessment of fetal maturity is often mandatory when there is a concern of preterm delivery, or when the fetus may have an abnormal delay in functional maturation, for instance, unknown gestational age, premature rupture of fetal membranes, pre-eclampsia, placenta praevia, twin pregnancy, isoimmunization or diabetic pregnancy.

The most important property of a maturity test is its predictive accuracy in detecting mature and immature lungs. Among a population of low fetal risk, the predictive value of a "mature" test is most critical. On the other hand, in a high risk population there is an additional requirement for a high predictive value of an "immature" test result. The test result preferably allows an estimate of the time required for maturation, evaluates the degree of immaturity and gives evidence on postmaturity.

Thus, the requirements for fetal maturity tests include: (a) high predictive accuracy of detecting both lung maturity or immaturity; (b) resistance to blood, meconium or vaginal secretions, that frequently contaminate the amniotic fluid (AF) specimens; (c) simple, fast, precise, and reproducible test performance; and (d) low cost.

Amniotic fluid (AF) contains phospholipids (PL), including lecithin (L), sphingomyelin (S), phosphatidylinositol (PI), phosphatidylglycerol (PG) and phosphatidilethanolamine (PE), some enzymes of the pathways of PL synthesis, lamellar bodies, and lung specific apoproteins. The amount of these substances in AF changes towards the end of gestation in a manner related to fetal lung maturity.

Since the introduction of the amniocentesis, a number of tests have been suggested for fetal maturity. They can be divided (Hallman 1984) in:

1. *Non-surfactant tests* (alfa-fetoprotein, bilirubin, creatinine, cortisol, fat cell count)
2. *Indirect surfactant tests* (optical density at 650 nm, fluorescent polarisation, thromboplastic activity)
3. *Biophysical methods* (the shake test, the foam stability index, surface tension measurement, bubble clicking, capillary flow rate)
4. *The concentration of surfactant components* (total phospholipid, lecithin, palmitic acid, phosphatidic acid phosphohydrolase, surfactant phospholipid, phosphatidylglycerol, saturated lecithin, surfactant protein)
5. *The determination of surfactant components versus reference compound* (palmitate/stearate, phosphatidylinositol/phospholipid, saturated lecithin/total lecithin, phosphatidylglycerol/total phospholipid, lecithin/sphingomyelin)

Non-surfactant tests, and non-specific surfactant tests are not accurate, and should therefore not be used. The biophysical tests are often fast (not necessary simple) to perform, but the test result is influenced even by minor contamination of the amniotic fluid (Cosmi and Di Renzo 1983).

Analysis of PL could provide an index of fetal lung maturity and of the risk of RDS. Gluck et al. (1971) were the first to demonstrate that the AF lecithin to sphingomyelin ratio (L/S) is a reliable index of fetal lung maturity. To date, the L/S ratio is still the most accepted clinical test to assess fetal lung maturity. Subsequently, a number of additional indices of lung maturity based on the determination of surfactant constituents in the AF have been proposed. It is generally agreed that an L/S ratio of 2 or greater indicates fetal lung maturity and negligible risk to RDS, whereas a ratio below 1.5 indicates lung immaturity and risk to RDS. With an intermediate ratio of 1.5 to 2.0 most, but not all, babies escape the disease. In practice, however, standards must be established and validated in each laboratory.

Phosphatidylglycerol (PG) is the second most abundant phospholipid in the surfactant system of the mature lung. The surfactant produced by immature lungs, however, contains PI in place of PG, and in several species lung maturation is accompanied by a developmental decline in PI that is associated with an increase of PG (Quirk and Bleasdale 1986). Although many indirect evidences support the view that PG may be essential for proper surfactant function, the biophysical and biochemical properties of PG and its physiological role are still uncertain (Di Renzo et al. 1986).

A number of investigators have found that preterm infants, delivered before the appearance of surfactant that is rich in PG are at increased risk of RDS. Conversely, it has been suggested that when PG is present in lipid extracts obtained from AF, the RDS does not occur even when the L/S ratio is less than 2 (Whittle et al. 1983). In addition, the assessment of amniotic fluid PG seems to provide additional information concerning fetal pulmonary maturation, especially in case of pregnancies complicated by diabetes mellitus.

L/S ratio and PG complement each other in assessment of lung maturity and may be the most accurate method. The "lung profile", i.e., the simultaneous determination of L/S ratio, PG, PI and PE in AF is essentially not a more difficult method than the L/S ratio (Kulovich et al. 1979; Feijen et al. 1982).

Surfactant contains specific proteins. Surfactant apoprotein A (34 kD), together with saturated lecithin and PG form a complex that is able to stabilize the alveoli (King et al. 1975). These lipids alone are rather ineffective. Immunoassays of apoprotein A have been described, and proposed for tests of lung maturity. Regular immunoassay (RIA, ELISA), are so far unsatisfactory, because the antibodies have a very low affinity. In order to improve the sensitivity, Hallman et al. (1987) introduced time resolved fluorometry for apoprotein A. In five cases of RDS AF apoprotein A was lower (12 % of the mean at term), whereas the 29 non-RDS cases had substantial amounts of apoprotein A (20 % of the mean at term). The verification of the predictive accuracy requires further study.

Determination of L/S ratio is by far the most widely used and accepted method. However, there is still controversy regarding the high incidence of false immature and the increased false mature values (from 1 %–15 %) in complicated pregnancies, especially with diabetes mellitus. According to several reports, an immature L/S ratio may predict RDS only in about 50 % of cases (Cosmi and Di Renzo, 1983). Attempts to improve predictability by modifying Gluck's original methods have not been successful.

The incidence of false immature L/S ratio as well as of other AF tests depends upon patient variability, method employed, threshold value taken for differentiating a normal from an abnormal condition, and on the fact that few authors report their results in terms of sensitivity and specificity. Furthermore, interpretation and comparison of results is hampered by differences within the methodology itself. All methods have more false immature than false mature results. In general, the larger the difference between a given value and the critical value, the greater the predictability.

Several factors may explain the discrepancy between neonatal outcome and results for immature values in AF tests, including: (a) mode of delivery; (b) fetal distress and/or chronic fetal hypoxia; (c) small-for-date infants; (d) prolonged rupture of fetal membranes; and (e) preterm delivery and birth asphyxia. Furthermore, abnormal maturation in most cases is only detectable during a critical time period, and not for instance throughout the third trimester.

The well-established diagnostic value of the L/S ratio had led to its widespread use in the obstetric practice. Nonetheless, the analysis of PG is very useful in certain instances, and its complementary value has been well documented.

When PG is present in the AF; it is very unlikely that the newborn baby will develop RDS, even in the presence of an L/S ratio less than 2 and despite certain pregnancy complications, such as diabetes mellitus or perinatal asphyxia. In our series of 104 babies, where clinical correlation between presence of PG and perinatal outcome was considered, none of the babies where PG was present in the amniotic fluid died from RDS, whereas 6 out of 12 babies where PG was absent in AF died from RDS. This opinion is not shared by others, who reported cases of RDS even in the presence of PG in amniotic fluid (Barnes et al. 1984). These findings support the contention that the presence of PG is not totally reliable in excluding the risk of a RDS. Apparently, the answer to the contradictory results lies in the fact that in many series the determination of PG was made using qualitative methods of analysis. Unfortunately, no reliable cut-off values indicative of lung maturity are yet available and this is probably the major short-coming of the PG determination (Di Renzo et al. 1986).

The importance of fetal maturity assessment in planned premature delivery is obvious. However, its indications in unplanned premature labor, or planned term delivery are more controversial. The benefits of knowing fetal maturity should be weighed against possible risks of amniocentesis and possible fetal risks of continuing the pregnancy with immature lungs. The economical issue also needs to be taken into consideration. Unfortunately, these factors have not been sufficiently analyzed.

In conclusion the utilization of the tests for assessing fetal lung maturity is recommended for: (a) timing for delivery prior to elective cesarean section; (b) complicated pregnancies, and (c) institution of pharmacologic prevention of RDS in utero or at delivery.

Prenatal Prevention

A wide spectrum of agents have been found to enhance the synthesis and/or secretion of lung surfactants, the most thoroughly investigated being glucocorticoids.

In the last ten years, studies have proceeded at an increasing pace toward the development of other drugs which when administered antenatally to the mother can accelerate the maturation of the fetal lung, thereby decreasing the risk of neonatal RDS.

Many reports indicate that the administration of corticosteroids to the mother decreases the incidence of RDS in the neonate delivered between 28 and 32 weeks and weighing less than 1,500 g (Hawkins and Di Renzo 1986). A full course of dexamethasone or preferably betamethasone therapy over 4 to 5 days is necessary to obtain the expected result, i.e., acceleration of fetal lung maturity, as judged by accurate fetal lung maturity test in AF. A compilation of many published-controlled double blind studies on the use of betamethasone or desamethasone confirms the efficacy of the drug in the prevention of RDS (Di Renzo et al. 1987). Modification of the glucocorticoid preparation, the dosage, plus time of administration may somewhat improve the results, though the incidence of potential side effects may increase. In addition, there remain a certain number of newborn babies (9 %–10 %) who will develop RDS despite the administration of corticosteroids to the mother.

It should be stressed that these drugs in the small clinical dose used do not cause or exacerbate maternal hypertension, and do not convey a worse prognosis if there is proteinuric hypertension (Crowley 1981). They are not contraindicated in most forms of heart disease, provided the patient is not subsequently subjected to fluid overload and electrolyte imbalance as often induced by the administration of β-sympathomimetic agents to inhibit uterine activity (Crowley 1981). In fact, the combination of β-mimetics and corticosteroids requires special caution as there is a synergistic effect on water retention, cardiac output and capillary permeability. Most cases of pulmonary edema have occurred after this combination therapy. Glucocorticoids are not contraindicated in patients with diabetes mellitus, provided blood sugar levels are closely monitored and controlled. They do not increase the chance of intra-uterine infection, as distinct from bacterial colonisation, in case of preterm rupture of membranes (Hawkins and Di Renzo 1986).

However, in view of the certain maternal and fetal side effects of the antenatal administration of corticosteroids, their use must be restricted to fetuses of gestational age between 28 and 32 weeks, when the delivery is imminent and/or unavoidable and when there are no contraindications. In addition to corticosteroid treatment, cesarean section is often indicated to prevent the asphyxia of birth. For fetuses of less than 28 weeks the success of any treatment is unpredictable in terms of perinatal outcome since the problems of fetal immaturity are related not only to the respiratory but also to other systems.

Besides glucocorticoids, other drugs capable of accelerating fetal lung maturity have been tested and even suggested for clinical use. Among these, ambroxol and aminophylline have been evaluated for the antenatal prevention of RDS.

Ambroxol has been shown to reduce significantly the incidence of RDS compared to placebo without causing important adverse effects either on mother or the baby (Wauer et al. 1982). A recent multi-centre prospective controlled study to compare the effectiveness of antenatal administration of ambroxol or betamethasone in the prevention of RDS has shown that ambroxol is as effective as betamethasone in reducing the incidence of RDS (Luerti et al. 1986).

Our experimental studies on pregnant rabbits have shown that aminophylline exerts only minor beneficial effect on lung maturation and surfactant production, depending on the dosage and the time of administration (Cosmi et al. 1986). One clinical study suggests that aminophylline may accelerate fetal lung maturity; however, its use for the prevention of neonatal RDS in the human needs further studies.

In attempts to minimise the risk of neonatal RDS, it is important to identify pregnant women at increased risk for preterm labor and to inhibit their labor if it is convenient for both the mother and the fetus. If delivery is imminent, intensive intrapartum monitoring is mandatory. Non-traumatic delivery including liberal use of cesarean section will reduce the possibility of birth asphyxia. An experienced neonatologist other than the anesthesiologist should be present in the delivery room well before birth in order to take appropriate care and resuscitation of the newborn. Specific guidelines should be established for the use of any prenatal drug to be administered for prevention of RDS (Quirk, 1986). Requirements are listed in Table 1.

Table 1. Indications for pharmacologic acceleration of fetal lung maturity

Preterm labor
Fetal membranes intact
Gestational age between 28 and 32 weeks
Pregnancy not complicated by infection, hemorrhage, or hypertension
Fetal pulmonary immaturity as judged by amniotic fluid tests
Labor can be inhibited safely for at least 48 h
Male fetus

Postnatal Prevention and Treatment

Supplementary surfactant derived from animal lung, human (amniotic fluid) or artificial surfactant instilled into the trachea at birth or when respiratory failure is present have shown to improve oxygenation of premature babies and reduce the severity of RDS both in the animal and the human baby.

After the first pioneering human trial of Fuijwara et al. (1980) several other trials have been performed with different surfactant preparations adopting various criteria of case entry and evaluation.

Much discussion concerning the use of surfactant replacement therapy has focused on the criteria that should be satisfied for clinical trials (Enhorning 1980). Preparations of surfactant either natural or synthetic should be well defined biochemically and biophysically (Table 2). Ability of surface films

Table 2. Suggested tests of clinically useful surfactants

Biophysical
Surface tension with reduced surface area
Adsorption degree and rate
Spreading rate
Respreading characteristics
Morphology and homogeneity of particles
Biochemical
Lipid analyses
Protein amount and characteristics
Animal testing
Lung morphology
Lung/water permeability
Lung mechanics/PV curves
Arterial blood pH and gas tensions
Distribution, pool locations, and half lives
Toxicity (e.g., membrane lysis)
Interaction with natural surfactants
Foreign protein
Resistance to oxidation and hydrolysis
Stability, storage, sterility characteristics

Table 3. Main complications of RDS

Patent ductus arteriosus
Bronchopulmonary displasia
Extrapulmonary extravasation of air
Intraventricular hemorrhage
Necrotizing enterocolitis

formed without solvent to lower surface tension upon compression on a Wilhelmy balance is a *sine qua non.* Complete protein and lipid analysis should be done to assess batch to batch variability as well as stability over time, under various conditions including temperature. Surfactant preparation should be tested in the animal (rabbit, sheep or monkey) newborn in order to evaluate important endpoints such as lung morphology, permeability, surfactant distribution, pool size, half-life, interaction with inhibitors. In addition lung compliance and volumes, blood gases and lung injury should be evaluated. Toxicity tests are of obvious importance.

Concerning the clinical application of supplementary surfactant several considerations should be made. Having a good surfactant preparation, successfully tested *in vitro* and *in vivo,* a consensus should be made to which candidate the surfactant might be administered. Experimental randomised clinical trials such as the one of Merritt et al. (1986) and the European Multicentre Trial (Robertson et al. 1986; Robertson 1987) have selected some restrictive criteria for the entry, which might be changed when final results will be available and a general consensus to a routine clinical use of surfactant will be obtained.

Moreover, the effects of surfactant therapy can be studied in two ways: short-term physiological studies on the acute effects before and after intervention do not require many babies, depending on the variable measured for judging the outcome. Long-term effect studies, particularly those related to complications and morbidity (Table 3), require larger numbers of babies. The need of multicentre trials is necessary because of the actual low incidence of prematurity in many centres (calculating an average of 2,000 deliveries per year in a centre, the incidence of babies weighing less than 1,200 g will be 40–45 babies per year of whom, on average, 20–22 will develop RDS).

Table 4. Postinatal prophylaxis and treatment of RDS with surfactant replacement therapy

	Prophylaxis	Treatment
Why	Prevention of RDS in very preterm babies with immature AF test	To reduce risks of mortality and morbidity associated with RDS
When	After intubation, under artificial ventilation	Within 2 h from initial symptoms; after intubation
Where	Only in intensive care unit	Only in intensive care unit
How	Endotracheal instillation	Endotracheal instillation

Table 4 summarizes the most important steps for postnatal prophylaxis and treatment of RDS with surfactant replacement. Finally it should be realised that RDS is a much more complex disease with many confounding factors than a simple deficiency of lung surfactant. Therefore, further research is needed to clarify the individual role of many factors responsible for the disease and the results of ineffective or partially effective surfactant replacement therapy.

Acknowledgements. This work was supported by grants from C.N.R., Italy, Target Programme "Preventive and Rehabilitative Medicine" and from the Ministry for Public Instruction (40 %). We are indebted to Ms. Annette Borzim for the secretarial assistance.

References

Barnes DA, Biskinis E, Tupulur I, Toot PJ (1984) Respiratory distress syndrome and presence of phosphatidylglycerol: report of three cases. Am J Obstet Gynecol 148:347–348

Cosmi EV, Di Renzo GC (1983) Diagnosis of fetal lung maturity. In: Cosmi EV, Scarpelli EM (eds) Pulmonary surfactant system. Elsevier, Amsterdam pp 75–94

Cosmi EV, Saitto C, Barbati A et al. (1986) Effect of aminophylline on lung maturation in preterm rabbit fetuses. Am J Obstet Gynecol 154:436–439

Crowley P (1981) Corticosteroids in pregnancy: the benefits outweigh the costs. J Obstet Gynaecol 1:147–149

Di Renzo GC, Marangi M, Anceschi MM, Cosmi EV (1986) The measurement of amniotic fluid phosphatidylglycerol for the assessment of fetal lung maturity. In: Vignali M, Cosmi EV, Luerti M (eds) Diagnosis and treatment of fetal lung immaturity. Masson, Milano pp 64–71

Di Renzo GC, Anceschi MM, Guidetti R, Cosmi EV (1987) Requirements of perinatal prevention and treatment of RDS. Eur J Respir Dis [Suppl] (in press)

Enhorning G (1980) Artificial surfactant to prevent and treat neonatal respiratory distress syndrome. Pediatrics 66:799–800

Feijen HWH, Di Renzo GC, Nederstigt J, Houx PCW, Eskes T KAB (1982) Evaluation of the fetal lung profile including the two dimensional L/S ratio for the establishment of fetal lung maturation. Gynecol Obstet Invest 14:142–150

Fujiwara T, Chida S, Watabe Y et al. (1980) Artificial surfactant therapy in hyaline membrane disease. Lancet 1:55

Gluck L, Kulovich MV, Borer RC, Brenner PH, Anderson GG, Spellacy WM (1971) Diagnosis of the respiratory distress syndrome by amniocentesis. Am J Obstet Gynecol 109:440

Hallman M (1984) Antenatal diagnosis of lung maturity. In: Robertson B, Van Golde LMG, Batenburg JJ (eds) Pulmonary surfactant. Elsevier, Amsterdam, pp 419–439

Hallman M, Arjomaar P, Mizumoto M, Akino T (1987) Surfactant proteins in the diagnosis of fetal lung maturity. I Predictive accuracy of the 35KD protein, the L/S ratio and phosphatidylglycerol. Am J Obstet Gynecol (in press)

Hawkins DF, Di Renzo GC (1986) Preterm premature rupture of membranes. In: Di Renzo GC, Hawkins DF (eds) Perinatal medicine: problems and controversies. Raven, New York, pp 157–162

King RJ, Ruch J, Gikas EG et al. (1975) Appearance of apoproteins of pulmonary surfactant in human amniotic fluid. J Appl Physiol 39:735–741

Kulovich MV, Hallman M, Gluck L (1979) The lung profile in normal pregnancy. Am J Obstet Gynecol 135:57–62

Luerti M, Corbella E, Casolati E et al. (1986) The effect of ambroxol versus corticosteroids in the prevention of respiratory distress syndrome. In: Cosmi EV, Di Renzo GC (eds) Selected topics in perinatal medicine. CIC, Rome, pp 149–164

Merritt TA, Hallman M, Bloom BT et al. (1986) Prophylactic treatment of very premature infants with human surfactant. N Engl J Med 315:785–790

Quirk JG (1986) Risks and benefits of steroid therapy for the prevention of hyaline membrane disease. In: Vignali M, Cosmi EV, Luerti M (eds) Diagnosis and treatment of fetal lung immaturity. Masson, Milan, pp 109–115

Quirk JG, Bleasdale JE (1986) Fetal lung maturation in the pregnancy complicated by diabetes mellitus. In: Di Renzo GC, Hawkins DF (eds) Perinatal medicine: problems and controversies. Raven, New York, pp 109–116

Robertson B (1987) International multicentre trial of surfactant replacement in neonatal respiratory distress syndrome. In: Cosmi EV, Di Renzo GC, Hawkins DF (eds) Advances in perinatology: Gordon and Breach, London

Robertson B, Noack G, Bevilacqua G et al (1986) In: Vignali M, Cosmi EV, Luerti M (eds) Diagnosis and treatment of fetal lung immaturity. Masson, Milan, pp 193–197

Wauer RR, Schmalisch G, Heurel K et al. (1982) The antenatal use of ambroxol (bromexine metabolite VIII) to prevent hyaline membrane diseases. A controlled double blind study. Int J Biol Res Pregnancy 3:84–91

Whittle MJ, Wilson AI, Whitfield CR (1983) Amniotic fluid phosphatidylglycerol: an early indicator of fetal lung maturity. Br J Obstet Gynaecol 90:134–137

Diagnosis of Surfactant Defects in Newborn, Children and Adults in the Era of Surfactant Therapy*

M. Hallman, J. L. Abraham, O. Pitkänen, and P. Arjomaa

Introduction

The risk of the respiratory distress syndrome (RDS) in the newborn is accurately evaluated by studying the lung effluent phospholipids (Cosmi 1987; Hallman 1984; Hallman et al. 1986a). In adult respiratory distress syndrome (ARDS) there is also a severe defect among the bronchoalveolar lavage phospholipids (Hallman et al. 1982). Surfactant defects appear early in respiratory failure, and the surface activity of the lung effluent deteriorates concurrent with the appearance of respiratory failure and correction of the surfactant defect ameliorates the respiratory failure (Lachmann et al. 1987). Although the surfactant system appears to be of central importance in the pathogenesis of respiratory failure, there are inherent limitations in obtaining the lung effluent and to study surfactant function and metabolism *in situ*. It has been claimed that the airway specimen, recovered during partial lung lavage, may not be representative of alveolar fluid. Secondly, there may be no reliable methods to quantify the surfactant pool. In this report it is proposed that surfactant diagnostics are of value when surfactant therapy is contemplated.

Methodology

Recovery of Lung Effluent

Lung effluent may be obtained during routine suctioning of the airways. In most cases the catheter is not wedged, but instead advanced to the main bronchus, and a small quantity of saline (0.5–10 ml depending on the size of the patient) is introduced. Shortly thereafter, the central airways are suctioned, and the small volume airway specimen is collected into a trap.

For bronchoalveolar lavage (BAL) a bronchoscope or a catheter is wedged to a lung segment. Thereafter, five to sixty milliliters of normal saline is injected to lung periphery. Fluid from the airways is then suctioned through the

* Supported by the Finnish Academy, the Sigrid Jusélius Foundation, and the Foundation for Pediatric research in Finland.

bronchoscope and collected to a container. The procedure may be repeated. This technique is similar to the quantitative BAL that involves lavage of the whole lung through the trachea.

Measurement of Surfactant and Its Activity

There are several possibilities for surfactant analysis in lung effluent. In the following brief discussion we point out the various approaches. Some of them may not be accurate. *Post mortem* analysis of surfactant is not discussed here.

Concentration or Amount of Surfactant Components. In emphysema the concentration and the recovery tend to decrease, possibly reflecting a decrease in lavageable alveolar area but not necessarily a deterioration in the surfactant system. We believe that measurements of the concentration or the amount of surfactant are not suitable for accurate diagnosis of the surfactant status (Hallman et al. 1982).

Concentration in Alveolar Epithelial Lining Fluid. The concentration of surfactant components in alveolar epithelial lining fluid may be estimated using an internal standard that readily crosses cellular membranes and therefore has an uniform concentration in extracellular fluids. The volume of epithelial lining fluid is obtained on the basis of the Fick principle by comparing concentration of urea in BAL fluid to that in serum. However, some urea diffuses into BAL fluid during multiple lavage, leading to an overestimate of the volume and underestimate of the concentration of surfactant component (Marcy et al. 1987).

Reference Compounds. The expression of surfactant component or surface activity on the basis of the reference compound is currently the method of choice for diagnostic purposes (Hallman 1984). The possible references include sphingomyelin, total phospholipid, albumin or total protein. The frequently used indexes include the saturated phosphatidylcholine/sphingomyelin ratio (the L/S ratio), phosphatidylglycerol/total phospholipid (PG), and minimum surface activity adjusted to a constant phospholipid concentration.

Fractionation of Lung Effluent. The acellular sample may be processed according to techniques used for subcellular fractionation. The phospholipids may be separated from the protein-rich fraction by centrifugation. The density gradient centrifugation allows the isolation of the membranous fraction, enriched with the surface active complex (Kankaanpää and Hallman 1982; Hallman et al. 1982).

External markers. Specific markers, attached to exogenous surfactant, may be used to study both the turnover of the complex and to estimate the surfactant pool size (Hallman et al. 1986b).

Interpretation of Results

According to animal studies, the L/S ratio in a small volume airway specimen is very similar to the L/S ratio in quantitative BAL. The L/S ratio correlates with the pool size of lavageable surfactant both in primary and secondary surfactant defects (Hallman et al. 1985). In RDS during the 4–10 neonatal hours, the quantitative L/S ratio correlated significantly with the surfactant pool size (Hallman et al. 1986b).

Low PG and low phosphatidylinositol (PI) illustrates lack of surfactant phospholipids (Hallman and Epstein 1982). Lack of PG in itself does not deteriorate alveolar stability, provided that there is sufficient quantity of PI-rich surfactant. Inositol is responsible for the suppression of surfactant PG and for the increase in surfactant PI. Inositol promotes the beneficial effect of glucocorticoid on the lung (Hallman and Bry 1987b). In a randomized, double-blind study, inositol supplementation to immature infants decreased the severity of RDS and increased the L/S ratio (Hallman et al. 1987a). Therefore this sugar, perhaps together with glucocorticoid, may prove to be an effective drug in the treatment of primary and secondary surfactant defects.

Lactosylceramide among the lipids of lung effluent is apparently derived from granulocytes. Upon phagocytic stimulation, neutrophils release lactosylceramide into the extracellular space. This glycolipid (Rauvala and Hallman 1984) as well as other granulocyte lipids disturb surfactant function. These cells additionally release elastase (Merritt et al. 1983), collagenase, phospholipase A_2 (Hallman et al. 1982), that destroy lung parenchyma and break down surfactant. The burst of free oxygen radicals may additionally damage alveolar epithelium (Haagsman and Van Golde 1985; McGuire et al. 1982). Since substantial quantities of lactosylceramide in lung effluent additionally is almost invariably associated with severe respiratory failure, this glycolipid may prove to be an accurate marker of destructive lung inflammation (Rauvala and Hallman 1984; Hallman et al. 1987c).

Respiratory Distress Syndrome

In near-term babies with RDS or persistence of fetal circulation, the "maturation", the surfactant indices, the L/S ratio and PG, takes place within 2–7 days, parallel to the recovery. In small preterm infants of less than 32 weeks of gestation, who recover without chronic lung disease, the L/S ratio and PI increase during the first few days, whereas PG remains low until the conceptional age of about 32 weeks (Hallman et al. 1986b). In infants whose respiratory course is complicated by bronchopulmonary dysplasia, the L/S ratio increases slowly, and lactosylceramide is strikingly elevated during the first neonatal week (Hallman et al. 1987c).

Surfactant Inhibitors

The indices that are based upon analysis of individual surfactant phospholipids do not always explain the large variation in the severity of the respiratory failure and why some infants develop bronchopulmonary dysplasia. In addition, surfactant inhibitors strikingly modify the severity of RDS, or occasionally precipitate RDS in a small preterm infant with "mature" lung surfactant (Kankaanpää and Hallman 1982). The inhibitors may originate from plasma, damaged alveolar epithelium or from inflammatory cells.

The leak of plasma through capillary-alveolar barrier has been studied in preterm animals, undergoing ventilation and oxygen treatment. The bidirectional protein leak is indirectly proportional to the gestational age, and increases after various injuries (mechanical shear forces and hypoxic-hyperoxic injury) (Jobe 1986). In RDS, alveolar epithelial permeability has been studied using ^{99}Tc-DPTA aerosols, and found to be high (Jefferies et al. 1984). Many proteins and lipids inhibit the surface activity. Fibrin monomer is perhaps the most potent surfactant inhibitor (Seeger et al. 1985).

To study the role of surfactant inhibitors in RDS, the cell-free lung effluent may be fractioned into "surfactant", supernatant, and "membraneous non-surfactant" fractions using density gradient centrifugation. According to Ikegami et al. (1984), in RDS a surfactant fraction lowers the minimum surface tension close to 0 mN/m. However, the protein-rich supernatant fraction contains surfactant inhibitors.

Serum-derived surfactant inhibitors are predominant in the soluble fraction of lung effluent. According to our recent study the specific activity of the soluble surfactant inhibitor in a small volume airway specimen was always high during days 1 to 3 in RDS. On the other hand, the surface activity of AS was strictly dependent upon the amount of surfactant in relation to the amount of soluble protein. The ratio between the amount of surfactant phospholipid in the membraneous fraction and the amount of protein in the soluble fraction, the phosphatidylcholine/protein ratio, correlated closely with the severity of RDS (Hallman et al. unpublished).

Adult Respiratory Distress Syndrome

In ARDS there is an increase in the epithelial and endothelial permeability regardless of etiology (Staub 1978). There is additionally a striking abnormality in BAL phospholipids that identifies these patients with high sensitivity and specificity (Petty et al. 1979; Hallman et al. 1982). Increases in serum-derived inhibitory proteins and granulocyte-derived glycolipids in BAL (Rauvala and Hallman 1984) are additional characteristic features that in part explain the abnormally high minimum surface tension. In present report we briefly discuss mechanisms of surfactant system damage and describe new features that may explain the role of the surfactant and alveolar epithelial cell in early pathogenesis of ARDS.

Mechanism of Surfactant System Damage

The mechanisms of the surfactant system damage may involve the synthesis, intracellular transport, secretion and maintenance of surfactant function in epithelial lining fluid, alveolar clearance or catabolism. A number of injuries, including hyperoxia, ozone, free oxygen radical-generating substrates or enzymes, radiotherapy, nitrogen dioxide, paraquat, bleomycin, N-nitroso-N-methylurethane, butylated hydroxytoluene, and other chemicals cause acute alveolar damage (Haagsman and Van Golde 1985).

Analysis of surfactant damage following hyperacute, progressive lung injury revealed a sequence of events that appear similar to other severe lung injuries, regardless of the etiology. Immediately after induction of high permeability edema there was an inhibition of surfactant function by plasma-derived proteins. Shortly thereafter the lavagable surfactant pool decreased and membraneous components entered the alveolar lining causing a decrease in the L/S ratio (Lachmann et al. 1987). These hyperacute disturbances were not caused by abnormal synthesis or secretion of surfactant. Following injection of N-nitrose-N-methylurethane (NMU) a severe lung injury developed within several days.

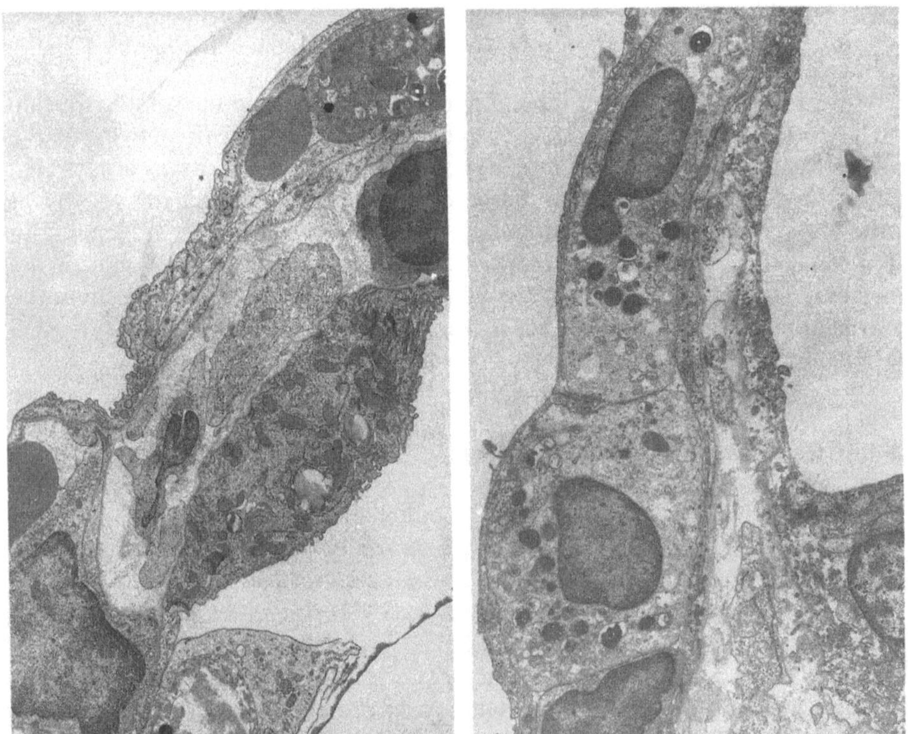

Fig. 1. The alveolar lining of adult rabbits 8 days after N-nitroso-N-methylurethane with no supplemental inositol *(left)* or inosital supplementation *(right)* as described previously (Hallman and Epstein 1982)

There was a gradual decrease in lavagable surface active material, and increase in membraneous lavagable non-surfactant material. Dietary inositol decreased these NMU-induced changes in the surfactant system (Hallman and Epstein 1982). As shown in Fig.1, the lamellar inclusion bodies were decreased in rabbits that were given NMU eight days earlier (7.5 mg/kg body weight). Furthermore, the pinocytic vesicles on the surface of type II cells decreased. Inositol supplementation preserved the morphology of the type II cells, and retarded the decrease in the surfactant pool.

According to a recent proposal, inositol is required for maintenance of the phosphoinositide-dependent signal transmission, synthesis and secretion of surfactant in "immature" or "damaged" type II cells (Hallman and Bry 1987b). The hormones and inositol may therefore prove to be effective in prevention and treatment of surfactant deficiency syndromes.

Appearance of Surfactant Defect in Acute Lung Injury in Adults

The relationship between the damage of the surfactant system and the onset of respiratory failure has not been previously studied in humans. We have recently approached this question by measuring BAL constituents before and after unilateral lung irradiation, given as a therapy of mesothelioma in the pleural space. BAL was obtained from four individuals once a month until there was advanced unilateral radiation pneumonitis. The phospholipids, the 35 kD surfactant protein, surface activity and surfactant inhibitors were analyzed. Following features became evident: 1. Proteinous surfactant inhibitors appeared shortly after the six-week course of radiotherapy. The concentration of both 35 kD surfactant protein and PG decreased during the asymptomatic phase of radiation injury. Finally, there was a decrease in the L/S ratio before development of unilateral pneumonitis (Hallman et al. unpublished).

According to these findings and previous studies of individuals with a lung disease that was less severe than ARDS (Hallman et al. 1982), a defect in the surfactant appears before the development of full-blown respiratory failure. The functional consequences of surfactant defect become evident when the surfactant system is not sufficient to "neutralize" surfactant inhibitors.

Genetic Predisposition to ARDS

Lysinuric protein intolerance (LPI) is a recessively inherited disease involving a defect in intestinal, renal, and hepatic transport of diamino acids. The defect resides at the basolateral membrane of epithelial cells (Desjeux et al. 1980). The massive renal loss of diamino acids causes depletion of the urea cycle intermediates, hyperammonemia and malnutrition. These defects are apparently corrected by dietary supplementation or arginine, ornithine or citrulline.

Out of the 41 Finnish LPI patients three have developed ARDS; two of them died. During ARDS the lung effluent phospholipids characteristically had low L/S ratio and undetectable PG. The possible association between the extrapulmonary epithelial transport defect and ARDS prompted us to study the non-proteinous amino acids in BAL.

Two asymptomatic patients with LPI were studied. They did not have respiratory failure, and the lung function studies were normal. Analysis of BAL revealed a striking increase in percentage distribution of lysine, arginine and ornithine (4–20-fold). The concentrations of these amino acids in epithelial lining fluid were similarly increased, whereas these amino acids in plasma were normal. There was no detectable abnormality in surfactant phospholipids.

The present findings suggest that the alveolar epithelial cells possess an amino acid transport similar to the intestinal and renal tubular epithelium. A defect in the diamino acid transport increases these amino acids in alveolar epithelial lining fluid and in alveolar epithelial cells. A consequence of this single gene mutation is an increased incidence of ARDS. Its pathogenesis remains exlusive (Hallman et al., in press).

Conclusions

Lavage of the airways using a small volume of saline yields phospholipids that are qualitatively strikingly similar to those obtained by quantitative BAL. Therefore, the present technique of the in situ lung lavage yields accurate qualitative information about the surfactant system. Exogenous surfactant substitution additionally allows to study the pool size and the turnover of surfactant (Hallman et al. 1986b).

The major determinants of the surfactant function, evaluated on the basis of a small volume bronchoalveolar lavage, are: (a) the pool size, (b) the quality of the lipid-protein complexes, and (c) the quantity and quality of the surfactant inhibitors. The performance of a surfactant system additionally depends on the area of the air-liquid interphase, the amount of the alveolar lining fluid, the interfascial pressures, degree of curvature of the air spaces, and the tidal volumes.

A serious limitation in postnatal evaluation of the surfactant system is the lack of accurate and fast techniques. Several possible indices have been investigated. We propose that an accurate test, useful in evaluation of the dose-requirement of surfactant substitution is the ratio between the phosphatidylcholine and the total protein in the lung effluent. It is possible to quantitate this ratio within one hour.

Surfactant supplementation will barely remain the sole form of pharmacologic intervention. The intensive care may create hormonal and nutritional deficiencies that affect the surfactant system and that can be detected using a detailed analysis of the lung effluent. Some of the future drugs may be given systemically and may simultaneously improve other critical functions as well. Exogenous surfactant may further serve as a carrier of drugs that need to be evenly distributed along the peripheral airway lining.

References

Cosmi EV (1987) (This book)
Desjeux J-F, Rajantie J, Simell O, Dumontier A-M, Perheentupa J (1980) Lysine fluxes across the jejunal epithelium in lysinuric protein intolerance. J Clin Invest 65:1382–1387
Haagsman HP, Van Golde LMG (1985) Lung surfactant and pulmonary toxicology. Lung 163:275–303
Hallman M, Epstein BL (1982) Myoinositol decreases N-nitroso-N-methylurethane induced lung surfactant deficiency. Life Sci 31:175–180
Hallman M, Spragg R, Harrell JH et al. (1982) Evidence of lung surfactant abnormality in respiratory failure. Study on bronchoalveolar lavage phospholipids, surface activity, phospholipase activity, and plasma myoinositol. J Clin Invest 70:673–683
Hallman M (1984) Antenatal diagnosis of lung maturity. In: Robertson B, Van Golde LMG, Batenburg JJ (eds) Pulmonary surfactant. Elsevier, Amsterdam, pp 419–448
Hallman M, Arjomaa P, Tahvanainen J, Lachmann B, Spragg R (1985) Endobronchial surface active phospholipids in various pulmonary diseases. Eur J Respir Dis [Suppl] 142:67:37–47
Hallman M, Arjomaa P, Romu M, Tahvanainen J (1986a) The lung profile in diagnosis of fetal and neonatal surfactant defects. In: Vignali M, Cosmi EV, Luerti M (eds) Diagnosis and treatment of fetal lung immaturity. Masson, Milano, pp 41–49
Hallman M, Merritt TA, Pohjavuori M, Gluck L (1986b) Effect of surfactant substitution on lung effluent phospholipids in respiratory distress syndrome: evaluation of surfactant phospholipid turnover, pool size, and the relationship to severity of respiratory failure. Pediatr Res 20:1228–1235
Hallman M, Arjomaa P, Hoppu K (1987a) Inositol supplementation in respiratory distress syndrome: relationship between serum concentration, renal excretion, and lung effluent phospholipids. J Pediatr 110:604–610
Hallman M, Bry K (1987b) Phosphoinositide cycle in differentiation of type II alveolar cells. Pediatr Res 21:227
Hallman M, Pitkänen O, Rauvala H, Merritt TA (1987c) Glycolipid accumulation in lung effluent in bronchopulmonary dysplasia. Pediatr Res 21:454
Ikegami M, Jobe A, Jacobs H, Lam R (1984) A protein from the airways of premature lambs that inhibits surfactant function. J Appl Physiol 57:1134–1142
Jefferies AL, Coates G, O'Brodovich H (1984) Pulmonary epithelial permeability in hyaline-membrane disease. N Engl J Med 311:1075–1082
Jobe A (1986) Surfactant and the developing lung. In: Thibeault DW, Gregory GA (eds) Neonatal pulmonary care, 2nd edn. Appelton-Century-Crofts, Norwalk, pp 75–100
Kankaanpää K, Hallman M (1982) Respiratory distress syndrome in very low birth weight infants with occasionally normal surfactant phospholipids. Eur J Pediatr 139:31–34
Lachmann B, Hallman M, Bergmann K-C (1987) Respiratory failure following anti-lung serum: study on mechanisms associated with surfactant system damage. Exp Lung Res 12:163–180
Marcy TW, Merrill WW, Rankin JA, Reynolds HV (1987) Limitations of using urea to quantify epithelial lining fluid recovered by bronchoalveolar lavage. Am Rev Respir Dis 135:1276–1280
McGuire WW, Spragg RG, Cohen AB, Cochrane CG (1982) Studies on the pathogenesis of the ARDS. J Clin Invest 69:543–553
Merritt TA, Cochrane CG, Holcomb K et al. (1983) Elastase and α_1-proteinase inhibitor activity in tracheal aspirates during respiratory distress syndrome. Role of inflammation in the pathogenesis of bronchopulmonary dysplasia. J Clin Invest 72:656–666
Petty TL, Silvers GW, Paul GW, Stanford RE (1979) Abnormalities in lung elastic properties and surfactant function in adult respiratory distress syndrome. Chest 75:571–574
Rauvala H, Hallman M (1984) Glycolipid accumulation in bronchoalveolar space in adult respiratory distress syndrome. J Lipid Res 25:1257–1262
Seeger W, Stöhr G, Wolf HRD, Neuhof H (1985) Alteration of surfactant function due to protein leakage: special interaction with fibrin monomer. J Appl Physiol 58:326–338
Staub NC (1978) Pulmonary edema due to increased microvascular permeability to fluid and protein. Circ Res 43:143–151

Assessment of Surfactant Function by Measurement of the Pulmonary Clearance of Inhaled Tc-99m-DTPA

P. Wollmer, E. Evander, and B. Jonson

Introduction

The transfer of water soluble substances across the alveolo-capillary barrier has been studied mainly in efforts to measure the permeability of the alveolar epithelium. A clinical technique has evolved (Jones et al. 1982) which entails administration of a radiolabeled tracer substance in the form of an aerosol and measurement of the pulmonary clearance of the tracer by external detection of radioactivity over the thorax. Recently, it has become clear that the transfer of water soluble tracer molecules across the alveolo-capillary barrier thus measured reflects not only the permeability of the alveolar epithelium but may also to a great extent reflect the properties of the pulmonary surfactant system. This paper describes the technique for measuring the alveolo-capillary transfer of water soluble tracers and reviews some experimental and clinical information relevant to surfactant function.

Methodological Aspects

Preparation and Administration of Tracer Substance. The most commonly used tracer substance is diethylene-triamine-pentaacetic acid (DTPA), a hydrophilic chelating agent which readily binds Tc-99m. This radioactive isotope is very frequently used in clinical nuclear medicine and can easily be obtained from a radionuclide generator. The labeling of DTPA with Tc-99m can be accomplished with commercially available kits. Quality control is important, since some preparations may contain significant amounts of Tc-99m-O_4^- and Tc-99m-O_2, which have characteristics quite different from Tc-99m-DTPA.

The Tc-99m-DTPA is administered to the patient in the form of an aerosol. The generation of the aerosol requires attention, since the particle size largely determines the site of deposition in the airways and hence the site of transfer to the blood. In normal subjects, inhaled particles with a diameter of less than about 2 μm are mainly deposited in the peripheral airways and alveoli by sedimentation or diffusion (Brain and Valberg 1979). Larger particles are to a great extent deposited in central airways by inertial impaction. The breathing pattern during aerosol inhalation also influences the site of deposition, high flow rates causing increased impaction of particles in large airways. In patients with obstructive airways disease, there is frequently increased deposition of particles by impaction in central airways.

The relation between the site of deposition of Tc-99m-DTPA and the rate of clearance has not been systematically studied. In normal subjects, the clearance rate of Tc-99m-DTPA administered as an aerosol with a mean particle diameter of 6 μm is of the same order of magnitude as if the compound is administered in an aerosol with particles in the sub-micron range (Elwood et al. 1983; Jones et al. 1980). In our own experience (unpublished) the site of deposition of aerosol does not heavily influence the clearance rate except in case of heavy localized deposition in large airways, when the clearance rate tends to be reduced. If the purpose of the measurement is to study the properties of the surfactant system, peripheral aerosol deposition should obviously be aimed at and an aerosol with sub-micron particles used. Several suitable air jet nebulizers are commercially available (Dahlbäck et al. 1986). Ultrasonic nebulizers should be avoided, since the Tc-99m-DTPA complex may be degraded in the nebulizer (Waldman et al. 1987). The administration of Tc-99m-DTPA generally requires a few minutes of tidal breathing of aerosol.

Clearance Measurements. Tc-99m is ideally suited for detection with standard nuclear medicine equipment. A single detector may be used to measure the count rate over part of one lung. Alternatively, the count rate over the entire thorax can be recorded with a gamma camera. A single detector system has the advantage of being mobile and possible to use bed-side. Limitations of the technique are that no information about the initial distribution of activity is obtained and that the clearance curve measured represents only one region of the lung. With a gamma camera connected to a computer, a sequence of images of the thorax is obtained. Changes in count rate with time can then be analyzed in freely selected regions over the lungs. The distribution of the tracer is thus known, and clearance curves can be measured over different lung regions (Fig. 1).

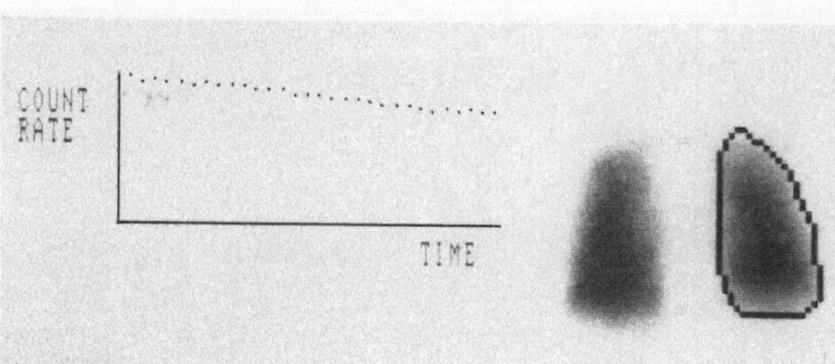

Fig. 1. Gamma camera image *(right)* of the lungs of a normal subject after inhalation of Tc-99m-DTPA aerosol. A series of 30 1-min frames were obtained. A region of interest has been placed over the right lung, and the time-activity curve thus obtained is shown *(left)*

After being absorbed from the lung to the blood, Tc-99m-DTPA is filtered in the kidney and excreted in the urine. In subjects with normal glomerular filtration, the background radiation from recirculating Tc-99m-DTPA in the chest wall and pulmonary blood volume is therefore small. It is possible to correct for extrapulmonary activity by measuring a separate time-activity curve over the thigh (Jones et al. 1980) or outside the lungs in gamma camera studies (O'Doherty et al. 1985). An intravenous injection of Tc-99m-DTPA is performed at the end of the clearance measurement. This results in a sudden increase in count rate over the lung and background region, the relative size of which provides information about the amount of non-pulmonary activity in the two regions. The correction is small and may not always be essential.

The clearance of Tc-99m-DTPA from a normal lung follows a monoexponential course. The clearance measurement is conveniently analyzed by fitting the equation

$$N = M_0 e^{-kt} \tag{1}$$

to the experimental data (Fig. 2), where N is the count rate at any time t and N_0 is the count rate at t = 0. Clearance is thus described by the rate constant k or by the half-time ($t_{1/2}$), i. e. the time at which $N/N_0 = 0.5$. The rate constant is related to the half-time by the equation

$$t_{1/2} = \frac{\ln 2}{k} \tag{2}$$

In normal non-smokers, the $t_{1/2}$ is approximately 70 min. In disease states, pulmonary clearance of Tc-99m-DTPA is sometimes bi-exponential. In these

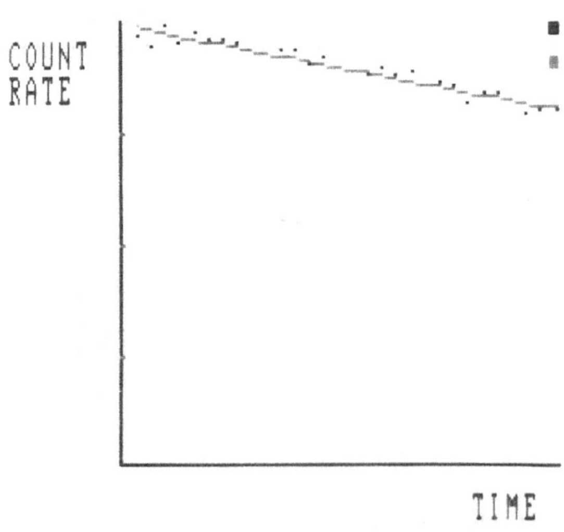

Fig. 2. Exponential function fitted to the data shown in Fig. 1. The activity over the right lung declines with a $t_{1/2}$ of 99 min

cases, two exponential functions can be fitted to the experimental data to accurately describe the clearance. The two components of such curves may be interpreted to represent lung compartments with different disease activity.

Interpretation of Clearance Measurements

When Tc-99m-DTPA is absorbed across the alveolo-capillary barrier, the tracer traverses the surfactant lining of the alveolus with its aqueous hypophase, the alveolar epithelium, the basement membrane and the capillary endothelium. Clearance of Tc-99m-DTPA is also dependent on pulmonary blood flow. Rizk et al. (1984) found no effect on the pulmonary clearance of Tc-99m-DTPA after bronchial artery occlusion. After pulmonary artery occlusion, clearance was reduced, but not absent, on the occluded side, but unaffected on the contralateral side where flow was increased. Coates and O'Brodovich (1986) showed that lymphatic drainage accounts for a very small fraction of the clearance. The pulmonary clearance of Tc-99m-DTPA thus seems to be primarily dependent on the transfer of the tracer from the alveolus to the blood.

The rate of pulmonary clearance of Tc-99m-DTPA is often claimed to reflect the permeability of the alveolar epithelium to water soluble molecules. In disease states, pulmonary clearance of Tc-99m-DTPA often increases dramatically. Thus, Jones et al. (1982) reported a mean $t_{1/2}$ of 17 min in a group of healthy smokers compared to 59 min in a group of non-smokers. Although changes in alveolar epithelial permeability are certain to affect the clearance of Tc-99m-DTPA, we found it difficult to envisage such large differences in clearance rates reflecting different permeability of the alveolar epithelium only, without this being reflected in clinical characteristics.

The surfactant system is often neglected in discussions about the clearance of Tc-99m-DTPA. The surfactant layer consists mainly of phospholipids with the hydrophobic part of the molecule oriented towards the air space. According to the classical theory, an aqueous sub-phase separates the hydrophilic part of the molecule from the alveolar epithelium (Goerke and Clements 1986). Hills and Bryan-Brown (1983) have recently challenged this opinion by suggesting that surfactant is absorbed directly onto the epithelium. According to this model, the small amounts of fluid present in the alveoli would be confined to "pools" in the alveolar corners. When a minute droplet of water containing dissolved Tc-99m-DTPA is introduced into an alveolus it thus encounters a hydrophobic surface. The rate of penetration of the tracer through the surfactant layer may well be dependent on the properties of the surfactant and could conceivably be very different in health and disease. The theory of Hills and Bryan-Brown is especially attractive, since in this case, the rate of transfer of Tc-99m-DTPA may reflect the available diffusion area, i. e. the size of the "corner pools", or a transformation from a normal dry state to a pathological wet state.

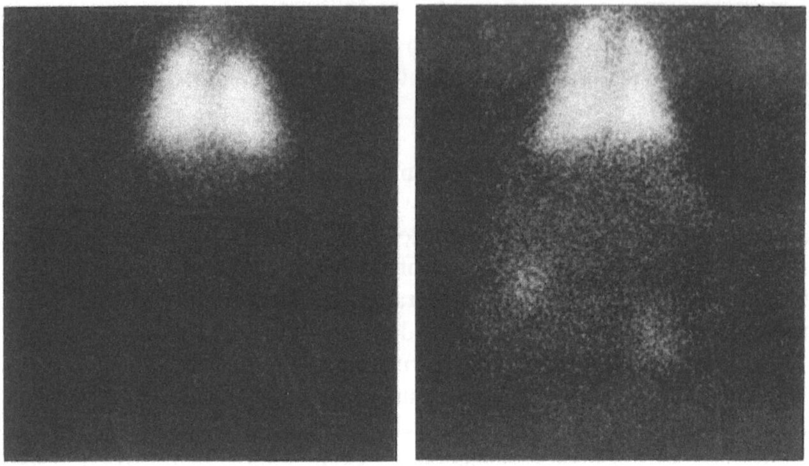

Fig. 3. Gamma camera images of rabbits obtained immediately after administration of Tc-99m-DTPA aerosol. In the control animal *(left)*, there is uniform activity in the lungs, and no tissue background is seen. In the animal studied after lung lavage *(right)*, there is also uniform distribution of activity in the lungs, but in this case, radioactivity is also observed in the kidneys, indicating increased rate of clearance of Tc-99m-DTPA from the lungs. (Reproduced, with permission, from the Journal of Applied Physiology)

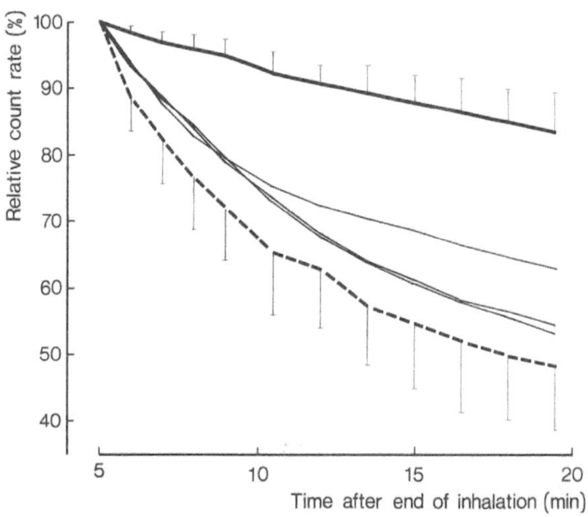

Fig. 4. Time-activity curves obtained from the right lung in a group of six control rabbits *(heavy continuous line)* and a group of six rabbits studied after lung lavage *(interrupted line)*. Three rabbits, which were studied after lung lavage and surfactant replacement, are each indicated by *thin lines*. Values are means ± SD. (Reproduced, with permission, from the Journal of Applied Physiology)

Pulmonary Clearance of Tc-99m-DTPA in States of Surfactant Dysfunction

Surfactant Depletion by Lung Lavage. We have measured the pulmonary clearance of Tc-99m-DTPA in two animal models of surfactant dysfunction. In the first study (Evander et al. 1987), repeated lung lavage was performed to remove surfactant. If isotonic saline is instilled into the airways under a moderate pressure, surfactant can be removed efficiently (Lachmann et al. 1980). A state of respiratory distress ensues, characterized by severly impaired gas exchange and reduced compliance of the respiratory system. Special modes of artificial ventilation are required to avoid structural damage to the lung (Lachmann et al. 1982).

When an aerosol containing Tc-99m-DTPA was administered to rabbits after lung lavage, activity appeared in the kidneys within the first few minutes, before any tissue background activity was seen in control animals not subjected to lavage (Evander et al. 1987; Fig. 3). Time-activity curves over the lungs clearly demonstrated a faster clearance of Tc-99m-DTPA in lavaged animals than in controls (Fig. 4).

In three animals, we studied the effect of surfactant replacement after lung lavage. A bolus of natural surfactant was instilled via the tracheal tube and the pulmonary clearance of Tc-99m-DTPA subsequently measured. There are many reasons why this treatment cannot be expected to completely reverse the effects of lung lavage, e. g. the biological effect of surfactant being adversely affected by the extraction procedure and the crude mode of administration by tracheal instillation. Nevertheless, we observed a clear tendency for the clearance of Tc-99m-DTPA to be slower in treated animals studied than in animals after lavage only (Fig. 4).

Surfactant Dysfunction Induced by Detergent. In an effort to study more subtle derangement of surfactant function, we measured clearance of inhaled Tc-99m-DTPA after administration of a synthetic detergent, dioctyl sodium sulfosuccinate, in the form of an aerosol (Jonson et al. 1986; Evander et al. 1988). This detergent has been shown to interfere with pulmonary surfactant function in vitro (Nieman and Bredenberg 1985). When administered in aerosol form, the detergent is thought to displace normal surfactant from the alveolar surface. Nieman and Bredenberg (1985) treated dogs for 30–45 min with an aerosol produced from a 1 % solution of dioctyl sodium sulfosuccinate in a vehicle of ethanol and saline. Immediately after the administration period, they observed reduced lung compliance and increased venous admixture. The surface tension of lung extracts became abnormal and pulmonary oedema developed 2 hours after the end of the administration period. The authors did not observe any structural abnormalities of the alveolar walls, and have subsequently demonstrated (Bredenberg et al. 1986) that the detergent does not increase pulmonary microvascular permeability to proteins.

In our studies with the dioctyl sodium sulfosuccinate aerosol, we wanted to measure the effects of an acute challenge to the surfactant system. We therefore used a higher concentration of detergent (5 % solution) and shorter administra-

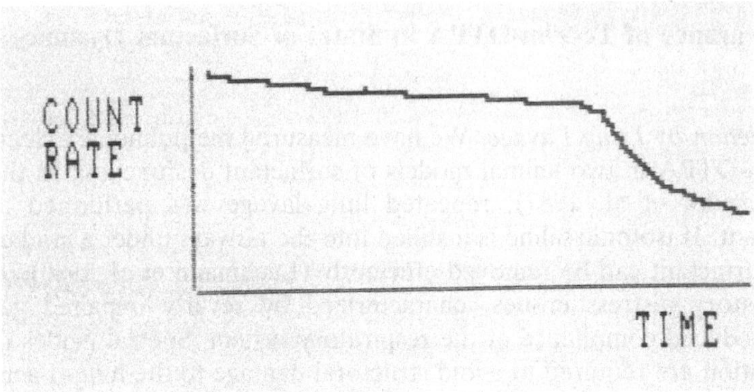

Fig. 5. Time-activity curve obtained from the right lung of a rabbit after administration of Tc-99m-DTPA aerosol. The initial part of the curve shows the normal clearance rate. After 20 min, administration of an aerosol containing dictyl sodium sulfosuccinate was started. This immediately resulted in dramatically increased clearance of Tc-99m-DTPA from the lung

tion periods (maximum 5 min). Figure 5 shows an experiment in which a Tc-99m-DTPA aerosol was administered to a rabbit and the radioactivity monitored over the lungs. The count rate slowly declines to reach 85 % of the initial value after 20 min. At this time, administration of detergent aerosol is started, resulting in an immediate, rapid reduction in count rate over the lungs, corresponding to greatly increased clearance of Tc-99m-DTPA. Administration of the detergent aerosol (obtained from a 5 % solution of dioctyl sodium sulfosuccinate) for 5 min invariably results in dramatically increased rate of clearance of Tc-99m-DTPA (Evander et al. 1988). Up to 30 min after the administration of detergent aerosol, we could not detect any systematic changes in compliance of the respiratory system or venous admixture.

The prompt effect of dioctyl sodium sulfosuccinate on the pulmonary clearance of Tc-99m-DTPA in the absence of any detectable changes in lung compliance and gas exchange suggests that the clearance measurement is a very sensitive indicator of the functional integrity of the surfactant system. It also shows that the functional disturbance induced by dioctyl sodium sulfosuccinate is not immediately associated with a substantial increase in surface tension.

Infantile Respiratory Distress Syndrome. The infantile respiratory distress syndrome (IRDS) is known to result from surfactant deficiency. Jefferies et al. (1984) measured pulmonary clearance of Tc-99m-DTPA during the first days of life in infants suffering from IRDS. They found biphasic clearance curves with extremely rapid initial clearance rate ($t_{1/2} = 1.6 \pm 0.6$ min) in all infants with IRDS studied during the first 72 hours of life. During recovery, most infants showed monophasic clearance curves with much longer $t_{1/2}$. In the original article, the authors interpreted their findings in terms of alveolar epithelial permeability, but in a subsequent review article (O'Brodovich and Coates 1987) they seem to favour surfactant dysfunction.

Effects of Smoking on Pulmonary Clearance of Tc-99m-DTPA and the Surfactant System

There have been no studies comparing the pulmonary clearance of Tc-99m-DTPA with more traditional methods to assess surfactant function in disease states. One of the most striking findings in clinical studies of the pulmonary clearance of Tc-99m-DTPA has been the observation of dramatically increased clearance rate in otherwise normal smokers. In this section, we review the reports of effects of smoking on clearance of Tc-99m-DTPA and on the surfactant system.

Smoking and Tc-99m-DTPA Clearance. The first systematic study of the pulmonary clearance of Tc-99m-DTPA in smokers was published by Jones et al. (1980), and their finding has subsequently been confirmed by other groups (Huchon et al. 1984, Kennedy et al. 1984).

In an experimental study in rabbits Witten et al. (1985) demonstrated increased pulmonary clearance of Tc-99m-DTPA after acute exposure to cigarette smoke. No ultrastructural evidence of cellular injury or widening of intercellular junctions were found. In humans, pulmonary clearance of Tc-99m-DTPA increased after smoking 10 cigarettes daily for three days (Minty et al. 1984). The effects of cigarette smoke on clearance of Tc-99m-DTPA are caused by the particulate matter in the smoke, the gas phase alone having no effect (Minty and Royston 1985). Nicotine administered in a chewing-gum has no effect on clearance of Tc-99m-DTPA (Minty et al. 1984).

No clear cut exposure-response relation has been reported in smokers. Jones et al. (1983) and Minty et al. (1984) suggest a hyperbolic relation between $t_{1/2}$ and the concentration of carboxyhaemoglobine in the blood, whereas Witten et al. (1985) found no exposure-response relationship.

In subjects who give up smoking, the $t_{1/2}$ for pulmonary clearance increases already the first day after cessation of smoking (Minty et al. 1981). Further improvement is seen during the first week, after which $t_{1/2}$ appears to stabilize at a slightly sub-normal level.

Smoking and the Surfactant System. The features of the surfactant system studied in relation to cigarette smoke exposure include surfactant synthesis and composition, the amount of surfactant recoverable by lung lavage and the surface active properties of lung washings and extracts. There appears to be little effect of cigarette smoke exposure on the synthesis of surfactant, as judged by measurements of the incorporation of radiolabeled palmitic acid (Giammona et al. 1971; Hamosh et al. 1979). There is also relatively little difference in composition of surfactant recovered by bronchoalveolar lavage in non-smokers and smokers. Low et al. (1978) reported a reduced ratio of lipid phosphorous to protein in smokers, but Pré et al. (1980) found no difference in the ratio between phosphatidyl choline and protein in non-smokers and smokers.

Finley and Ladman (1972) demonstrated a substantially reduced yield of surfactant in the cell-free fraction of broncho-alveolar lavage fluid of smokers.

There was a seven-fold difference in the amount of phosphatidyl choline recovered in the supernatant. Lavage was repeated after cessation of smoking, in one subject after two weeks, in one subject after one month and in a third subject after five months. In all three cases, the yield of surfactant increased.

In an experimental study, Le Mesurier et al. (1981) exposed rats to cigarette smoke and showed the yield of surfactant to be reduced already after five days of exposure for one hour daily. The authors also found the exposure-response relation to be very complex. Reduced yield of surfactant always preceded ultrastructural changes, which were not seen prior to 25 days of exposure to cigarette smoke.

Several studies have shown that cigarette smoke affects the surface active properties of surfactant (Miller and Bondurant 1962; Giammona 1967; Webb et al. 1967). The changes in surface active properties can be attributed to the particulate matter in the smoke (Webb et al. 1967). Abnormal surface tension of lavage fluid has also been recorded in clinical studies of cigarette smokers (Cook and Webb 1966).

The effects of cigarette smoke on the pulmonary clearance of Tc-99m-DTPA and surfactant function thus show some interesting parallel features. In both cases, the functional abnormalities occur after a few days of exposure and precede any ultrastructural changes in the alveoli. In both cases, the functional effects can be attributed to the particulate matter in the smoke, and the exposure-response relationship in complex. Finally, the functional abnormalities are in both cases reversible with a time course of days to weeks. These similarities suggest that adverse effects of cigarette smoke on surfactant function could be the basis for increased pulmonary clearance of Tc-99m-DTPA. Clearly, further studies have to be made to verify this hypothesis.

Conclusion

Measurement of the pulmonary clearance of inhaled Tc-99m-DTPA is a simple and non-invasive procedure, which can be performed bedside with standard nuclear medicine equipment. The increase in the rate of clearance observed in lung diseases is almost certainly multi-factorial. A growing body of direct and indirect evidence suggests that increased pulmonary clearance of Tc-99m-DTPA may be related to abnormal surfactant function. Further studies are required to clarify the precise relation between pulmonary clearance of Tc-99m-DTPA and surfactant function. Possible future applications of clearance measurements include studies of the role of the surfactant system in the pathogenesis of lung diseases, assessment of artificial surfactant preparations and evaluation of surfactant replacement therapy.

Acknowledgements. This work was supported by the National Association Against Chest and Heart Diseases and by Swedish Medical Research Council Grants 02872 and 04732.

References

Brain JD, Valberg PA (1979) Deposition of aerosol in the respiratory tract. Am Rev Respir Dis 120:1325–1373

Bredenberg CE, Nieman GF, Paskanik AM, Hart AKE (1986) Microvascular membrane permeability in high surface tension pulmonary edema. J Appl Physiol 60:253–259

Coates G, O'Brodovich HM (1986) The contribution of lymphatic drainage to the clearance of inhaled ^{99m}Tc-DTPA from the lungs. Clin Invest Med 9:15–20

Cook WD, Webb WR (1966) Surfactant in chronic smokers. Ann Thorac Surg 2:327–333

Dahlbäck M, Nerbrink O, Arborelius M, Hansson HC (1986) Output characteristics from three medical nebulizers. J Aerosol Sci 17:563–564

Elwood RK, Kennedy S, Belzberg A, Hogg JC, Paré PD (1983) Respiratory mucosal permeability in asthma. Am Rev Respir Dis 128:523–527

Evander E, Wollmer P, Jonson B, Lachmann B (1987) Pulmonary clearance of inhaled ^{99m}Tc-DTPA: effects of surfactant depletion by lung lavage. J Appl Physiol 62:1611–1614

Evander E, Wollmer P, Jonson B (1988) Pulmonary clearance of inhaled ^{99m}Tc-DTPA: effect of the detergent dioctyl sodium sulfosuccinate in aerosol. Clin Physiol 8:105–111

Finley TN, Ladman AJ (1972) Low yield of pulmonary surfactant in cigarette smokers. N Engl J Med 286:223–227

Giammona ST (1967) Effects of cigarette smoke and plant smoke on pulmonary surfactant. Am Rev Respir Dis 96:539

Giammona ST, Tocci P, Webb WR (1971) Effects of cigarette smoke on incorporation of radioisotopically labeled palmitic acid into pulmonary surfactant and on surface activity of canine lung extracts. Am Rev Respir Dis 104:358–367

Goerke J, Clements JA (1986) Alveolar surface tension and lung surfactant. In: Macklem PT, Mead J (eds) The Respiratory System. American Physiological Society, Bethesda, pp 247–261 (Handbook of Physiology, vol 3)

Hamosh M, Shechter Y, Hamosh P (1979) Effect of tobacco smoke on the metabolism of rat lung. Arch Environ Health 34:17–23

Hills BA, Bryan-Brown CW (1983) Role of surfactant in the lung and other organs. Crit Care Med 11:951–956

Huchon GJ, Russell JA, Barritault LG, Lipavsky A, Murray JF (1984) Chronic air-flow limitation does not increase respiratory epithelial permeability assessed by aerosolized solute, but smoking does. Am Rev Respir Dis 130:457–460

Jefferies AL, Coates G, O'Brodovich H (1984) Pulmonary epithelial permeability in hyaline-membrane disease. N Engl J Med 311:1075–1080

Jones JG, Minty BD, Lawler P, Hulands G, Crawley JCW, Veall N (1980) Increased alveolar epithelial permeability in cigarette smokers. Lancet 1:66–68

Jones JG, Minty BD, Royston D (1982) The physiology of leaky lungs. Br J Anaesth 54:705–721

Jones JG, Minty BD, Royston D, Royston JP (1983) Carboxyhaemoglobine and pulmonary epithelial permeability in man. Thorax 38:129–133

Jonson B, Evander E, Lachmann B, Wollmer P (1986) Pathophysiological considerations on special modes of ventilation in severe respiratory distress syndrome. In: Gasparetto A (ed) International resuscitation days. Excerpta Medica, Amsterdam, pp 145–152

Kennedy SM, Elwood RK, Wiggs BJR, Paré PD, Hogg JC (1984) Increased airway mucosal permeability of smokers. Am Rev Respir Dis 129:143–148

Lachmann B, Robertson B, Vogel J (1980) In vivo lung lavage as an experimental model of the respiratory distress syndrome. Acta Anesthesiol Scand 24:231–246

Lachmann B, Jonson B, Lindroth M, Robertson B (1982) Modes of artificial ventilation in severe respiratory distress syndrome. Crit Care Med 10:724–732

Le Mesurier SM, Stewart BW, Lykke AW (1981) Injury to type-2 pneumocytes in rats exposed to cigarette smoke. Environ Res 24:207–217

Low RB, Davis GS, Giancola MS (1978) Biochemical analyses of bronchoalveolar lavage fluids of healthy human volunteer smokers and non-smokers. Am Rev Respir Dis 118:863–875

Miller D, Bondurant S (1962) Effects of cigarette smoke on the surface characteristics of lung extracts. Am Rev Respir Dis 85:692–696
Minty BD, Jordan C, Jones JG (1981) Rapid improvement in abnormal pulmonary epithelial permeability after stopping cigarettes. Br Med J 282:1183–1186
Minty BD, Royston D, Jones JG, Hulands GH (1984) The effect of nicotine on pulmonary epithelial permeability in man. Chest 86:72–74
Minty BD, Royston D (1985) Cigarette smoke induced changes in rat pulmonary clearance of ^{99m}Tc-DTPA. Am Rev Respir Dis 132:1170–1173
Nieman GF, Bredenberg CE (1985) High surface tension pulmonary edema induced by detergent aerosol. J Appl Physiol 58:129–136
O'Brodovich H, Coates G (1987) Pulmonary clearance of ^{99m}Tc-DTPA: a noninvasive assessment of epithelial integrity. Lung 165:1–16
O'Doherty MJ, Page CJ, Croft DW, Bateman NT (1985) Lung ^{99m}-Tc-DTPA transfer: a method for background correction. Nucl Med Commun 6:209–215
Pré J, Baladier D, Battesti JP (1980) An improved method for the determination of the lecithin content of human broncho-alveolar lavages: values of smokers and non-smokers. IRCS Med Sci 8:225
Rizk NW, Luce JM, Hoeffel JM, Price DC, Murray JF (1984) Site of deposition and factors affecting clearance of aerosolized solute from canine lungs. J Appl Physiol 56:723–729
Waldman DL, Weber DA, Oberdörster G, Drago SR, Utell MJ, Hyde RW, Morrow PE (1987) Chemical breakdown of Technetium-99m DTPA during nebulization. J Nucl Med 28:378–382
Webb WR, Cook WA, Lanius JW, Shaw RR (1967) Cigarette smoke and surfactant. Am Rev Respir Dis 95:244–247
Witten ML, Lemen RJ, Quan SF, Sobonya RE, Roseberry H, Stevenson JL, Clayton J (1985) Acute cigarette smoke exposure increases alveolar permeability in rabbits. Am Rev Respir Dis 132:321–325

Pneumotachographic Measurements of Lung Function in Preterm Infants with Respiratory Distress Syndrome*

E. Kattner, K. Kosack, and M. Obladen

Introduction

Measurements of Lung Mechanics

In newborn infants requiring artificial ventilation, the ventilator therapy usually is controlled and directed to obtain normal blood gases in the infant. Methods to ascertain lung function include the use of body plethysmography and measurement of esophageal pressure [2, 11, 25, 30], flow measurements by hot-wire anemometry [12], and pneumotachographic measurements [5, 11, 15, 22, 24, 26, 27]. Few data concerning compliance and resistance of the lung in small infants measured by pneumotachography are available. Due to variations in this technique there are often problems in comparing results. The aim of our study was to apply this objective method on term and preterm infants suffering from different lung problems particularly from respiratory distress syndrome and to gain data during the course of disease. Later this technique was also used to study lung mechanics in infants after surfactant replacement therapy.

Lung function in ventilated newborn infants

The first step of the study was to gain information about lung function parameters (i. e. tidal volume, dynamic compliance) in ventilated infants with different pulmonary problems. Normal values should be established, and changes in compliance during the course of disease were documented.

Patients

We studied 28 newborns and preterm infants with various lung diseases (Table 1) which were classified according to Hjalmarson [10]: Pulmonary maladaptation (i.e. wet lung), pulmonary insufficiency of prematurity and respiratory distress syndrome, grade I to IV diagnosed by X-ray [8]. All infants were intubated with an uncuffed endotracheal tube and ventilated with a

* Supported by Deutsche Forschungsgemeinschaft, Ob 43/5-1.

Table 1. Clinical data and measurement of expiratory tidal volume per kilogram bodyweight (V_T spec) and compliance of the respiratory system (C_{RS}) in ventilated infants with different types of respiratory insufficiency (Hjalmarson 1974) on the 1st or 2nd day of life

Type of respiratory problem	*n*	Birthweight[a] (g)	Gestational age[a] (weeks)	Sex (male/female)	V_T spec[a] (ml/kg)	C_{RS}[a] (ml/cm H_2O/kg)
No lung pathology	4					
Term infants		2923 (2500–3180)	38 (37–40)	2/2	5.0 (4.0–6.8)	0.74 (0.41–0.99)
Preterm		1055 (770–1500)	29 (27–35)	1/3	4.9 (4.4–5.6)	0.61 (0.44–0.72)
PMA	5	2144 (1250–3160)	35 (30–38)	5/0	5.8 (4.4–6.9)	0.37 (0.26–0.46)
PIP	2	660–770	26	0/2	4.5–7.1	0.21–0.28
RDS I, II	4	1535 (850–1999)	31 (26–35)	1/3	5.1 (4.3–7.1)	0.38 (0.27–0.46)
RDS III, IV						
Survivors	4	1382 (1170–1550)	30 (29–31)	2/2	5.5 (4.8–6.4)	0.26 (0.19–0.32)
Non-survivors	5	1672 (1090–2390)	30 (27–35)	5/0	5.7 (4.5–6.8)	0.16 (0.14–0.18)

PMA, pulmonary maladaptation; PIP, pulmonary insufficiency of prematurity; RDS, respiratory distress syndrome.

[a] Mean (range).

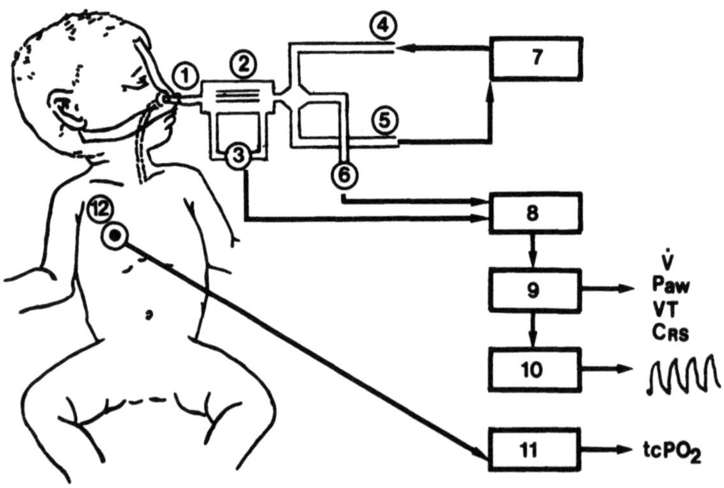

Fig. 1. Equipment and set-up for pneumotachographic measurements of lung mechanics in ventilated newborn infants. *1,* endotracheal tube; *2, 3,* pneumotachograph and transducer; *4, 5, 7,* ventilator and tubes; *6,* pressure transducer; *8,* preamplifier; *9,* computer (Buxo model 6); *10,* recorder; *11, 12,* transcutaneous oxygen sensor and monitor; $\dot{V}$, gas flow; P_{aw}, airway pressure; *VT,* tidal volume, $tcP0_2$, transcutaneous partial pressure of oxygen. For details see text

pressure limited time-cycled ventilator (Bourns BP 200, Bear 2001, humidifier VH 820, Bear Medical Systems, Inc., Riverside, CA, USA; heated tubes). The infants were not paralyzed and were weaned to intermittent positive pressure ventilation according to blood gas analysis. The usual initial ventilator settings were: rate 50–60 per minute, inspiratory pressure 20–25 cm H_2O, positive endexpiratory pressure 2–3 cm H_2O, oxygen concentration 50–100 %. All infants were studied continuously by pneumotachography throughout the course of disease. During the first week of life data were obtained at least every second day. The last measurement was done before extubation or shortly before death.

Methods

Pneumotachography

Measurement of lung mechanics was carried out in the set-up shown in Fig. 1. Flow of respiratory gases was measured by means of a heated pneumotachograph (A. Fleisch 00, Metabo, Epaepalinges, Switzerland) inserted in series to the endotracheal tube. For amplification of inspiratory and expiratory flow it was connected to a differential pressure transducer (Validyne MP 45, Hugo Sachs Elektronik, Hugstetten, FRG). Airflow was integrated electronically to tidal volume.

Airway pressure changes (Δ p) were measured proximal of the pneumotachograph using a Validyne transducer DP 45–24 (Hugo Sachs Elektronik, Hugstetten, FRG). Simultaneous registration of tidal volume, flow and airway pressure was performed with a four channel strip chart recorder (Schwarzer US 266, Cambridge Instruments GmbH, München, FRG).

Calibration of the volume: After flow zero adjustment preamplifier was calibrated by injection of two defined volumes of air with a syringe through the pneumotachograph. For pressure calibration a water manometer was used (O. H. Gauer, Hugo Sachs Elektronik, Hugstetten, FRG) set to 8 cm H_2O across the pressure transducer.

Parameters of Lung Mechanics

Tidal volume (V_T) was measured by integrating gas flow. *Specific* tidal volume was defined as tidal volume per kilogram bodyweight (V_t spec).

Dynamic compliance of the total respiratory system (C_{RS}) was defined as the ratio of expiratory tidal volume and the airway pressure difference at zero flow (Δ p) measured at points of no flow at peak inspiration and peak expiration of continuous respiratory cycles:

$$C_{RS} = V_T / (\Delta p) \text{ (ml/cm } H_2O)$$

Each value for Δ p and V_T represents the mean of 10 single values taken from continuous breathing cycles when the infant was quiet, laminar flow and clear inspiratory plateau of pressure could be obtained and spontaneous breathing did not interfere with ventilator action. In some infants leakage was present and was detected by a difference between in- and expiratory volumes (i. e. leaks around the endotracheal tube, at the humidifier, or in case of pneumothorax) (Fig. 2). If the leak exceeded 20 % of the inspiratory volume or the flow was not laminar, the measurement was not evaluated.

The compliance of the respiratory system (C_{RS}) was normalised by dividing the C_{RS} by kilogram bodyweight [10, 15].

Reproducibility

To test reproducibility of the method, two infants were measured six times during one day. They were suffering from severe RDS and were in stable condition.

The tidal volume was determined ten times during a short period of measurement (20 min), where each value was the mean of ten single breaths.

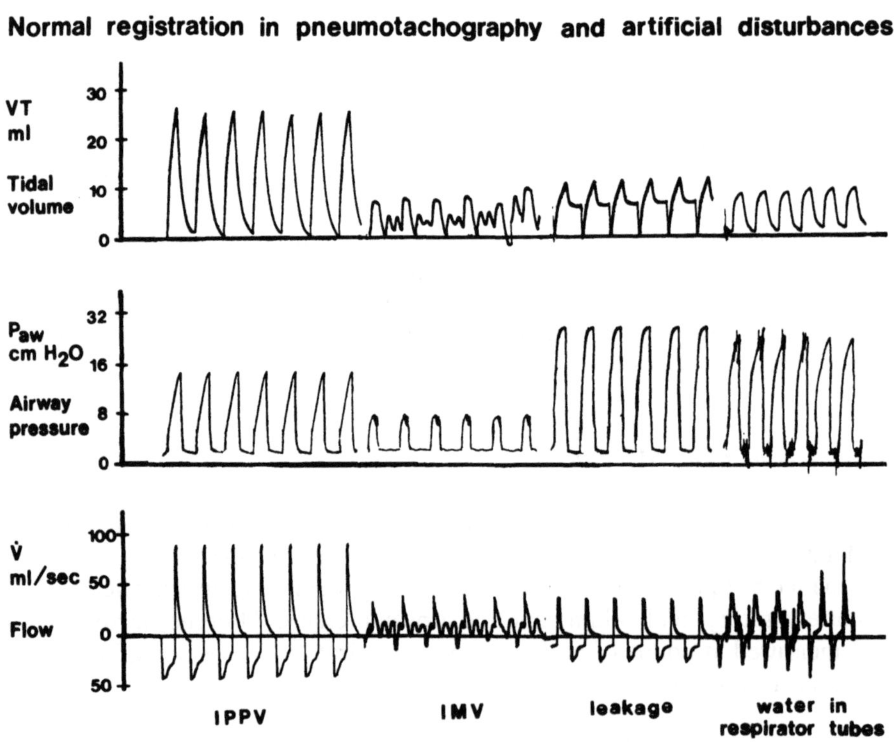

Fig. 2. Typical recording obtained by pneumotachography during intermittent positive pressure ventilation *(IPPV)*, intermittent mandatory ventilation *(IMV)*, air leak, and with water in ventilator tubing. For other abbreviations see legend to Fig. 1

Statistics

To show the correlation between tidal volume and bodyweight a linear regression analysis was performed.

Differences between specific tidal volumes in sick infants at the beginning of the disease and immediately before extubation were tested by Student's t-test. The homogenity of variances was proven.

In repeated measurements of tidal volume in one infant mean and standard deviation were calculated. Differences between means were tested by variance analysis. Intervals of confidence were calculated.

Results

Limits of the Method

Measurement of lung mechanics by pneumotachography showed no signs of irritation even in very small infants. No decrease in transcutaneously monitored partial oxygen pressure or increase in partial carbon dioxide pressure was documented. Dead space added to the breathing circuit was 2.3 ml.

The measurements are disturbed by leakage or water in ventilatory tubes because of non-laminarflow. Measurements in infants fighting the ventilator were not evaluated, as no regular breathing curve could be obtained. In most cases it was possible to investigate these infants at the same day, for example after feeding.

Reproducibility

Repeated measurements of specific tidal volume (n = 10) in two infants resulted in 2.38 ± 0.16 ml/kg and 3.16 ± 0.12 ml/kg (mean ± SD). The interval of confidence was determined with [2.0; 2.8] and [2.86; 3.46] respectively. Variance analysis showed no significant differences between repeated measurements.

Repeated measurements in two infants for the compliance (normalized per kg bodyweight) resulted in 0.15 ± 0.02 and 0.22 ± 0.005 ml/cm H_2O/kg respectively.

Tidal Volume

Figure 3 shows the correlation of expiratory tidal volume and birthweight. The correlation coefficient was determined with r = 0.90. The same result was obtained when tidal volume at time of extubation was used (r = 0.88). The mean specific tidal volume during the first two days of life and at the time of extubation was determined with 5.4 ± 1 ml/kg and 5.8 ± 1 ml/kg respectively.

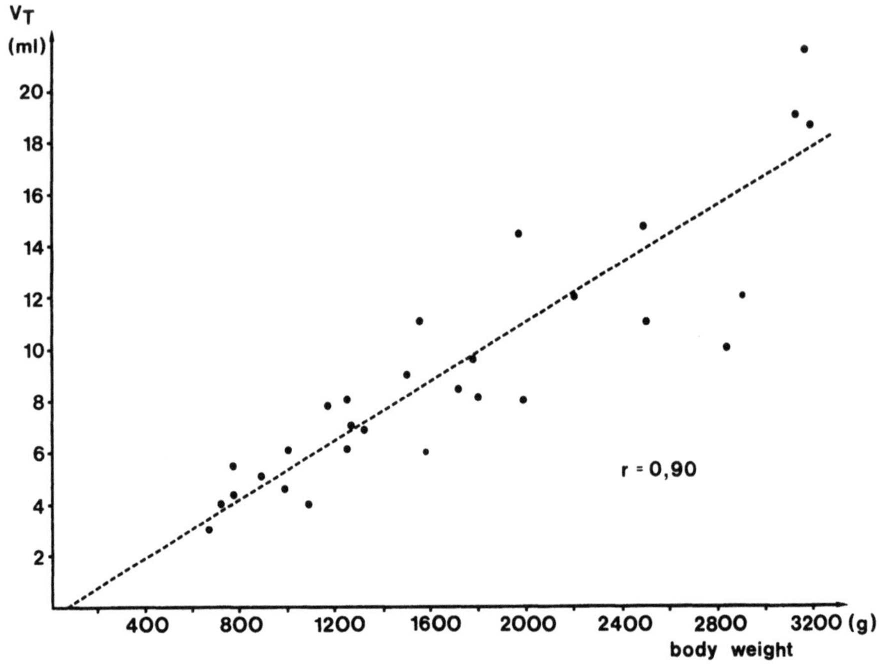

Fig. 3. Correlation of expiratory tidal volume *(V_T)* and birthweight. Measurements on 1st or 2nd day of life in infants with different respiratory problems

No statistical differences were found ($p = 0.25$). Homogenity of variances could be proven ($F = 1.38$).

The *specific* tidal volume in infants with different lung problems is rather constant (Table 1), although no statistics could be applied because of the small number of cases. The values ranged between 5.1 and 5.8 ml/kg bodyweight.

Dynamic Compliance of Respiratory System

Infants with Different Lung Problems. The values of the dynamic compliance of the respiratory system obtained on the first two days of life are shown in Table 1. Normal values in intubated infants were determined with 0.74 ml/cm H_2O/kg for term and 0.61 ml/cm H_2O/kg for preterm infants. In contrast to the values of tidal volume those of the compliance show differences between various lung diseases. The lowest compliance was found in infants who suffered from severe RDS, who developed bronchopulmonary dysplasia and died. Infants who recovered had higher compliances in the beginning.

Although pulmonary maladaptation (PMA) is a transitory problem, a clear reduction in compliance was seen. All these infants could be extubated at least at the third day of life. The two very tiny infants suffering from pulmonary insufficiency of prematurity (PIP) had low values, similar to those observed in RDS. One of these infants died, one recovered well.

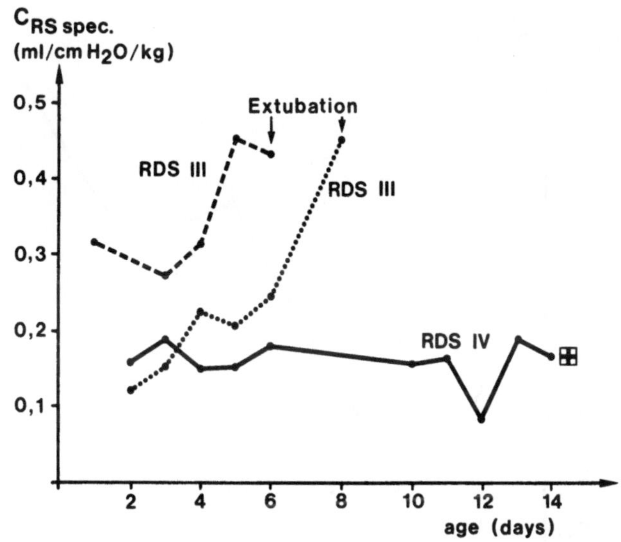

Fig. 4. Examples of course of RDS in preterm infants. C_{RS} dynamic compliance of the respiratory system. Patient data: ■, 1170 g birthweight, 30 weeks of gestation, RDS III (●---●); ■, 1540 g birthweight, 29 weeks of gestation, RDS III (●·····●); ■, 1300 g birthweight, 28 weeks of gestation, RDS IV (●—●)

Infants with Respiratory Distress Syndrome. The course of RDS was studied in thirteen infants, the results are shown in Fig. 5. The infants with RDS I or II were extubated within the first week of life, those who recovered from RDS III or IV within two weeks. Lower degrees of RDS were correlated with higher values of compliance throughout the whole course of disease. It was noticed that the compliance in those infants decreased around the third day and increased quickly thereafter.

In severe RDS leading to death nearly no changes of compliance could be observed (Figs. 4, 5). The values in those infants who recovered rose around

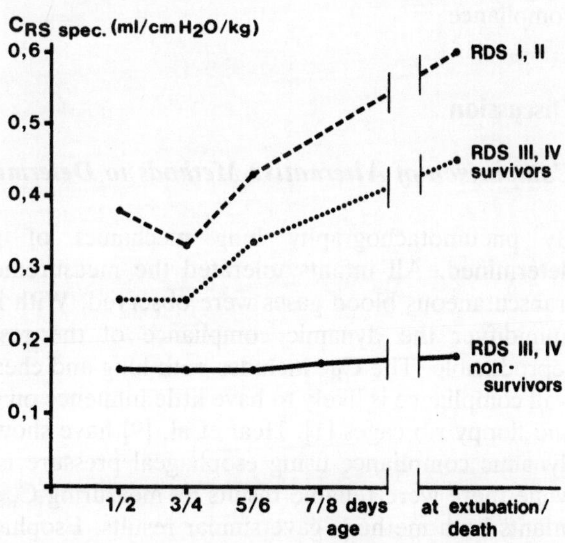

Fig. 5. Dynamic compliance (C_{RS}) of the respiratory system in infants with different degrees and outcome of RDS during the course of disease: infants with RDS I and II ($n = 4$; ●---●); infants with RDS III and IV who recovered ($n = 4$; ●····●); infants with RDS III and IV who died ($n = 5$; ●—●). Last measurements at the day of extubation or shortly before death

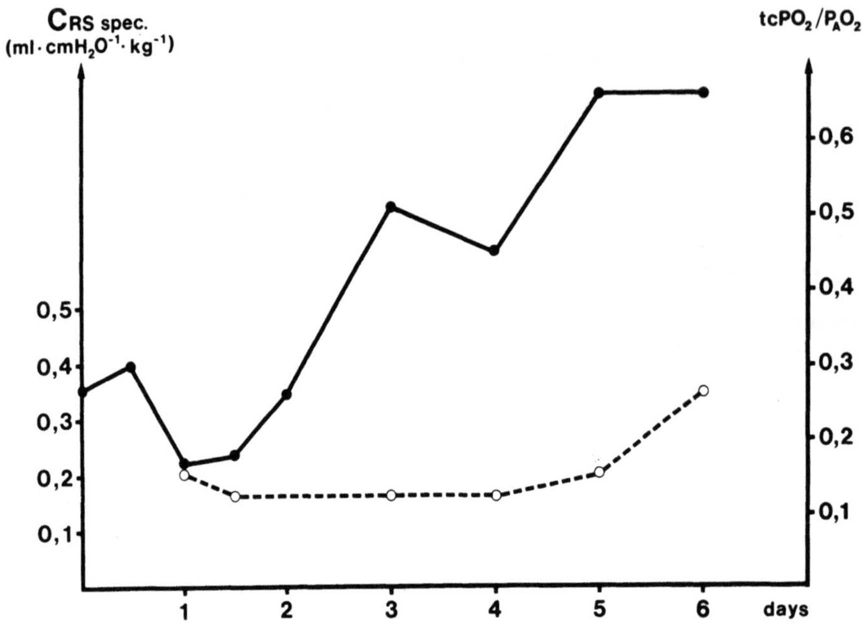

Fig. 6. Compliance and oxygenation in a low-birthweight infant with RDS III. C_{RS}, dynamic compliance of the respiratory system (●—●); $tcPO_2/P_AO_2$, transcutaneous/alveolar oxygen tension ratio (○---○)

the fourth day. Usually, the rise of compliance preceded the time when ventilator therapy could be reduced. Figure 6 shows the much earlier increase in transcutaneous/alveolar oxygen tension ratio than in compliance in an infant with RDS III. Obviously, extubation is possible when compliance is still lower than normal. In some cases the infant had to be reintubated a few days later because of apnoea or pneumonia, only in the latter case with reduced compliance.

Discussion

Comparison of Alternative Methods to Determine Lung Function

By pneumotachography lung mechanics of preterm infants could be determined. All infants tolerated the measurement well and no changes in transcutaneous blood gases were observed. With identical tubing, heating and humidifyer the dynamic compliance of the respiratory system (C_{RS}) was reproducible. The C_{RS} includes both lung and chest wall compliance, but chest wall compliance is likely to have little influence on C_{RS} in infants with stiff lungs and floppy rib cages [1]. Heaf et al. [9] have shown that measurements of the dynamic compliance using esophageal pressure is not reliable in sick infants while there were reliable results by measuring C_{RS} in those infants. In healthy infants both methods gave similar results. Esophageal pressure measurements by balloon technique was often used to assess lung function in newborns, but rib cage distortion and maldistribution of transpulmonary pressure may make this technique invalid [7, 14, 29]. Noninvasive flow measurements have also been proposed with hot-wire-anemometry [13].

Correlation of Data for Lung Function to Body Weight

As shown, there is a good correlation of tidal volume and bodyweight. Dividing results for tidal volume and compliance through bodyweight allows to compare results in infants of different bodyweight. This may lead to problems in hypotrophic or hypertrophic babies. Correlation to body length as proposed for measurements in the body plethysmography will give less exact results because it is difficult to measure exact body length in newborn infants.

Tidal Volume and Dynamic Compliance of Newborns

The tidal volume per kg bodyweight was 5.8 ml/kg in infants immediately before extubation and was not significantly lower in the very sick infants. Similar results are found in healthy, spontaneously breathing newborns by plethysmographic methods [6, 23]. In spontaneously breathing infants with respiratory disease these authors found a reduced tidal volume of 3 to 1 ml/kg. Due to increased breathing rate however, minute volume remained in the normal range [10, 23].

Inspite of markedly reduced compliance, tidal volume is normal in ventilated infants due to high inspiratory pressure during artificial ventilation.

Normal values for dynamic compliance in healthy newborns are rare and differ considerably [3, 4, 10, 11, 16, 17, 19, 21, 24]. Most of the values derived from body plethysmographic measurements are higher than our results. Polgar and Lacourt [20] found the static compliance of the respiratory system to be 1.94–0.70 ml/cm H_2O/kg. Olinsky et al. [18] determined the dynamic compliance of the respiratory system in healthy spontaneously breathing infants with 1.56–0.77 ml/cm H_2O/kg. Our own results for infants without lung pathology are in the lower range of these data. This may be explained by reports that dynamic compliance is lower with rising respiratory rates and rising ventilator rates and increased ratio of inspiration to expiration time [12].

The C_{RS} in infants with severe lung disease was low, similar to those reported by Thomson et al. [28, 29]: 0.3–1.2 ml/cm H_2O, increasing with recovery. Those infants who required prolonged mechanical ventilation had very low C_{RS} and this seems to indicate poor prognosis as suggested by Cunningham and Desai [5] and Le Souef et al. [15]. Cunningham and Desai [5] found C_{RS} of about 1 ml/cm H_2O in infants less than 1000 g with modest ventilator settings. The authors reviewed the results of several investigators who determined the compliance by pneumotachography in very low birthweight infants. The values ranged from 0.21 up to 1.6 ml/cm H_2O/kg. Our lowest compliance was measured in infants with lethal RDS. We think that the measurement around the fourth to fifth day may have a prognostic value and rise in those infants who recover. The decrease in C_{RS} on the second day observed in some infants with less severe disease might be caused by alteration in the immature surfactant system secondary to ventilator therapy.

Advantages of the Method

We found pneumotachography a promising non-invasive technique to investigate lung mechanics even in very sick and very low birthweight infants. No irritation of the infants could be observed during measurement. This method may help to classify severity of disease and to individualize ventilator setting. Monitoring pulmonary function in artificial ventilated infants will lead to more knowledge and understanding about lung mechanics in various diseases, during recovery or relapse. We think this method is a good technique to measure lung function under exogenous surfactant therapy and to control effectiveness of the treatment.

References

1. Avery ME, Cook CD (1961) Volume-pressure relationships of the lungs in fetal and adult goats. J Appl Physiol 16:1034–1038
2. Beardsmore CS, Helms P, Stocks J, Hatch DJ, Silvermann M (1980) Improved esophageal balloon technique for use in infants. J Appl Physiol 49(4):735–742
3. Cook CD, Sutherland JM, Segal S, Cherry RB, Mead J, McIllroy MB, Smith CA (1957) Studies of respiratory physiology in the newborn infant: III measurements of the mechanics of respiration. J Clin Invest 36:440
4. Cook CD, Helliesen PJ, Agathon S (1958) Relation between mechanics of respiration, lung size and body size from birth to young adulthood. J Appl Physiol 13:349–352
5. Cunningham MD, Desai NS (1986) Methods of assessment and findings regarding pulmonary function in infants less than 1000 grams. Clin Perinatol 13(2):229–313
6. Doershuk CF, Fisher BJ, Matthews LW (1975) Pulmonary physiology of the young child. In: Scarpelli EM (ed) Pulmonary physiology of the fetus, newborn and child. Lea and Febiger, Philadelphia
7. Gerhardt T, Bancalari E (1980) Chestwall compliance in full-term and premature infants. Acta Paediatr Scand 69:359–364
8. Giedion A, Haefliger A, Dangel P (1973) Acute pulmonary X-ray changes in hyaline membrane disease treated with artificial ventilation and positive endexpiratory pressure. Pediatr Radiol 1:145
9. Heaf DP, Turner H, Stock J, Helms P (1986) Lung compliance measurements in convalescent and sick intubated infants. In: Wolf P (ed) Neonatal physiological measurements. Butterworths, London, pp 249–255
10. Hjalmarson O (1974) Epidemiology and classification of acute neonatal respiratory disorders. Acta Paediatr Scand 70:773–783
11. Lachmann B (1981) Neonatal pulmonary mechanics during spontaneous ventilation. In: Scarpelli EM, Cosmi EV (eds) Reviews in perinatal medicine, vol 4. Raven, New York: 381–419
12. Lachmann B, Grossmann G, Freyse J, Robertson B (1981) Lung-thorax compliance in the artificially ventilated premature rabbit neonate in relation to variation in I:E ratio. Pediatr Res 15:833–838
13. Leidig E, Noller F, Mentzel H (1986) Nichtinvasive kontinuierliche Überwachung der künstlichen Beatmung Früh- und Neugeborener durch fortlaufende Messung von

Atemminutenvolumen, Sauerstoffaufnahme und Kohlendioxidabgabe. Klin Padiatr 198:321–325

14. Le Soeuf PN, Lopes JM, England SJ et al. (1983) Influence of chest wall distortion on esophageal pressure. J Appl Physiol 55:353–364
15. Le Soeuf PN, England SJ, Bryan Ch (1984) Passive respiratory mechanics in newborns and children. Am Rev Respir Dis 129:552–556
16. Mortola JP, Fisher JT, Smith B (1982) Dynamics of breathing in infants. J Appl Physiol 52:1209
17. Nightingale DA, Richards CC (1964) Volume-pressure relations of the respiratory system of curarized infants. Anesthesiology 26(6):710–714
18. Olinsky A, Bryan AC, Bryan MH (1976) A simple method of measuring total respiratory system compliance in newborn infants. S Afr Med J 50:128–130
19. Polgar G (1961) Airway resistance in the newborn infant. J Pediatr 59(6):915–921
20. Polgar J, Lacourt G (1972) A method for measuring respiratory mechanics in small newborn (premature) infants. J Appl Physiol 32(4):555–559
21. Richards CC, Bachmann L (1961) Lung and chestwall compliance of apneic paralysed infants. J Clin Invest 40:273
22. Rüttimann UE, Galioto FM jr, Franke JR, Rivera O (1981) Measurement of tracheal airflow in newborns by a differential flow system. Crit Care Med 9(11):801–804
23. Sandberg K, Dagbjartson A, Hjalmarson O, Olsson T, Sjoquist BA (1982) Distribution of ventilation in healthy newborn infants and infants with transient tachypnea.
24. Simbruner G, Coradello H, Lubec G, Pollak A, Salzer H (1982) Respiratory compliance of newborns after birth and its prognostic value for the course and outcome of respiratory disease. Respiration 43:414–423
25. Stocks J, Levy NM, Godfrey S (1977) A new apparatus for the accurate measurement of airway-resistance in infancy. J Appl Physiol: 43(1):155–159
26. Swyer PR, Reiman RC, Wright JJ (1960) Ventilation and ventilatory mechanics in the newborn. J Pediatr 56:612–622
27. Thomson A, Silverman M (1985) Single-breath measurement of lung mechanics in VLBW-infants. Crit Care Med 13(1):4–8
28. Thomson A, Silverman M. Single breath lung mechanics during mechanical ventilation of very low birthweight infants. In: Wolf P (ed) Neonatal physiological measurements. Butterworths, London, pp 268–274
29. Thomson A, Elliott J, Silverman M (1983) Pulmonary compliance in sick low birthweight infants. How reliable is the measurement of oesophageal pressure? Arch Dis Child 58:891–896
30. Warren RA, Alderson SH (1985) Calibration of computer-assisted (respicomp) respiratory inductive plethysmography in newborns. Am Rev Respir Dis 131:564–567

Part III
Surfactant Replacement in Newborn Infants

Factors Affecting the Response to a Postnatal Single Dose of a Reconstituted Bovine Surfactant (Surfactant TA)

T. Fujiwara, M. Konishi, S. Chida, and H. Maeta

Exogenous surfactant preparations of various types have now been used in treatment of established RDS [1–4] and to prevent development of RDS [5–9]. However, the beneficial results have not been consistent [10]. In addition, the reasons for the variation in response and the identity of the non-responsive nature in some infants have not been clarified.

This report describes our method of treatment and observations regarding the responses to surfactant TA of 96 infants treated at our nursery.

Patients and Surfactant Administration

Ninety-six severe RDS infants treated with a single dose (120 mg/kg) of surfactant TA are the subjects of this study. Birth weights ranged from 520 to 2400 g (25 patients < 1000 g; 44, 1000–1499 g; 19, 1500–1999 g; 8, 2000–2400 g); 63 were male, 33 female; 47 were outborn and 49 inborn. RDS was diagnosed by clinical and radiological criteria, and confirmed by a microbubble stability test of the gastric juice obtained at birth for the evaluation of surfactant deficiency. Excluded from this report are those infants with congenital heart diseases, neurologic disorders and primary infection.

All infants were ventilated according to the guidelines of Reynolds [11] with a time-cycled, pressure-limited respirator. The respirator was initially set at a peak inspiratory pressure of 20–30 cm H_2O, a positive end-expiratory pressure of 4 cm H_2O, and a respiratory rate of 20–30 cpm with an inspiratory time of 1.0 s; no muscle relaxants were used. Subsequent alterations in the respirator settings were based on blood gas analysis (PaO_2 60–80 torr and $PaCO_2 < 50$ torr).

Following blood gas stabilization, infants who still required more than 7 cm H_2O mean airway pressure and 40 % oxygen were enrolled.

Surfactant TA (Tokyo Tanabe C, Japan; 120 mg of lyophilized surfactant lipid in a vial) was suspended in 4 ml saline by sonication to give a lipid concentration of 30 mg/ml.

This surfactant has been characterized previously by Fujiwara [12], Tanaka et al. [13], and Taeusch et al. [14], and testing of multiple batches in an animal model has shown it to be very effective.

Surfactant suspension was warmed to 37° C and passed through a 26 gauge needle immediately prior to administration. In 5 divided doses, 4 ml/kg of this suspension was instilled into the lungs through an endotracheal tube via a catheter that was passed to the tip of the endotracheal tube. To promote uniform distribution and to avoid hypoxia, repeated instillation of surfactant, 0.5–1.0 ml each, was followed by manual bagging with 100 % oxygen while changing the patient's position. The respirator pressures were maintained at 4 cm H_2O higher than the previous settings, while the respirator rate was kept at 60 cpm with a 1 : 1 inspiratory/expiratory ratio to ensure that transcutaneous PO_2 ($tcPO_2$) was not less than 60 torr. Upon completion of the instillation, which took about 5 min, the patient was promptly returned to the respirator with the settings used prior to treatment. 15 min post-treatment, subsequent adjustments were made first for FiO_2, and then for peak inspiratory pressure and rate to maintain the normal blood gas values. The mean age at which surfactant was given was 6 h.

Evaluation of Response

All patients were evaluated at set intervals using treatment age as "time zero" during the first 120 h after treatment in order to determine the initial response and its duration. Assessment consisted of physical examination including vital signs, arterial blood gases, arterial blood pressures, respirator settings, mean airway pressure (MAP, Pneumogard), FiO_2, and chest radiograms.

Each patient was also evaluated for the presence of a patent ductus arteriosus (PDA) by clinical, radiographic, and echocardiographic criteria. Doppler- and contrast echocardiographies were also used for evaluating the presence of tricuspid regurgitation, and right-to-left shunting through fetal pathways.

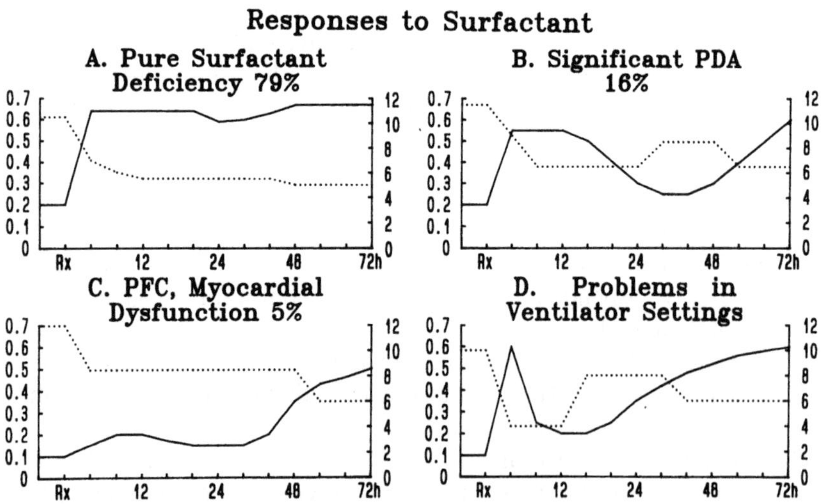

Fig. 1. Response patterns in terms of changes in a/APO_2 *(continuous line)* and MAP *(dotted line)* of RDS infants following surfactant treatment. *Left axis* is a/APO_2 ratio, *right axis* MAP (cm H_2O). For details see text and Tables 1, 2

Responses to Surfactant

The magnitude of initial response was evaluated by the composite changes in arterial-alveolar PO_2 ratio (a/APO_2) and mean airway pressure (MAP) values. The a/APO_2 is an index of oxygenation or physiologic intra- and extrapulmonary shunt, and the MAP is an index of non-compliant lung disease.

Based on the magnitude of initial response to a single dose of surfactant TA and its duration, the response patterns were classified into 4 categories as follows: (a) immediate and sustained response; (b) immediate response, relapse and recovery; (c) poor or no response; and (d) transient response.

The response patterns are schematically illustrated in Fig. 1. Table 1 lists summary data pertaining to the factors affecting responses. Table 2 shows the response patterns of infants weighing < 1000 g and outcome.

Immediate and Sustained Response

This was an increase in a/APO_2 ≥ 0.2 combined with a fall in MAP ≤ 2 cm H_2O from the pre-surfactant values, and these improvements were maintained during the subsequent 120 h after treatment.

This type of response was the one seen in 79 % (76/96) of the surfactant recipients, and may represent "pure RDS" in which surfactant deficiency is the primary factor underlying their disease. This group included 16 infants weighing less than 1000 g.

Chest x-ray films cleared rapidly in most cases. From the radiologic standpoint, aeration was nearly complete when evaluated at 6 h following surfactant administration, because the small focal areas of non-aerated lung are too small to be seen on the radiograph. A slight but not significant fall in a/APO_2 from the baseline present during remission occurred between 24 and 48 h after surfactant treatment, but this may be attributed to mild to moderate lung edema due to a PDA.

RDS is characterized by markedly increased pulmonary permeability [15];

Table 1. Factors affecting responses

Responses	*n*	BW[a] (g)	Factors	Mortality (*n*)
Immediate-sustained	76 (79%)	1326 (640–2400)	Mild-moderate PDA	0 (0%)
Relapse-recovery	15 (16%)	1351 (520–2320)	Significant PDA	5 (33%)
Poor or no response	5 (6%)	882 (540–1580)	Cardiogenic shock Myocardial dysfunction PFC	2 (40%)
Total	96			7 (7%)

[a] Mean (range).

Table 2. Response to surfactant in infants < 1000 g

Responses	*n* (%)	Mortality (*n*)
Immediate-sustained	16 (64%)	0 (0%)
Relapse-recovery	5 (20%)	3 (60%)
Poor or no response	4 (16%)	1 (25%)
Total	25	4 (16%)

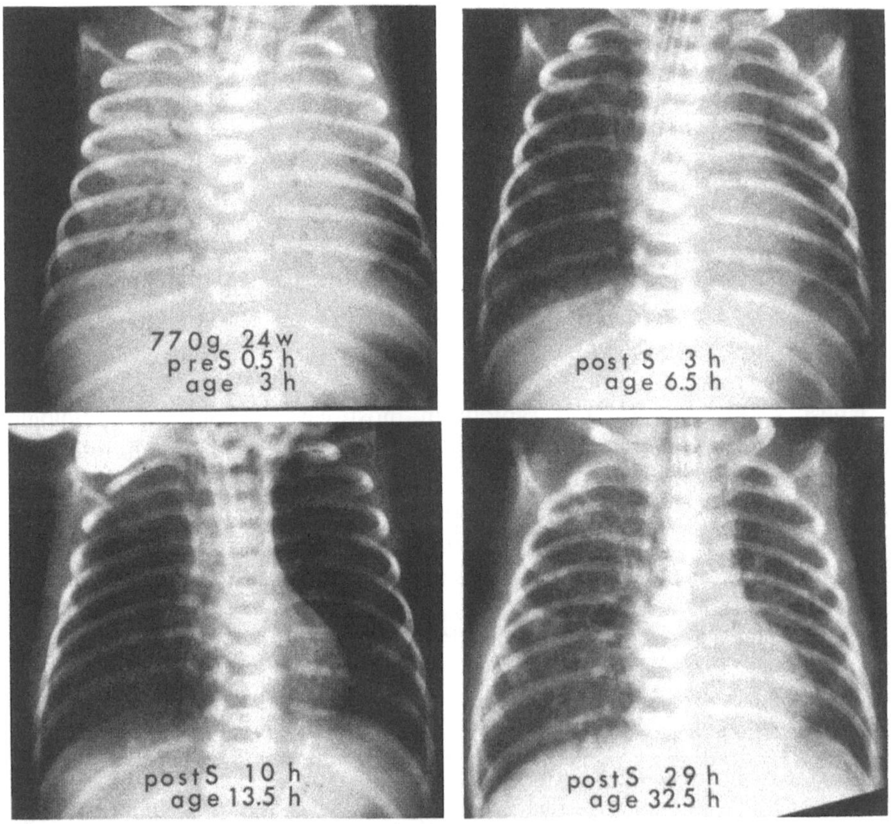

Fig. 2. Serial chest radiographs of a 24-week, 770-g infant with severe RDS and hemodynamically significant PDA. The radiograph on the *upper left* was taken at age 3 h after birth, prior to surfactant administration; that on the *upper right,* was taken 3 h following surfactant administration; that on the *lower left,* taken 10 h post-surfactant, corresponds to the clinical improvement; that on the *lower right,* taken at age 32.5 h, is consistent with a PDA. Serial echocardiographs showed the presence of a left-to-right ductal shunt at age 6–12 h. Left systolic time interval (LSTI) decreased from 0.33–0.38 to 0.26–0.28 between 24 and 48 h of age, suggesting a large ductal shunt; concurrently, a/APO_2 decreased from the baseline value of 0.45 to 0.25 (44 % reduction). Because of the reopening of the PDA after 6 doses of mefenamic acid, the patient underwent surgical ductal ligation at age 5 days and had an uncomplicated recovery

this increased permeability enhances the significance of the increased blood flow from a PDA even before overt cardiac failure occurs [16–18]. In those infants in whom the findings of RDS have cleared following surfactant treatment, the radiographic findings are the first to alert one to the presence of an underlying ductal shunt [1, 19].

This is the most important to appreciate and, in our nursery, many of the surfactant recipients who subsequently developed the PDA demonstrated positive radiographic findings (Fig. 2) before the murmur of the PDA was clearly audible clinically.

Changes present in serial echocardiographs and radiographs often preceded the appearance of physical signs of PDA. Many of these infants were given prophylactically mefenamic acid [12], a potent prostaglandin synthetase inhibitor to close the PDA; consequently, none of the infants developed a significant PDA during the first week of life. However, 9 infants weighing < 1000 g underwent surgical ductal ligation afterwards (at 10–13 days of life), because of PDA-reopening. None of the infants weighing > 1000 g required surgical PDA ligation.

A representative case of this category is shown in Fig. 3. The sustained response seen in this category of infants may represent the maintenance of an intra-alveolar surfactant pool size sufficient to counteract inhibitors present in the airways, and to prevent the progression of lung disease, thereby decreasing the protein leak into the airways. This possibility is supported by a recent animal study by Jobe et al. [20] which demonstrated that surfactant treatment in preterm lambs has a beneficial effect on decreasing the protein leak into the airways, and also by recent in *in vitro* studies showing that the higher the ratio of surfactant to protein the less serious the inhibition [21–23].

An immediate and sustained improvement with a high-dose surfactant treatment may be important in the light of recent data implicating pulmonary instability with intraventricular hemorrhage [24].

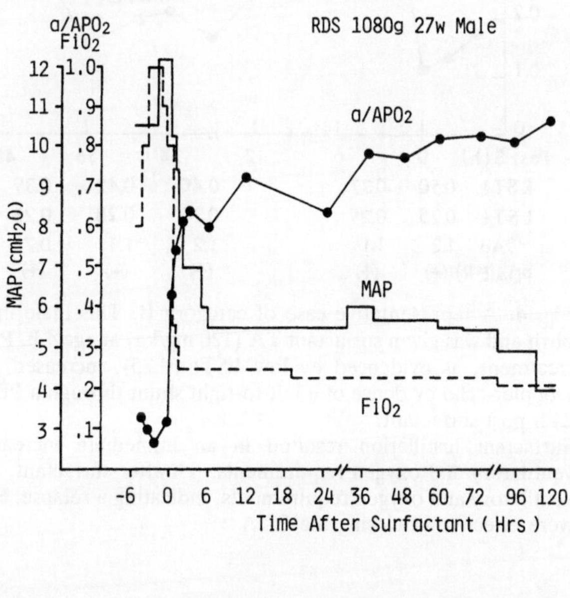

Fig. 3. Changes in FiO_2, MAP, and a/APO_2 in an infant with severe RDS before and after surfactant treatment

Immediate Initial Response, Relapse and Recovery

The magnitude of initial response and duration are similar to those seen in the category A until about 24 h following surfactant administration when there was a gradual decline of a/APO_2 accompanied by an increase in MAP by approximately 50 % from each of the baseline present during remission. This category of infants comprised 16 % (15/96) of the surfactant recipients and may represent a hemodynamically significant PDA.

This group of infants did not significantly differ from the category A of infants in birth weight or gestational age, but were significantly more asphyxiated at birth. Asphyxia is known to be an important determinant of decompensation due to a left-to-right shunt through a PDA [25]. Many of these infants recovered after intervention with mefenamic acid. Of these, one infant weighing < 1000 g underwent surgical ligation of the duct at age 5 days. Three of the 5 infants weighing < 1000 g had either contraindications to pharmacological intervention of PDA or were too ill to undergo surgical

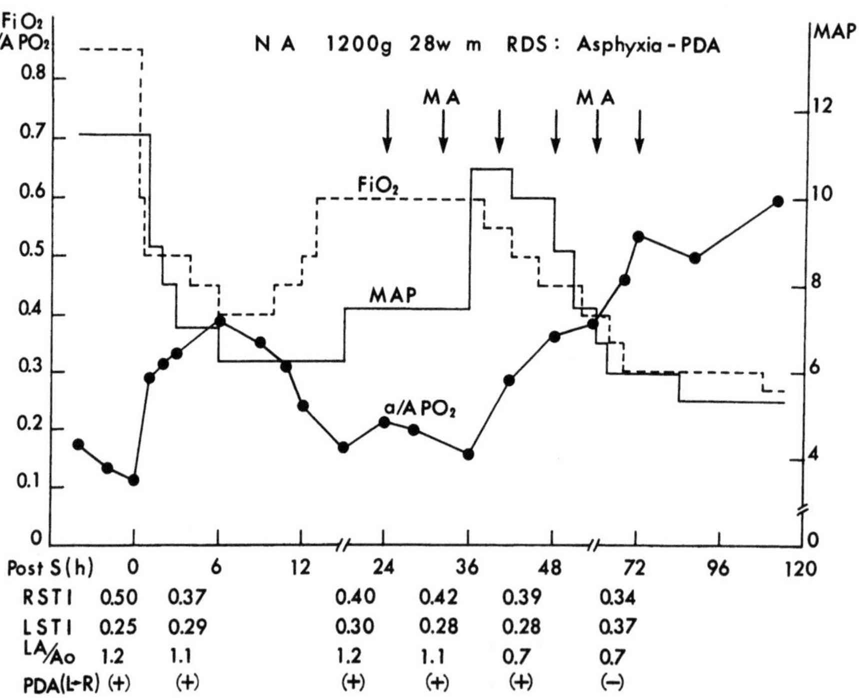

Fig. 4. A representative case of category B. This male infant was severely asphyxiated at birth and was given surfactant TA (120 mg/kg) at age 4 h. PDA was present before surfactant treatment, as evidenced by low LSTI (0.25), increased LA/AO ratio (1.2), and pulsed-Doppler-echo evidence of a left-to-right shunt through a PDA. No murmur was audible until 24 h post-surfactant.
Surfactant instillation resulted in an immediate increase in a/APO_2, with decreasing ventilatory and oxygen requirements; 6 h after surfactant, a/APO_2 declined, with increasing ventilatory and oxygen requirements, indicating a relapse. Six doses of mefenamic acid *(MA)* were effective in closing the PDA

ligation; all died of IVH. Five of 15 infants in this category died. A representative case of this category is shown in Fig. 4.

When the infants in this category were combined with those in the category A, surfactant treatment was successful in 95 % of the infants with 93 % survival rate in this series.

Similar results have been obtained in our recent three series of multicenter, nonrandomized and randomized trials in a total of 336 infants.

Poor or No Response

A poor response was defined as an increase in a/APO_2 < 0.2, and no response was defined as an increase in a/APO_2 < 0.1 when evaluated at 6 h post-surfactant, combined with little change in respiratory requirements. This category of infants was the one seen in 6 % (5/96) of the surfactant recipients. All of these infants had severe perinatal asphyxia.

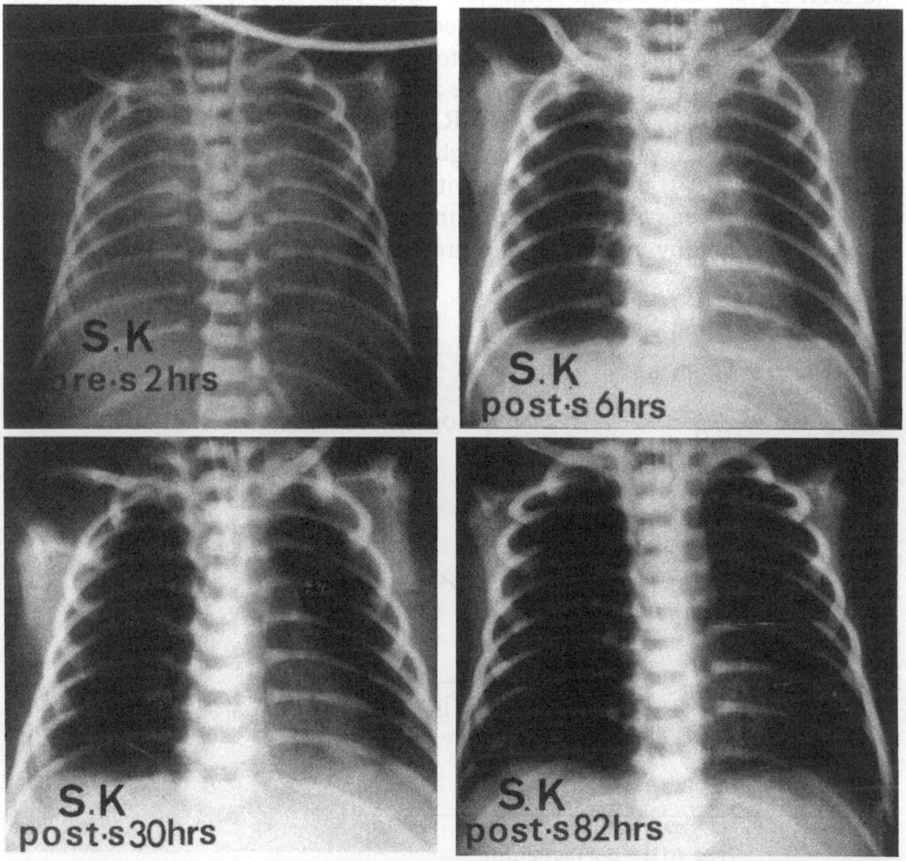

Fig. 5. Serial chest radiographs from a RDS-PFC patient (birth weight 1450 g, gestational age 30 weeks) before and after treatment with surfactant TA

RDS is not simply characterized by surfactant deficiency and, therefore, it is not surprising that there are always some unresponsive categories of infants.

Cardiogenic shock [26] or myocardial dysfunction [27] or persistent fetal circulation [28–30] are the other cardiopulmonary conditions presenting early in the neonatal period, signs and symptoms of respiratory distress, which develop resembling RDS. Five infants had myocardial dysfunction, one of whom had Doppler-detectable tricuspid regurgitation, and two larger infants had PFC. These patients were treated with cardiotonic agents; three recovered (3/5 = 60 %).

Report of a Case of RDS-PFC. A severely asphyxiated infant (1450 g, gestational age 30 weeks) was admitted at 2 hours of age. The initial chest radiograph in this infant was compatible with grade 3 RDS (Fig. 5). The microbubble stability test revealed lack of surfactant in the gastric juice. At age 4 h he was given surfactant (120 mg/kg). Although the MAP was reduced from the initial value of 12 cm H_2O to 10 cm H_2O after surfactant instillation, and there was a clear improvement in the aeration of the lungs as shown in Fig. 5, the a/APO_2 did not change until 48 h after surfactant (Fig. 6). Infusion of calcium gluconate or dopamine resulted in a temporary increase in a/APO_2 (Fig. 7). Interruption of these cardiotonic agents resulted in a decrease in a/APO_2. The RSTI (right systolic time interval) was high, indicating elevated pulmonary vascular resistance (PVR) [28–30]. At this point, contrast echocardiography revealed a large right-to-left shunt via foramen ovale. A decrease in the PVR was associated with a rising a/APO_2.

Thus, in this case, surfactant treatment resulted in clearing of the lung parenchyma, thereby unmasking the presence of the associated PFC.

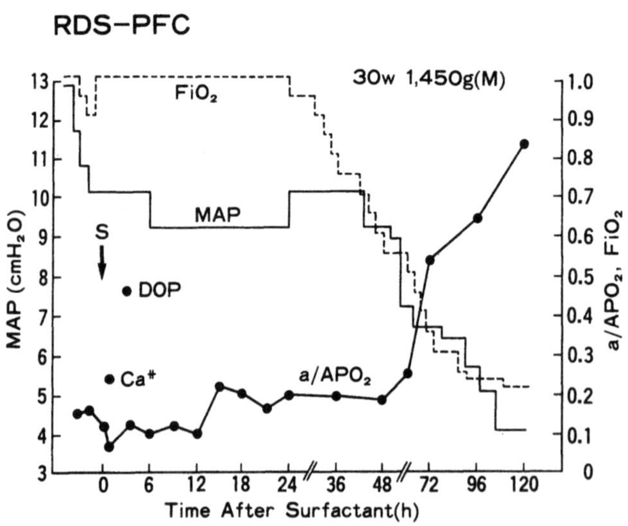

Fig. 6. Clinical course of a patient with RDS-PFC. *S*, surfactant treatment; *Ca*, infusion of calcium gluconate; *DOP*, infusion of dopamine. For details see text

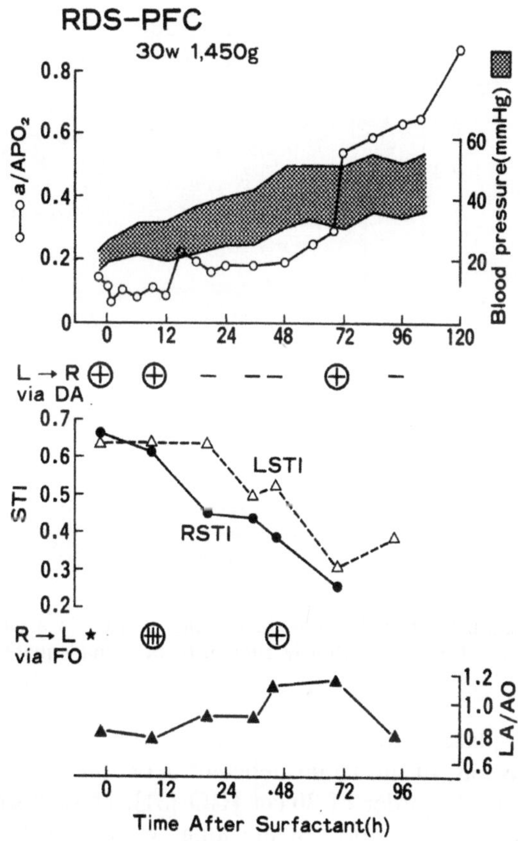

Fig. 7. Hemodynamic changes in the patient considered in Fig. 6. *RSTI,* right systolic time interval; *LSTI,* left systolic time interval; *R-L,* right-to-left shunt; *FO,* foramen ovale; *DA,* ductus arteriosus; *LA/AO,* left-atrial to aortic-root dimension ratio

Transient Response due to Inadequate Respirator Settings

While this was not seen in this series, in our multicenter uncontrolled trial we encountered a small number of infants who had a transient response that appeared to be related to inadequate respirator settings. Because of the fear of acute barotrauma, some attendant physicians in that study did not follow the study protocol and employed higher rates and lower peak inflating pressures. In some extreme cases, the pre-surfactant settings used were: PIP 12–15 cm H_2O, frequency 80–100 cpm, inspiratory time 0.1–0.3 s, PEEP 2–10 cm H_2O, FiO_2 1.0, and the blood gases were: PaO_2 40–50 torr, $PaCO_2$ 50–80 torr, and pH 7.1–7.2.

Surfactant instillation produced a rapid increase in PaO_2, but this was not sustained as shown in Figs. 1D, 8. The initial settings used in these patients do not seem to be adequate for opening up the closed lung units and displacing the fluid from the airways.

To provide adequate aeration of the liquid-filled lung at birth, the inflating pressure must be high enough to overcome capillarity in the finer conduction airways. Furthermore, this opening pressure must be applied for a period of time which is long enough to overcome the viscosity of the fluid that has to be

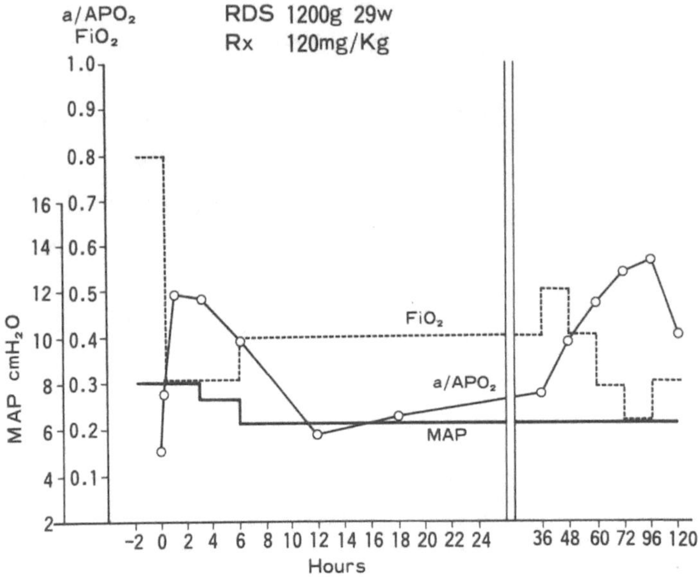

Fig. 8. Changes in FiO_2, MAP, and a/APO_2 in a RDS patient in whom the presurfactant ventilator settings shortly after birth were inadequate (an example of category D)

displaced from the airways as the air-liquid interface moves towards the alveoli. In surfactant-deficient neonatal rabbit lungs, this critical pressure level is in the order of 30 cm H_2O [31]. Most normal newborn infants accomplish complete aeration of the lungs within the first breaths of life. On the other hand, there is another group of infants who are so weak and depressed at birth

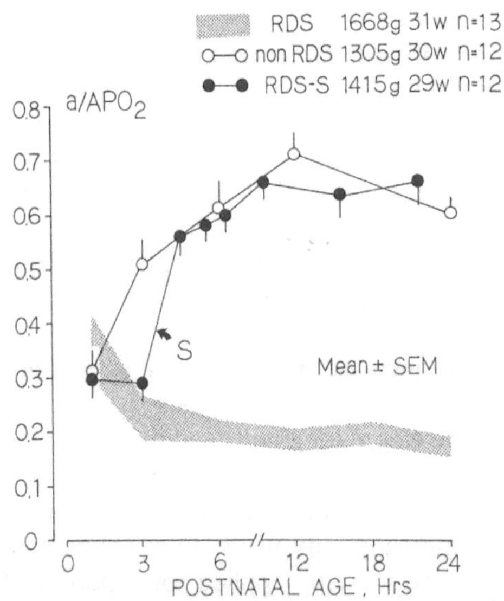

Fig. 9. Changes in a/APO_2 in infants with RDS and transient respiratory distress of the newborn (TRDN). The a/APO_2 levels shortly after birth were similar in the three groups. A brief initial stabilization with higher inflating pressure (25 cm H_2O) and longer inspiratory time (1 s) resulted in great improvement in patients with TRDN, while those with RDS remained stable. *RDS-S,* surfactant-treated RDS; *non RDS,* TRDN; *S,* surfactant treatment

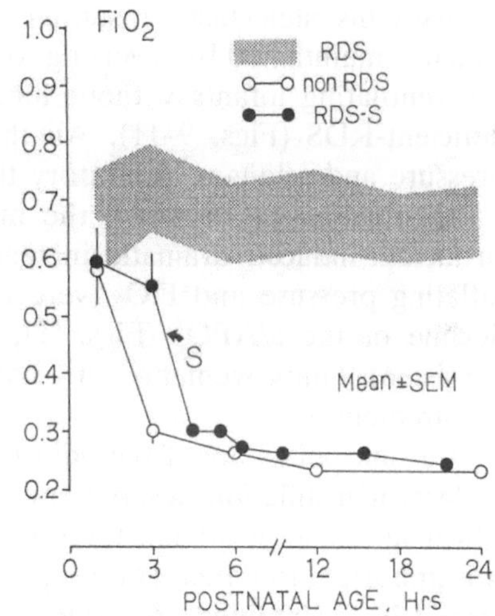

Fig. 10. Changes in FiO_2 in infants with RDS and TRDN. See Fig. 7

that they are unable to sustain normal respiratory activity. In particular, transient respiratory distress of the newborn (TRDN), or wet lung, commonly produced respiratory distress; many of these infants probably have been diagnosed and treated as cases of RDS [32]. If an infant with TRDN is initially ventilated with a pressure of only 12–15 cm H_2O with a shorter inspiratory time, i.e., 0.1–0.3 s, he may not be able to accomplish complete alveolar recruitment, because an endogenous surfactant release is delayed; the resorption of fetal pulmonary fluid from the airspaces is delayed, and pulmonary function is impeded; the infant would require a higher FiO_2. Sometimes a brief period of stabilization with a higher inflating pressure and longer inspiratory time

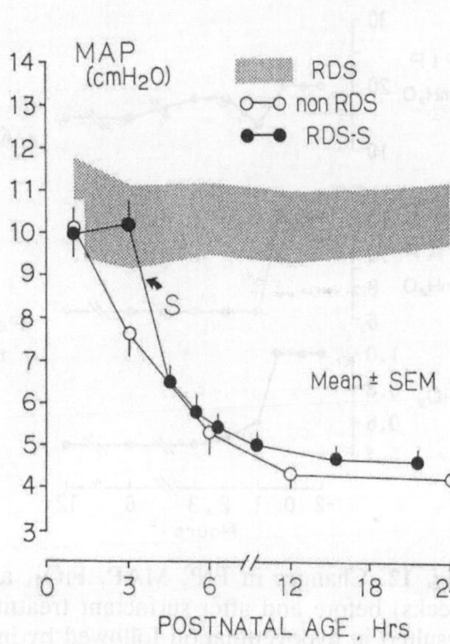

Fig. 11. Changes in MAP in infants with RDS and TRDN. See Fig. 7

reverses his atelectasis, resulting in a marked improvement in oxygenation which is maintained by reverting to much lower pressures similar to those used for ventilating infants without lung disease. Such infants are not surfactant deficient-RDS (Figs. 9–11). All this suggests that initially a higher inflating pressure and a longer inspiratory time should be the preferred settings.

In other cases in which the initial settings were adequate, there was a surfactant-induced dramatic increase in PaO_2. At this point both the peak inflating pressure and FiO_2 were reduced too rapidly, resulting in a gradual decline of the a/APO_2 (Figs. 1D, 12). Adjustment of the settings led to a consistent improvement in oxygenation with decreasing ventilatory requirements.

Certain points about the adjustment of respirator settings during and after surfactant instillation deserve emphasis. It must be noted that the manner in which surfactant is administered is simply to deliver liquid to the baby's lungs. It is important to note that during instillation of surfactant the inflating pressure should be maintained at a few cm H_2O higher than the previous settings to promote uniform distribution of surfactant in the lung and to prevent fluid plugging up the airways. In addition, the settings used prior to treatment should be maintained to produce good air entry into the lung, adequate oxygenation and ventilation. Lachmann and his associates [31] demonstrated that when surfactant-treated preterm animals were ventilated with a fixed peak inflating pressure of 25 cm H_2O and a frequency of 40 cpm, a longer inspiratory time (0.9 s) resulted in better alveolar recruitment and compliance characteristics compared to a shorter inspiratory time (0.3 s).

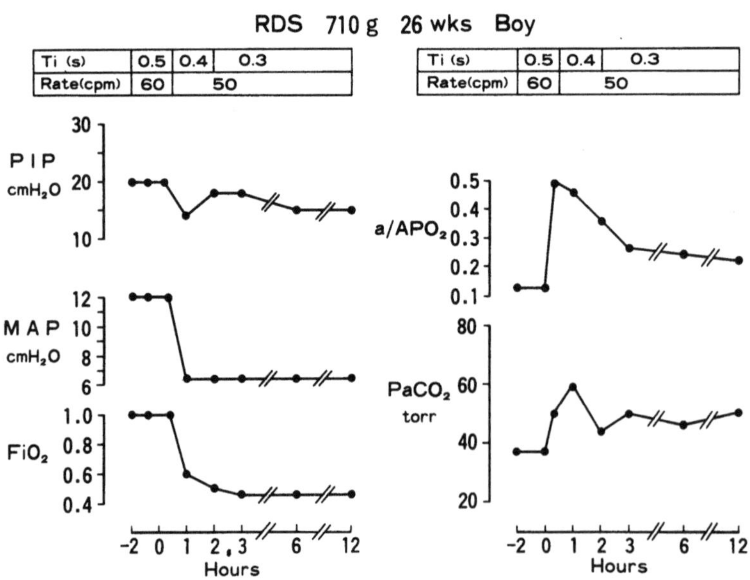

Fig. 12. Changes in PIP, MAP, FiO_2, a/APO_2, and $PaCO_2$ in a RDS patient (710 g, 26 weeks) before and after surfactant treatment. A too rapid reduction in both FiO_2 and PIP resulted in hypoventilation followed by increasing physiologic shunt (see text)

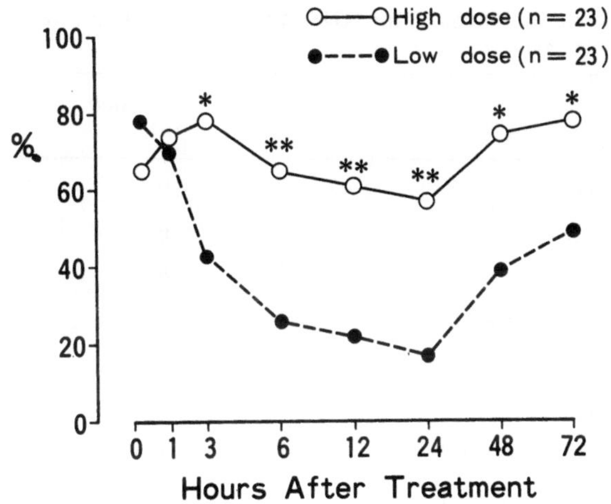

Fig. 13. Proportion of patients meeting criterion for clinical improvement following surfactant treatment. An increase in a/APO_2 > 0.2 from the pretreatment value was defined prior to analysis as clinical improvement. *, p 0.05; **, $p < 0.01$ (Fisher exact test)

Once a clear improvement in the gas exchange has occurred, FiO_2 should be first lowered and the inflating pressure should be carefully reduced afterwards.

High vs Low Dose of Surfactant

Besides the timing of treatment, either at birth [5–9] or early vs late [33, 34], the dosage of surfactant is also important. In both the rescue and prophylactic trials with human surfactant [2, 8] there was a high recurrence rate of respiratory distress; however, in these trials, a lower dose (60 mg/kg) of surfactant was used. In a prospective, randomized, controlled trial comparing the efficacy of two doses (60 mg vs 120 mg/kg) of surfactant TA in severe RDS-patients weighing 1000–1500 g [35], we have demonstrated that both treatments improved oxygenation with decreased ventilatory requirements, the high-dose surfactant having a more beneficial effect in prolonging the response. The striking and acute improvement in oxygenation and ventilation achieved with the low-dose treatment suggest that initially this dose of surfactant was at least sufficient to neutralize the inhibitors that were already present in the airways before treatment. However, this initial beneficial effect was not sustained in a significant proportion of low-dose infants (Fig. 13). This finding is in accord with that of both the rescue and prophylactic trials with a lower-dose of human surfactant [2, 8]. Our study showed further that infants in the high-dose group had significantly less incidence of both IVH and BPD [35] (Table 3).

Thus, the use of an inadequate dose of surfactant may account not only for the variablity in response but also for the lack of a substantial reduction in the incidence of IVH in the series reported by others.

Table 3. Outcomes for RDS infants treated with low-dose (n = 23) or high-dose (n = 23) surfactant

	Low dose (*n*)	High dose (*n*)	p*
ND	0	1[b]	0.5000
BPD	6	0	0.0108
IVH	6	1	0.0479
BPD or IVH	11[a]	1[b]	0.0008
PIE or PNTX	1	1	0.7555
Significant PDA	9	10	0.5000
Ventilation for 30 days	7	1	0.0235
O_2 30% for 30 days	10	3	0.0236

* Fisher exact test.
[a] One patient had a combination of BPD and IVH.
[b] Same infant.

Abbreviations: ND, neonatal death; IVH, intraventricular hemorrhage; BPD, bronchopulmonary dysplasia; PIE, pulmonary interstitial emphysema; PNTX, pneumothorax; PDA, patent ductus arteriosus.

Mortality

Overall neonatal mortality was 7 % (7/96), 16 % (4/25) for infants weighing < 1000 g, and 4 % (3/71) for those > 1000 g. If we restricted the analysis to babies weighing 750–1000 g, the mortality was 10.5 % (2/19).

None of the 76 infants in the category A died; 5 (33 %) of the 15 infants in the category B, and 2 (40 %) of the 5 infants in the category C died, respectively (Tables 1 and 2). The causes of death were 6 from IVH and one from intractable renal failure.

Toxicity

One concern about using bovine surfactant is the presence of a small amount of very hydrophobic protein (about 1 %) in the preparation; however, we have been unable to detect any antibodies against surfactant TA in more than 600 sera from the 204 patients treated with surfactant TA [36]. This proteolipid apoprotein (MW = 5 kilodalton) is a functionally important constituent of surfactant TA [37–39].

Conclusion

In conclusion, surfactant replacement therapy unmasks the relative contributions of other mechanisms to the whole spectrum of RDS which include a hemodynamically significant PDA, PFC, cardiogenic shock, transient myocardial dysfunction with or without tricuspid regurgitation associated with severe perinatal asphyxia, etc.

In assessing the effectiveness of surfactant treatment we should consider possible effects of these underlying abnormalities that may complicate the interpretation of the response.

Serial echo- and Doppler echocardiographic examinations are extremely useful in identifying these underlying abnormalities.

With early recognition, management, and possible prevention of these cardiac involvements, it may be possible to increase the significance of surfactant treatment further, and facilitate recovery.

Appendix

Our recent multicenter, randomized, controlled trial of a postnatal single dose of surfactant TA in 91 severe RDS infants weighing 750–1750 g confirms the above observations and shows that surfactant treatment leads to a significant reduction in major morbidity of RDS such as PIE/PNTX, BPD and IVH when compared to conventional treatment.

References

1. Fujiwara T, Maeta H, Chida S, et al. (1980) Artificial surfactant therapy in hyaline membrane disease. Lancet 1:55–59
2. Hallman M, Merritt A, Jarvenpaa A-L, (1985) Exogenous human surfactant for treatment of severe respiratory distress syndrome: a randomized prospective clinical trial. J Pediatr 106:963–969
3. Gitlin JD, Soll RF, Parad RB, et al. (1987) Randomized controlled trial of exogenous surfactant for the treatment of hyaline membrane disease. Pediatrics 79:31–37
4. Raju TNK, Vidyasagar D, Bhat R, et al. (1987) Double-blind controlled trial of single-dose treatment with bovine surfactant in severe hyaline membrane disease. Lancet 1:651–655
5. Enhorning G, Shennan A, Possmayer F, et al. (1985) Prevention of neonatal respiratory distress syndrome by tracheal instillation of surfactant: a randomized clinical trial. Pediatrics 76:145–153
6. Kwong MS, Eagan EA, Notter RH, Shapiro DL (1985) Double-blind clinical trial of calf lung surfactant extract for the prevention of hyaline membrane disease in extremely premature infants. Pediatrics 76:585–592
7. Shapiro DL, Notter RH, Morin FC III, et al. (1985) Double-blind randomized trial of a calf lung surfactant extract administered at birth to very premature infants for prevention of respiratory distress syndrome. Pediatrics 76:593–599
8. Merritt A, Hallman M, Bloom BT, et al. (1986) Prophylactic treatment of very premature infants with human surfactant. N Engl J Med 315:785–790
9. Morley C (1987) Ten centre trial of artificial surfactant (artificial lung expanding compound) in very premature babies. Br Med J 294:991–996

10. Avery ME, Taeusch HW, Floros J (1986) Surfactant replacement. N Engl J Med 315:825–826
11. Reynolds EOR (1975) Management of hyaline membrane disease. Br Med Bull 31:18–24
12. Fujiwara T (1984) Surfactant replacement in neonatal RDS. In: Robertson B, Van Golde LMG, Batenburg JJ (eds) Pulmonary surfactant. Elsevier, Amsterdam, pp 479–503
13. Tanaka Y, Takei T, Aiba T, Masuda K, Kiuchi A, Fujiwara T (1986) Development of synthetic lung surfactant. J Lipid Res 27:475–485
14. Taeusch HW, Keough KMW, Williams M, et al. (1986) Characterization of bovine surfactant for infant with respiratory distress syndrome. Pediatrics 77:572–581
15. Jefferies AL, Coates G, O'Brodovich H (1984) Pulmonary epithelial permeability in hyaline-membrane disease. N Engl J Med 311:1075–1080
16. O'Brodovich HM (1986) Pulmonary edema in unresolved neonatal lung injury. Report of the 90th Ross conference on pediatric research. Ross Laboratories, pp 69–76
17. Dudell GG, Gersony WM (1984) Patent ductus arteriosus in neonates with severe respiratory disease. J Pediatr 104:915–920
18. Reller MD, Lorenz JM, Kontagal UR et al. (1985) Hemodynamically significant PDA: an echocardiographic and clinical assessment of incidence, natural history, and outcome in very low birth weight infants maintained in negative fluid balance. Pediatr Cardiol 6:17–24
19. Edwards DK, Hilton SVM, Merritt TA, et al. (1985) Respiratory distress syndrome treated with human surfactant: radiographic findings. Radiology 157:329–334
20. Jobe A, Ikegami M, Jacobs H, et al. (1983) Permeability of premature lamb lungs to protein and the effect of surfactant on that permeability. J Appl Physiol 55:169–176
21. Ikegami M, Jacobs H, Jobe A (1983) Surfactant function in respiratory distress syndrome. J Pediatr. 102:443–447
22. Fuchimukai T, Fujiwara T, Takahasi A, Enhorning G (1987) Artificial pulmonary surfactant inhibited by proteins. J Appl Physiol 62:429–439
23. Holm BA, Notter RH, Finkelstein JN (1984) Surface property changes from interactions of albumin with natural lung surfactant and extracted lung lipids. Chem Phys Lipids 38:287–298
24. Perlman JM, McMenahim JB, Volpe JJ (1983) Fluctuating cerebral blood-flow velocity in respiratory distress syndrome. N Engl J Med 369:204–209
25. Cotton RB (1984) Patency of the ductus arteriosus-its etiologic and pathogenetic relationship to respiratory distress syndrome. In: Stern L (ed) Hyaline membrane disease. Grune-Stratton, Orlando, pp 175–210
26. Cabal LA, Devaskar U, Siassi B (1980) Cardiogenic shock associated with perinatal asphyxia in preterm infants. J Pediatr 96:705–710
27. Bucciarelli RL, Nelson RM, Eagan EA, et al. (1977) Transient tricuspid insufficiency of the newborn: A form of myocardial dysfunction in stressed newborns. Pediatrics 59:330–337
28. Riggs T, Hirschfield S, Fanaroff A, et al. (1977) Persistence of fetal circulation syndrome: an echocardiographic study. J Pediatr 91:626–631
29. Johnson GL, Cunningham MZD, Desai NS, et al. (1980) Echocardiography in hypoxemic neonatal pulmonary disease. J Pediatr 96:716–720
30. Valdes-Cruz LM, Dudell GG, Ferrara A (1981) Utility of M-mode echocardiography for early identification of infants with persistent pulmonary hypertension of the newborn. Pediatrics 68:515–525
31. Lachmann B, Berggren P, Curstedt T, et al. (1982) Combined effects of surfactant substitution and prolongation of inspiration phase in artificially ventilated premature newborn rabbits. Pediatr Res 16:921–927
32. Swischuk LE (1984) Radiology of the newborn and young infant. Williams and Wilkins, Baltimore, pp 58
33. Fujiwara T, Konishi M, Chida S, et al. (1984) Exogenous surfactant therapy in infants with RDS: comparison of early vs late treatment. Pediatr Res 18:353A
34. Maeta H, Vidyasagar D, Raju TNK, et al. (1987) Comparison of early and late surfactant treatments in baboon HMD model. Pediatrics (in press)

35. Konishi M, Fujiwara T, Naito N, et al. (1987) Surfactant replacement therapy in neonatal respiratory distress syndrome: a multicenter, randomized clinical trial: comparison of high- vs low-dose of surfactant TA. Eur J Pediatr (in press)
36. Kawashima T, Chida S, Fujiwara T (1987) Measurements of anti-surfactant proteolipid-5-kDa apoprotein antibodies in sera from HMD-patients treated with surfactant TA using an ELISA. Perinatal Med [JPN] 17:473–476
37. Takahashi A, Fujiwara T (1986) Proteolipid in bovine lung surfactant: its role in surfactant function. Biochem Biophys Res Commun 135:527–532
38. Whitsett JA, Ohning BL, Ross G, et al. (1986) Hydrophobic surfactant associated protein (SAP 6–14) in whole lung surfactant and its importance for biophysical activity in lung surfactant extracts used for replacement therapy. Pediatr Res 20:460–467
39. Yu S, Possmayer F (1986) Reconstitution of surfactant activity by using the 6 kda apoprotein associated with pulmonary surfactant. Biochem J 236:85–89

A Multicenter Randomized Trial of Surfactant Replacement with Calf Lung Surfactant Extract: Effects of Pre- and Post-Ventilatory Instillation and of Multiple Doses

J. W. Kendig, R. H. Notter, J. M. Davis, A. Bartoletti, H. S. Dweck, H. M. Risemberg, C. Cox, and D. L. Shapiro

Introduction

This paper is concerned with the clinical therapy of lung disease, particularly the treatment of pathology which involves a functional deficiency (quantitative or qualitative) of pulmonary surfactant. As with many other kinds of medical therapies, however, the efficacy of treatment of surfactant-deficient lung disease has increased directly with the understanding of the underlying pathophysiology. Thus, although clinical aspects are stressed, the discussion is initiated by considering some basic concepts of lung surfactant (LS) and its influence on pulmonary physiology.

Pulmonary Surfactant and Its Actions in Respiration

Pulmonary surfactant is a mixture of specific phospholipids and apoproteins, produced and stored in the type II pneumocyte of the mammalian lung, prior to secretion into the alveolar hypophase and adsorption to the air-hypophase interface. This adsorbed film of lung surfactant (LS) is then compressed and expanded during breathing and can lower the alveolar surface tension to minimum values of the order of one dyne/cm. The specific biophysical properties of LS allow it to reduce the work of breathing (increase compliance), stabilize alveoli, and decrease the driving force for pulmonary edema. Indeed, pulmonary surfactant dominates respiratory mechanics, and the presence of active LS at the air-hypophase interface in alveoli is crucial for normal lung function and physiology. The direct coupling of specific biophysical properties of natural LS, and the generation of beneficial effects on respiration, have been reviewed in detail recently by Notter and Finkelstein [19] and Notter and Shapiro [21], and are listed in Table 1

Lung Surfactant in RDS, Lung Injury, and ARDS

The discovery in the mid-1950s by Pattle in England, and by Clements in the United States, that lungs contained powerful surfactants with direct effects on respiration was an exciting and important finding. However, an additional significant event, critical for clinical medicine, occurred in 1959, when Avery and Mead [1] demonstrated that a deficiency in lung surfactant correlated with

Table 1. Biophysical properties and physiological effects of active lung surfactants

Biophysical Properties
Ability to lower surface tension well under dynamic compression (i. e. order of 1 dyne/cm)[a]
Ability to adsorb rapidly to values near the equilibrium limit of phospholipids (near 20–24 dynes/cm)
Ability to respread after dynamic compression past collapse
Ability to vary surface tension as surface area changes
Resultant Physiological Effects
Decreased work of breathing and lung recoil forces
Increased compliance
Decreased atelectasis and increased inflation uniformity
Decreased driving force for pulmonary edema

The rationale for the biophysical properties listed, and their coupling with beneficial physiological effects in lungs, are given in detail elsewhere [e.g. 19, 21, for review].

[a] There are no theories which are precise enough to fix a single value on "required" minimum surface tension, or adsorption rapidity and final value of surface tension reached. Moreover, the resultant physiological effects of these properties are also not fixed, with some surfactants having "better activity" in lungs. Although biophysical-physiological correlates for surfactants are not exact, however, the general validity of the relationships here seems to be borne out by a variety of interdisciplinary studies [5, 19, 21, 24]. See text for further discussion.

the occurrence of hyaline membrane disease, now generally called respiratory distress syndrome (RDS) in premature infants. Specifically, Avery and Mead showed that the surface tension vs. surface area (γ-A) characteristics of lung washings from infants with RDS showed higher minimum surface tension, and reduced γ-A hysteresis, compared to lung washings from infants without RDS. Since their initial report [1], an array of studies using biophysical, physiological, biochemical and cell biological methods have confirmed that RDS is a disease of primary LS deficiency due to a lack of LS-producing type II pneumocytes at birth [e.g. 5, 19, 21, 24]. As expected from the physiologic effects of active LS in Table 1, the typical pathological features seen initially in RDS are decreased lung compliance, increased work of breathing, increased end-expiratory atelectasis and diffuse (but usually not fulminant) pulmonary edema [e.g. 5].

Although an absolute LS deficiency initiates neonatal RDS, additional pulmonary pathophysiology often becomes operative during the course of the illness. This is due in part to the requirements of affected infants for higher inspired oxygen levels, mechanical ventilation with positive endexpiratory pressure (PEEP) and abnormally high mean airway pressure. Complications of RDS that are related to high oxygen and mechanical ventilation include bronchopulmonary dysplasia (BPD), as well as progression to lung injury features reminiscent of the "adult" Respiratory Distress Syndrome (ARDS). Although denoted as adult RDS, this pathology can also affect pediatric patients. As reviewed by Hopewell and Murray [13] and Rinaldo and Rogers [23], a major pathologic feature of typical ARDS lung injury is high molecular weight pulmonary edema, resulting from damage to the alveolo-capillary membrane with increased permeability to solutes. Decreased lung compliance and functional residual capacity (FRC), together with intrapulmonary shunting, also are included in most definitions of the ARDS lung injury syndrome [13, 23].

Although ARDS is not a disease of primary surfactant deficiency, infants with neonatal RDS, particularly those ventilated for a significant period, can exhibit closely related pulmonary pathology. The practical significance of this perspective for neonatologists is that the timing of therapy for neonatal RDS can be critical. Treatment early in the course of neonatal RDS can provide quite specific therapy for a primary surfactant-deficiency, while later treatment may be complicated by a superimposed lung injury with potential LS inactivators such as plasma proteins present in alveolar edema e.g. [9–13, 23]. This phenomenon was directly responsible for some of the questions to be addressed later in terms of comparing the efficacy of pre-ventilatory vs post-ventilatory surfactant replacement therapy for RDS.

Rationale for Lung Surfactant Replacement Therapy for RDS

Since the discovery that a deficiency in natural lung surfactant was the primary cause of RDS [1], many studies have been directed toward developing effective exogenous surfactant replacement therapy for affected infants. This therapeutic approach was initially reported to have questionable efficacy in early clinical studies in the 1960s which used dipalmitoyl phosphatidylcholine (DPPC) as the exogenous surfactant. However, although DPPC is the primary phospholipid component of natural LS, it does not by itself exhibit all of the biophysical properties (Table 1) needed for an active surfactant in lungs. Specifically, DPPC gives LS its extraordinary surface tension lowering ability when compressed in a surface film, but DPPC has very poor adsorption facility, and also poor dynamic respreading. Thus, DPPC alone cannot adsorb to the air-water interface, or remain there after successive compression cycles passed monolayer collapse. It is now known that specific hydrophobic LS apoproteins (Mr = 6000 Daltons) contribute to this necessary adsorption facility [e.g. 22, 26].

Because the reason behind the failure of DPPC in early replacement therapy in infants [21 for review] was not immediately appreciated, the establishment of its potential benefits had to be proved by a variety of multidisciplinary studies. Perhaps most important here were studies showing the direct benefits of natural LS instilled into surfactant-deficient animal lungs, and correlations of these results with biophysical activity studies to define essential surface properties [19, 20 for review]. The consistently positive effects of biophysically active surfactants in animal models of RDS led to the resumption of clinical trials in infants at the start of the 1980s.

Representative Clinical Trials for Exogenous Surfactant Replacement in RDS

Overview

It has already been shown that exogenous surfactant replacement therapy has extensive clinical benefits in infants with RDS, providing that a biophysically active surfactant preparation is administered (i.e. an exogenous surfactant with the biophysical properties in Table 1). A summary of such studies, reviewed by Notter and Shapiro [21] is presented in Table 2. The tabulation in Table 2 shows, for example, that three randomized, controlled clinical trials, all published in 1985 [4, 15, 25], demonstrated significant beneficial effects over controls when given a single pre-ventilatory dose of a calf lung surfactant extract (CLSE) preparation (90 mg CLSE/3 ml). In addition, other controlled trials using human surfactant harvested from term amniotic fluid have also recently proved to be successful [8, 17], as have two trials using another surfactant extract preparation, surfactant-TA [6, 7]. All these surfactants have been found to exhibit the four surface biophysical properties in Table 1, and also have shown substantial physiologic benefits in animal models prior to use in humans [6, 21, 24, 26]. In this context it is not surprising that these surfactants were successful in trials involving extremely premature infants of 24 or 25 weeks gestation to 28 or 29 weeks gestation.

In examining the successful controlled clinical trials in Table 2 it is noted that all are restricted to examining the physiological benefits of a single dose (except for Merritt et al. [17]). As will be described in detail later, the concept of multiple doses is one variable, along with *initial* pre- or post-ventilatory dosage and other factors, that are being studied in our current multicenter clinical trial group. Before going on to describe this multicenter trial, the results of selected studies in Table 2 will be examined to illustrate the kinds of effects typically found in reduced RDS severity, incidence, and/or survival after only a single dose of exogenous surfactant.

Trials Using Bovine (Calf) Lung Surfactant Extract

The discussion in this section refers to the studies of Kwong et al. [15] and Shapiro et al. [25] in Table 2. The results of the clinical trial of Enhorning et al. [4] are generally consistent with these other two studies [15, 25] and used a similar extract (from cows rather than calves). However, the patient groupings and scoring systems used by Enhorning are not easily compared with those of Kwong et al. [15] and Shapiro et al. [25], and appropriate elements from these latter studies are now described. Two doubleblind randomized trials of a single pre-ventilatory dose of CLSE in very premature infants were reported in 1985. The composition of CLSE is shown in Table 3. Kwong et al. [15] enrolled 27 infants of 24–28 weeks gestation at the Children's Hospital of Buffalo, New York and Shapiro et al. [25] enrolled 32 infants at 25–29 weeks gestation at the University of Rochester Medical Center, New York. At both centers, enrolled

Table 2. Some controlled trials of exogenous lung surfactant replacement in very premature infants with RDS (from Notter and Shapiro [21])

Study[a]	Exogenous surfactant[b]	Type and Time of doses[c]	No. of doses	Selected conclusions[d]
Chu et al. [2]	DPPC	Post-ventilatory (aerosol)		No beneficial effect
Morley et al. [18]	7 : 3 DPPC:PG (dry)	At birth (instilled)	1	Reduced mortality of RDS
Fujiwara [6]	Supplemented bovine LS extract (Surfactant-TA)	Post-ventilatory (instilled)	1	Reduced mortality of RDS, in agreement with an earlier reported uncontrolled trial
Wilkinson et al. [27]	7 : 3 DPPC:PG (dry)	Pre-ventilatory [1]	1	
		Post-ventilatory [2] (instilled)	(in either trial) [1] or [2]	No beneficial effect in either trial [1] or [2]
Hallman et al. [8]	HS (amniotic fluid)	Post-ventilatory (instilled)	1	Therapy reduced mortality and severity of RDS
Enhorning et al. [4]	Cow lung surfactant extract	Pre-ventilatory (instilled)	1	Therapy reduced mortality and severity, plus some complications, of RDS
Kwong et al. [15]	Calf lung surfactant extract (CLSE)	Pre-ventilatory (instilled)	1	Therapy reduced severity of RDS
Shapiro et al. [25]	CLSE	Pre-ventilatory (instilled)	1	Therapy reduced severity of RDS
Merritt et al. [17]	Human surfactant (HS) (amniotic fluid)	At birth and later (instilled)	multiple	Therapy decreased mortality and severity of RDS, and bronchopulmonary dysplasia
Gitlin et al. [7]	Surfactant-TA	Post-ventilatory (instilled)	1	Reduced severity of RDS

[a] The table shows selected studies with sufficient patients for statistical significance ($p < 0.05$).

[b] Surfactant mixtures under this heading are a cow lung surfactant extract which is essentially identical to the calf lung surfactant extract (CLSE) developed by Notter and co-workers [see 21 for review] and used clinically by Kwong et al. [15] and Shapiro et al. [25]; human lung surfactant harvested from amniotic fluid by Merritt et al. [17]; a surfactant extract from minced cow lungs, and then enriched with DPPC and other additives, now called Surfactant-TA [6]; and two synthetic surfactants: pure DPPC [2] and a mixture of dry DPPC: egg-phosphatidylglycerol (egg-PG) [18]. The word "dry" associated with this latter surface active mixture refers to its formulation and delivery as a dry powder, which is meant to increase its surface spreading.

[c] The type and time of dose refers to intratracheal instillation vs aerosolization of surfactant (*type* of delivery), and to surfactant given prior to the first breath vs post-ventilatory delivery (*timing* of dose). When the description "at birth" is used, surfactant replacement was given very early in life, but not necessarily before the first breath.

[d] Conclusions of the various studies are reported only in generalized form. The study authors listed in Table 2 used a variety of criteria, including subjective clinical criteria, as well as more objective physiological criteria related to lung function, to define the effects of surfactant replacement on the incidence and severity of RDS and its complications such as BPD, intraventricular hemorrhage (IVH), pneumothorax, etc. The point of the table is not to extensively define clinical outcomes, but rather to provide an overview of the success of current replacement therapy, and the types of clinical trials that have been reported. See text for details.

Table 3. Composition of calf lung surfactant extract

Component	%
Phosphatidylcholine (80% disaturated)	79
Phosphatidylglycerol	6
Phosphatidylinositol/phosphatidylserine	5
Sphingomyelin	2
Lysophospatidylcholine	< 1
Cholesterol and cholesterol esters	4
Protein	1

infants were immediately intubated in the delivery room and given 3 ml of CLSE or 3 ml of saline prior to ventilation. The baseline perinatal characteristics of the infants were similar at both centers, except for the fact that the Buffalo group excluded infants whose mothers had received betamethasone therapy for more than 24 hours before delivery. Both of these first generation clinical trials were based on the premise that the pre-ventilatory

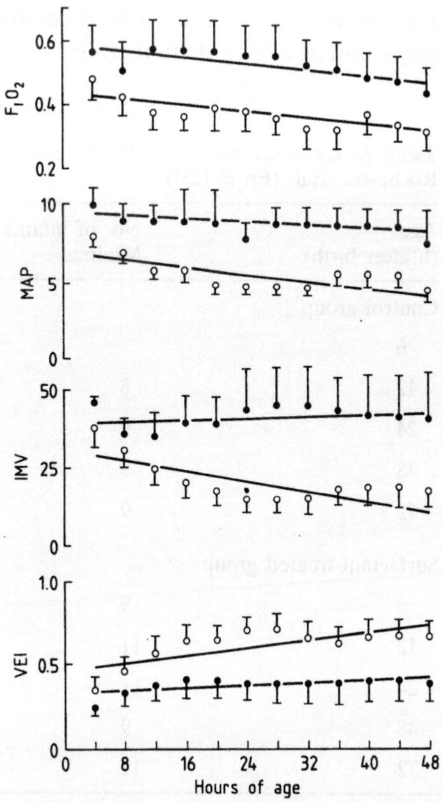

Fig. 1. Means and SEM for physiologic variables used to compare the groups are shown for saline-treated control infants (•) and calf lung surfactant extract-treated infants (○) for the first 48 h of life. *Lines* are linear regression by least squares method. *MAP*, mean airway pressure (mm Hg); *IMV*, intermittent mandatory ventilation (per minute); *VEI*, ventilator efficiency index (mL/mm Hg/kg/min). (From [15])

Table 4. Severity scale for respiratory distress

Minimal	Oxygen requirement < 40% and mean airway presure < 7 cm H_2O
Intermediate	Oxygen requirement of 40%–80% and/or mean airway pressure of 7–10 cm H_2O
Severe	Oxygen requirement of > 80% or mean airway pressure > 10 cm H_2O

administration would result in a homogeneous delivery of the CLSE to the alveoli during the process of fetal lung liquid absorption.

Both groups were able to demonstrate the efficacy of a single pre-ventilatory dose of CLSE in reducing the severity of RDS in very premature infants. Using a prospective clinical scoring system to define the incidence of RDS at 48 hours of age, the Buffalo group found a 14% incidence of RDS in the CLSE-treated group compared with a 54% incidence in the saline control group ($p = 0.033$ by Fisher's Extract Test). The Buffalo group also demonstrated (Fig. 1) that during the first 48 hours the CLSE-treated group had significantly lower inspired oxygen concentrations, mean airway pressures, and ventilator rates and higher ventilator efficiency indices ($p < 0.01$ to $p < 0.001$ by comparison at interceps of the regression lines using the 2-sided t test).

The Rochester study graded respiratory distress (Table 4) as minimal, intermediate, and severe based on inspired oxygen and mean airway pressure requirements. They demonstrated (Table 5) that a single pre-ventilatory dose of

Table 5. Incidence and severity of respiratory distress in control and treated infants in Rochester trial. (From [25])

Age (h after birth)	No. of infants with respiratory distress Minimal	Intermediate	Severe
Control group			
6	7	7	2
12	8	4	4
24	9	2	5
48	9	1	6
72	9	1	6
Surfactant-treated group			
6	9	7	0
12	11	4	1
24	11	4	1
48	9	3	4
72	11	1	4

CLSE produced a significant reduction in the incidence of severe RDS at 12 hours ($p < 0.01$) and at 24 hours ($p < 0.01$) after birth. The salutary effect, however, was not sustained, and no significant differences were found at 48 and 72 hours of age. It is not clear why the CLSE seemed to have a more sustained effect in the Buffalo study; the Rochester experience indicates that some infants may require multiple doses of CLSE. Wood et al. [28] have reported radiographic data to support the efficacy of CLSE in reducing the severity of RDS in premature infants.

The results reported in these two clinical trials are consistent with the previous in vitro and animal studies which were used to evaluate CLSE. The first generation clinical trials at Buffalo and Rochester have clearly demonstrated the efficacy of CLSE in reducing the severity of RDS in very premature infants of less than 30 weeks gestation. It soon became apparent that additional second generation clinical trials were necessary for the evaluation of different strategies for the optimization of surfactant therapy.

Currently Active Multicenter Trials with Calf Lung Surfactant Extract

Introductory Notes

A major goal of our current second generation randomized trial at the University of Rochester is to determine the best time to administer the first dose of exogenous CLSE (i.e. initial pre-ventilatory versus post-ventilatory). The trial also examines the effects of additional post-ventilatory doses of CLSE. Pre-ventilatory surfactant administration has the advantage of a more homogenous delivery to the alveoli during the process of fetal lung liquid absorption following birth. With this form of administration there is no opportunity to confirm the location of the endotracheal tube (by breath sounds or chest X-ray) and there is a risk of delivering the entire dose of CLSE into the right lung. A strategy of preventilatory administration also leads to the unnecessary treatment of the 20 to 30% of infants in the 24–29 weeks gestational range who do not develop RDS.

Post-ventilatory administration of CLSE at several hours of age permits the selective treatment of those infants with moderate or severe disease. The CLSE can be administered in the NICU under carefully controlled conditions following X-ray confirmation of endotracheal tube positions. Oxygenation can be monitored during the procedure. Our current randomized comparison trial examines the relative therapeutic efficacy of these two strategies for administration of CLSE. This trial is currently in progress at three tertiary medical centers in New York State (University of Rochester, Albany Medical College, and New York Medical College).

Overall Design and Description of Rochester Trial

The following protocol is currently being followed at all three centers. Infants between 24 and 29 weeks gestation, inclusive, are randomized prior-to-birth, for subsequent inclusion in pre-ventilatory and post-ventilatory groups (Fig. 2). All infants in the pre-ventilatory group are intubated immediately at birth and given a single dose of 90 mg of CLSE (in 3 ml of 0.15 M NaCl) by instillation into the endotracheal tube. By contrast, infants randomized to the post-ventilatory arm of the study receive an initial dose of CLSE in the neonatal intensive care unit at 2–3 hours of age if moderate to severe RDS has developed, i.e. if chest X-ray is characteristic for RDS, and assisted ventilation is required with an $FIO_2 \geq 0.40$ and/or a mean airway pressure ≥ 7.0 cm H_2O.

Infants in both the pre-ventilatory and post-ventilatory arms of the study may receive up to three additional doses of CLSE at 12–24 hour intervals if clinical criteria for moderate to severe RDS are present. Each post-ventilatory dose of CLSE is identical and contains 90 mg surfactant in 3 ml saline. The dose is administered in four aliquots of 0.75 ml each, while different lung fields are made dependent by positioning the infant. Between each aliquot, the infant is ventilated with hand bagging. Arterial blood gases are obtained at 15, 30, 45, and 60 minutes post CLSE administration, and ventilator settings are adjusted appropriately to maintain a PaO_2 of 50–70 mm Hg and a PCO_2 of < 50 mm Hg. A chest X-ray is obtained at 60 minutes post administration. Serial ventilator settings and blood gases are recorded at 2, 4, 6, 12, 24, 48, and 72 hours of age and entered into a computer database. The severity of the

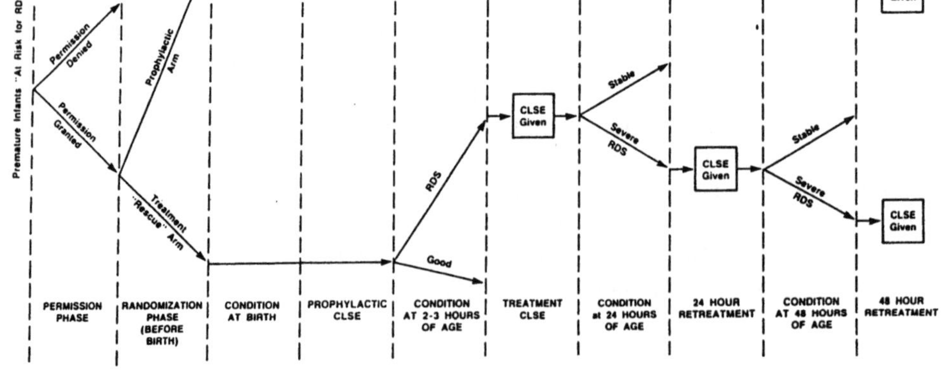

Fig. 2. Protocol for multi-center randomized trial of pre- vs post-ventilatory administration of CLSE

respiratory distress is graded at each time interval according to the scheme in Table 4. Over 200 patients have been enrolled in this multi-center randomized comparison trial, and an interim analysis is currently in progress. Outcome variables being evaluated include survival, physiological parameters and the incidence of neonatal complications such as bronchopulmonary dysplasia, intraventricular hemorrhage, and patent ductus arteriosus.

Selected Early Results

Following the enrolment of a total of 21 patients into each arm of the trial, we conducted a preliminary analysis [14] looking at the primary outcome variable of survival. We compared this group with the original first generation CLSE treated group to evaluate the effect of multiple doses on survival. Significantly, 20 out of the first 21 patients (95.2%) enrolled in the pre-ventilatory arm of the second generation trial survived. This represents a substantial improvement over the 70.6% survival rate found in our original trial following a single pre-ventilatory dose of CLSE ($p < 0.05$). The impact of multiple doses of CLSE on the incidence of neonatal complications such as bronchopulmonary dysplasia is currently under investigation.

An analysis of pre- and post-ventilatory strategies for the administration of the first dose of CLSE is currently in progress. We have found that post-ventilatory doses are tolerated surprisingly well. Figure 3 shows the chest radiographs of a 28 weeks gestation newborn with RDS before and 20 hours after the administration of a post-ventilatory dose of CLSE. The second X-ray shows a dramatic improvement in aeration.

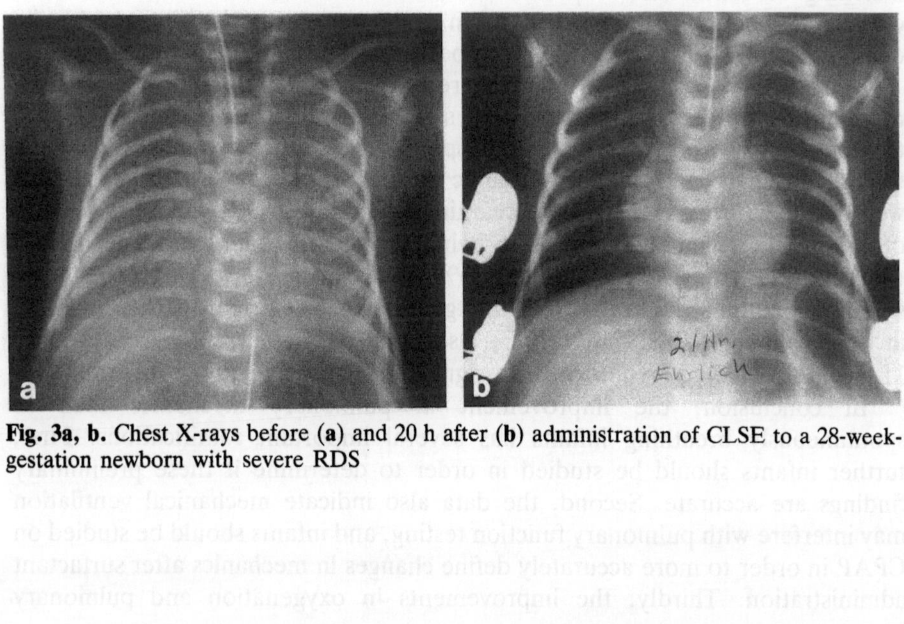

Fig. 3a, b. Chest X-rays before (**a**) and 20 h after (**b**) administration of CLSE to a 28-week-gestation newborn with severe RDS

Quantitative Analysis Enhanced by Computer-Generated Pulmonary Function and Mechanical Measurements and Calculations

We have developed a computer-based methodology to examine a broad spectrum of respiratory variables just before, and one hour after tracheal instillation of 90 mg of CLSE to each infant. Transpulmonary pressure [mouth minus esophageal (pleural) pressure] is determined by a differential transducer and balloon catheter, and inspiratory expiratory flows are directly measured by a pneumotachometer. Continuous, on-line microcomputer processing then allows calculation (for each complete breath) of additional variables including integrated volume (V_T), dynamic compliance (lung plus thorax, C_T), pulmonary resistance (R_T), total resistive work (Rw), and minute ventilation (MV).

We have reported [3] our preliminary experience using pulmonary function studies in newborns with the respiratory distress syndrome. Ten ventilated infants were studied before and after a single post-ventilatory dose of CLSE. Of the 10 infants studied, the birthweight was (mean ± SD) 1228 ± 366 grams and the gestational age was 29.1 ± 2.4 weeks. Required FiO2 decreased significantly after CLSE administration in all 10 infants studied from a mean ± SD of 0.61 ± 0.13 to 0.37 ± 0.13 ($p < 0.001$). These improvements in oxygenation indicate the large benefit in pulmonary function that occurs following surfactant replacement therapy in RDS. However, there were no significant changes in pulmonary mechanics demonstrated when infants were studied on the ventilator and only ventilator breaths were examined. Two factors might have been involved in this disparity between improvements in oxygenation and changes in pulmonary mechanics. Surfactant replacement therapy may be effective in infants with RDS by improving pulmonary blood flow and decreasing pulmonary ischemia. This would result in a better matching of ventilation to perfusion and a decrease in alveolar to arterial oxygen gradients. Little change in pulmonary mechanics would be demonstrated if this mechanism were occurring.

The second factor is that infants were studied only during positive pressure mechanical ventilation which uses significantly higher flow rates and transpulmonary pressures compared to spontaneous respiration. To determine if this factor significantly affects dynamic pulmonary mechanics, three patients were studied during spontaneous breathing on CPAP for brief sampling periods of 1–2 minutes, immediately after being studied on continuing mechanical ventilation. In these three infants, 33% increase in V_E was observed after surfactant administration (332–443 ml/kg/min) due primarily to a 66% increase in tidal volume (5.3–8.8 ml/kg). C_T also increased significantly from 4.35 to 11.60 ml/cm H_2O/kg. There was no significant change in R_T or W_R noted.

In conclusion, the improvement in pulmonary mechanics seen in spontaneously breathing infants has several important ramifications. First, further infants should be studied in order to determine if these preliminary findings are accurate. Second, the data also indicate mechanical ventilation may interfere with pulmonary function testing, and infants should be studied on CPAP in order to more accurately define changes in mechanics after surfactant administration. Thirdly, the improvements in oxygenation and pulmonary mechanics that are seen after surfactant administration may play a useful role in evaluating future strategies for surfactant replacement therapy.

Future Directions and Conclusions

New Synthetic Surfactants (Approaches and Development)

As described earlier, the clinical efficacy of naturally derived surfactants and surfactant extracts (CLSE, Surfactant-TA) has not yet been equaled by any "artificial" surfactant synthesized in the laboratory. However, from the standpoint of purity, sterility, and quality control, a synthetic surfactant of equivalent activity would be preferable if it were free from cellular toxicity and contained no animal lung protein which might be immunogenic in humans. However, it should be noted that lung surfactant extracts such as CLSE have many positive properties (are sterilized and have only tiny protein contents) plus *very* high efficacy. It will be a challenge for a synthetic surfactant to match or improve upon these characteristics.

There are several conceptual approaches which will be described here for the development of synthetic surfactant preparations of potentially high activity. All of these approaches are described in detail by Notter and Shapiro [21], and just an outline is given here. One approach that is of significant current interest for synthetic surfactant development involves the combination, in vitro, of synthetic phospholipids with isolated, purified SPA-6 apoprotein [22]. A second approach to be described is the formulation of mixtures of synthetic phospholipids with non-protein emulsifiers (e.g. perfluorocarbons) or other additives to enhance adsorption, and thus replace the need for protein in synthetic surfactant mixtures. A related approach to enhance the surface activity of phospholipid mixtures is by allowing them to attain specific physical states for better spreading, such as dehydrated phospholipids [18], or to attain specific microstructures in water for better adsorption, such as the Hex II phase or large unilamellar vesicles. Finally, a unique method for developing synthetic lung surfactants is through the *de novo* synthesis of analogs to biologic phospholipids which are designed to have specific hydrophobicity and resistance to enzymatic (phospholipases) transformability.

Development of Optimal Treatment Protocols

Additional clinical trials will be needed to explore the issues of dosage and the timing of administration. Our trials to date have used the same dose (90 mg) for all infants and a timing schedule that was constrained by a study design which attempted to limit the surfactant being given to infants. This limitation was based on concerns about biological toxicity or adverse clinical effects. To date, we have not found any significant toxicity or adverse clinical effects associated with the administration of surfactant.

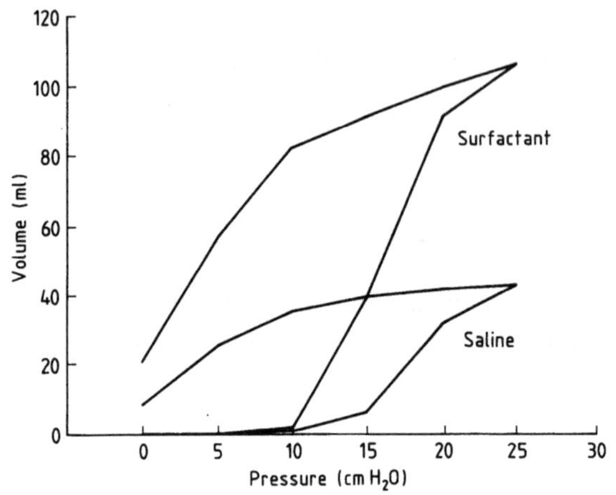

Fig. 4. Quasi-static open-chested lung pressure-volume curves for representative surfactant and saline rabbits. Both rabbits were exposed to 100% O_2 for 64 h followed by 24 h in room air. (From [16])

One of the problems in assessing dosage and timing issues is the ability to measure clinical response. Our current measures of oxygen need and ventilator settings are probably too crude. Our computer assisted pulmonary function testing system for the analysis of dynamic respiratory variables will be helpful in monitoring the clinical response to various treatment regimens.

Comparisons of Available Surfactants

Virtually all surfactants being tested in clinical trials have shown efficacy in reducing respiratory distress, and this has validated the therapeutic strategy of surfactant replacement. As new surfactant compounds become available, it is important that they be compared with known active surfactants such as CLSE rather than with untreated controls.

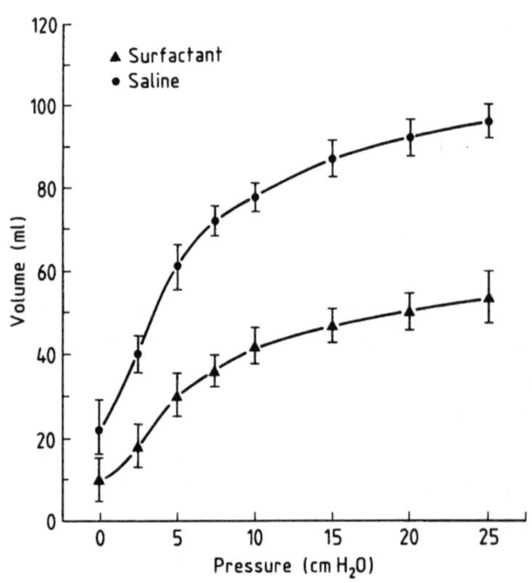

Fig. 5. Deflation limbs of quasi-static pressure-volume curves for surfactant- (n = 7) and saline- (n = 5) instilled rabbits. Each curve was calculated by averaging volume of each animal at indicated pressures. Values are means ± SE. (From [16])

Potential of Surfactant Replacement in Lung Injuries and ARDS

Earlier it was noted that severe lung injuries such as ARDS-like syndrome can develop a functional surfactant deficiency due, for example, to LS inactivation by plasma proteins. This has been shown conclusively to occur in severe hyperoxia by Holm, Notter, and co-workers [9, 11, 12], and also in the presence of hemorrhagic pulmonary edema [10]. However, Matalon et al. [16] have recently shown that this lung injury can be mitigated in vivo by instillation of sufficient CLSE (Figs. 4, 5). This will be an evolving area of research for those who have helped define exogenous surfactant replacement therapy for neonatal RDS.

References

1. Avery ME, Mead J (1959) Surface properties in relation to atelectasis and hyaline membrane disease. Am J Dis Child 97:517–523
2. Chu J, Clements JA, Cotton EK, et al (1967) Neonatal pulmonary ischemia: clinical and physiologic studies. Pediatrics 40:709–782
3. Davis JM, Veness-Meehan K, Shapiro DL (1987) Ventilatory response to exogenously administered calf lung surfactant extract (CLSE) in infants with respiratory distress syndrome. Pediatr Res 21:448A
4. Enhorning GE, Shennan A, Possmayer F, et al (1985) Prevention of neonatal respiratory distress syndrome by tracheal instillation of surfactant: a randomized clinical trial. Pediatrics 76:145–153
5. Farrell PM (ed) (1982) Lung development: biological and clinical perspectives, vols I, II. Academic, New York
6. Fujiwara T (1984) Surfactant replacement in neonatal RDS. In: Robertson B, van Golde LMG, Batenburg JJ (eds) Pulmonary surfactant. Elsevier, Amsterdam, pp 479–504
7. Gitlin JD, Soll RF, Parad RB, Horban JD, Feldman HA, Lucey JF, Taeusch HW (1987) Randomized controlled trial of exogenous surfactant for the treatment of hyaline membrane disease. Pediatrics 79:31–37
8. Hallman M, Merritt TA, Jarvenpaa A-L, Boynton B, Manniro F, Gluck L, Moore T, Edwards D (1985) Exogenous human surfactant for treatment of severe respiratory distress syndrome: a randomized, prospective clinical trial. J Pediatr 106:963–969
9. Holm BA, Notter RH (1986) Pulmonary surfactant effects in sublethal hyperoxic lung injury. In: Taylor AE, Matalon S, Ward P (eds) Physiology of oxygen radicals. APS Press, Bethesda, pp 71–86
10. Holm BA, Notter RH (1987) Effects of hemoglobin and cell membrane lipids on pulmonary surfactant activity. J Appl Physiol 63:1434–1442
11. Holm BA, Notter RH, Finkelstein JN (1985) Surface property changes from interactions of albumin with natural lung surfactant and extracted lung lipids. Chem Phys Lipids 38:287–298
12. Holm BA, Notter RH, Siegle J, Matalon S (1985) Pulmonary physiological and surfactant changes during injury and recovery from hyperoxia. J Appl Physiol 59:1402–1409
13. Hopewell PC, Murray J (1977) The adult respiratory distress syndrome. In: Shibel EM, Moser KM (eds) Respiratory emergencies. Mosby, St. Louis, pp 101–128
14. Kendig JW, Notter RH, Shapiro DL (1987) Improved survival in very premature infants treated with multiple doses of calf lung surfactant extract (CLSE). Pediatr Res 21:365A
15. Kwong MS, Egan EA, Notter RH, Shapiro DL (1985) A double blind clinical trial of calf lung lipid for the prevention of hyaline membrane disease in extremely premature infants. Pediatrics 76:585–592

16. Matalon S, Holm BA, Notter RH (1987) Mitigation of pulmonary hyperoxic injury by administration of exogenous surfactant. J Appl Physiol 62:756–761
17. Merritt TA, Hallman M, Bloom BT, Berry C, Beninschke K, Sahn D, Key T et al (1986) Prophylactic treatment of very premature infants with human surfactant. N Engl J Med 315:785–790
18. Morley CJ, Bangham AD, Miller N, Davis JA (1981) Dry artificial lung surfactant and its effects on very premature babies. Lancet 1:64–68
19. Notter RH, Finkelstein JN (1981) Pulmonary surfactant: an interdisciplinary approach. J Appl Physiol 57:1613–1624
20. Notter RH, Shapiro DL (1981) Lung surfactant in an era of replacement therapy. Pediatrics 68:781–789
21. Notter RH, Shapiro DL (1987) Lung surfactant for replacement therapy: biochemical, biophysical, and clinical aspects. Clin Perinatol 14:433–479
22. Notter RH, Shapiro DL, Ohning B, Whitsett JA (1987) Biophysical activity of synthetic phospholipids combined with purified lung surfactant 6,000 dalton apoprotein. Chem Phys Lipids 44:1–17
23. Rinaldo JE, Rogers RM (1982) Adult respiratory distress syndrome, changing concepts of lung injury and repair. N Engl J Med 15:900–909
24. Robertson B, van Golde LMG, Batenburg JJ (eds) (1984) Pulmonary surfactant. Elsevier, Amsterdam
25. Shapiro DL, Notter RH, Morin FC, Deluga KS, Golub LM, Sinkin RA, Weiss KI, Cox C (1985) A double-blind, randomized trial of a calf lung surfactant extract administered at birth to very premature infants for prevention of the respiratory distress syndrome. Pediatrics 76:593–599
26. Whitsett JA, Ohning BL, Ross G, Meuth J, Weaver T, Holm B, Shapiro DL, Notter RH (1986) Hydrophobic surfactant-associated protein (SAP 6000) in whole lung surfactant and its importance for biophysical activity in lung surfactant extracts for replacement therapy. Pediatr Res 20:460–467
27. Wilkinson A, Jenkins PA, Jeffrey JA (1985) Two controlled trials of dry artificial surfactant: early effects and later outcome in babies with surfactant deficiency. Lancet 2:287–291
28. Wood BP, Sinkin RA, Kendig JW, Notter RH, Shapiro DL (1987) Exogenous lung surfactant: effect on radiographic appearance in premature infants. Radiology 165:11–13

The European Multicenter Trial of Surfactant Replacement in Neonatal Respiratory Distress Syndrome

B. Robertson

Introduction

This paper summarizes the results of a large randomized multicenter trial, involving the use of a new type of surfactant (Curosurf), isolated from minced porcine lungs by a combination of washing, chloroform-methanol extraction, and liquid-gel chromatography. It contains 99% polar lipids, mainly phospholipids, and 1% hydrophobic, low molecular weight (< 15 kDa) proteins [1].

Trial design

The criteria for entry are given in Table 1 [2]. Babies were not randomized if there was evidence of prolonged rupture of the membranes (≥ 3 weeks; to avoid cases of pulmonary hypoplasia), intraventricular hemorrhage of grade III–IV [3], birth asphyxia as indicated by seizures, or major congenital anomalies. Our end-points were (a) to improve the quotient PaO_2/FiO_2 by 100% during six hours after surfactant replacement, (b) to reduce the period of artificial ventilation in survivors by 33%, and (c) to reduce mortality by 30%.

Those who entered the trial were allocated to the treatment or the control group by randomization with sealed envelopes, stratified for birth weight (700–1200 g, and 1201–2000 g). Surfactant was instilled into each main bronchus or into the lower trachea at a total dose of 2.5 ml/kg (80 mg/ml). Between and after the instillations, the baby was ventilated manually for a total of 2 min, using the same FiO_2 as before. Control babies received no material but were disconnected from the respirator system and ventilated manually for 2 min, using the same protocol as for the treated babies. After the period of manual ventilation the patients were reconnected to the respirator system and blood gases were analysed at pre-determined intervals. The respirator setting was modified with respect to the clinical response, in order to maintain adequate blood gases with the lowest possible levels of FiO_2 and insufflation pressure.

A number of parameters of gas exchange and lung mechanics were evaluated including PaO_2/FiO_2 ratio, right-to-left shunt, and mean airway pressure. We also recorded the incidence of intracerebral hemorrhage, persistent ductus arteriosus, pulmonary interstitial emphysema, pneumothorax, bronchopulmonary displasia (BPD), and fatal outcome.

Table 1. Criteria for entry

Birth weight 700–2000 g
Clinical and radiological findings typical of neonatal RDS [2]
Age at treatment 2–15 h
Requirement of artificial ventilation
$FiO_2 \geq 0.6$
No complicating disease

Results

A total of 146 babies were randomized in the trial, 77 surfactant-treated and 69 controls. The groups were well matched as regards gestational age (mean values 28.8 and 28.4 weeks; NS), birth weight (1246 and 1182 g; NS), male/female ratio (50/27 and 40/29; NS) age at randomization (8.6 and 8.3 h; NS), and FiO_2 before treatment (0.80 in both groups).

Treatment with surfactant resulted, within 5 min, in a prominent improvement of oxygenation, as reflected by a nearly threefold increase in the PaO_2/FiO_2 ratio (Table 2). The average therapeutic response was sustained, with statistically significant differences vs. the control group remaining for as long as 48 h. At the age of 6 h, there was a 98% improvement of the PaO_2/FiO_2 ratio in babies treated with surfactant, nearly satisfying the criterion for endpoint a. In spite of the striking improvement in oxygenation, mean airway pressure could be reduced but slightly in the surfactant-treated babies and statistically significant differences vs. the control group were observed only at the intervals 12 and 24 h after randomization (data not shown).

During the first four weeks after births, the median period of artificial ventilation among survivors was 35% shorter in surfactant-treated babies than

Table 2. PaO_2/FiO_2 (kPa, $\bar{X} \pm SD$) in surfactant-treated babies and controls at various intervals after randomization

Time	PaO_2/FiO_2		
	Surfactant-treated babies	Controls	$P <$
0[a]	10.3 ± 4.7	8.6 ± 4.6	NS
5 min	29.2 ± 16.2	10.6 ± 5.5	0.001
60 min	26.0 ± 13.1	10.6 ± 5.1	0.001
6 h	22.8 ± 11.2	11.5 ± 6.6	0.001
24 h	23.4 ± 11.2	12.8 ± 7.2	0.001
48 h	22.6 ± 11.2	16.9 ± 9.8	0.01

[a] At randomization.

in controls. The magnitude of this reduction matches end-point b, but there was a large variation in both groups and the difference is not statistically significant. The neonatal mortality ($\leq$ 4 weeks) was reduced in the surfactant-treated group (31% vs 51%; $P < 0.02$), satisfying the criterion for endpoint c. Treatment with surfactant also reduced the incidence of pulmonary interstitial emphysema (23% vs 39%; $P < 0.05$) and pneumothorax (18% vs 35%; $P < 0.05$) and the combined incidence of death and BPD of grade III–IV (45% vs 74%; $P < 0.001$). The incidence of intracerebral hemorrhage, patent ductus arteriosus, and pneumonia did not differ between the groups (data not shown).

Concluding Remarks

The results of the present and other randomized trials [4–7] indicate that treatment with surfactant improves survival rate, and reduces the incidence of air leaks and BPD in babies with moderate to severe RDS. In our study, the number of patients surviving without chronic lung disease was about twice as high in the surfactant-treated group as among controls (54% vs 26%; $P < 0.001$). These are clearly worthwhile effects that should encourage further use of surfactant replacement in neonatal RDS. To date, no long-term adverse effects have been reported after replacement therapy with heterologous surfactant in the neonatal period.

Details from the European Multicenter Trial including follow-up data will be published elsewhere [8].

Summary

The efficacy of replacement therapy with a new surfactant preparation (Curosurf) was tested in a randomized prospective clinical trial involving 146 patients with neonatal respiratory distress syndrome (RDS). At randomization, all patients required artificial ventilation with $FiO_2 \geq 0.6$ (average 0.8). Surfactant was instilled into the central airways at a dose of 200 mg phospholipids per kg body weight; no placebo was used in the control group. Within 5 min after surfactant replacement, there was a dramatic improvement of oxygenation as reflected by a threefold increase in the average PaO_2/FiO_2 ratio and a corresponding decrease in the right-to-left shunt. The therapeutic response was usually sustained, and statistically significant differences vs. the control group persisted until 48 h after randomization. Treatment with surfactant reduced the incidence of pulmonary interstitial emphysema and pneumothorax by 41% and 49%, respectively, and lowered early neonatal mortality by 39%. These data confirm the efficacy of surfactant replacement therapy in patients with severe neonatal RDS.

Acknowledgements. This work was supported by The Swedish Medical Research Council (Project No. 3351), The Swedish National Association against Heart and Chest Diseases, The "Expressen" Prenatal Research

Foundation, The General Maternity Hospital Foundation, and the Perinatal Trust Fund of Northern Ireland.

References

1. Noack G, Berggren P, Curstedt T et al (1987) Severe neonatal respiratory distress syndrome treated with the isolated phospholipid fraction of natural surfactant. Acta Paediatr Scand 76:697–705
2. Hjalmarson O (1981) Epidemiology and classification of acute, neonatal respiratory disorders. A prospective study. Acta Paediatr Scand 70:773–783
3. Papile LA, Burstein J, Burstein R, Koffler H (1978) Incidence and evolution of subependymal and intraventricular hemorrhage: a study of infants with birth weights less than 1,500 g. J Pediatr 92:529–534
4. Hallman M, Merritt TA, Jarvenpaa A-L et al (1985) Exogenous surfactant for treatment of severe respiratory distress syndrome: a randomized prospective clinical trial. J Pediatr 106:963–969
5. Gitlin JD, Soll RF, Parad RB et al (1987) Randomized controlled trial of exogenous surfactant for the treatment of hyaline membrane disease. Pediatrics 79:31–37
6. Raju TK, Vidyasagar D, Bhat R et al (1987) Double-blind controlled trial of single-dose treatment with bovine surfactant in severe hyaline membrane disease. Lancet 1:651–656
7. Fujiwara T, Konishi M, Ogawa Y et al (1987) Surfactant replacement for respiratory distress syndrome (RDS) – a multicenter clinical trial. Jpn J Pediatr 40:459–568
8. Collaborative European Multicenter Study Group (1988) An international randomized clinical trial of surfactant replacement therapy in severe neonatal respiratory distress syndrome. Pediatrics (in press)

Use of Surfactant in Very Low Birth Weight Infants with Severe Respiratory Distress Syndrome: Initial Experience and Problems

T. A. Clarke, S. Gormally, A. Bergin, M. King, and T. G. Matthews

It is now accepted that surfactant therapy is effective in prevention and amelioration of the respiratory distress syndrome (Avery et al. 1986). However, the question whether to await establishment of RDS or treat some or all infants at risk prophylactically at birth remains to be answered. The approach ultimately adopted may depend on the risk of mortality and morbidity from RDS in a particular perinatal unit and the efficacy of surfactant in routine clinical practice.

We report the use of porcine lung surfactant (Curosurf^R) in nine unselected very low birth weight (VLBW) infants with severe RDS. In addition we briefly review the mortality of VLBW infants in our hospital during the past two years as a background to provide guidelines for selection of infants for administration of surfactant at birth.

Materials and Methods

Infants of birth weight 700–1500 g with established severe RDS were considered for treatment. The decision to administer surfactant was made by a Consultant Paediatrician. A policy decision was made not to treat any infants prior to age 2 hours to allow confirmation of the diagnosis of RDS by chest X-ray. Infants were not excluded for reasons of perinatal asphyxia or of intraventricular haemorrhage (IVH). All infants had a cranial ultrasound prior to treatment.

Case Material

The infants birth weight ranged from 850–1335 g and gestational age from 24–29 weeks (Table 1). Pre treatment all required assisted ventilation with peak inspiratory pressures (PIP) 20–34 cm H_2O, FiO_2 = 0.8–1.0. The age at initial treatment varied from 2–6 hours. All treated infants received the isolated phospholipid fraction of porcine surfactant (Curosurf^R). The total surfactant dose was 2.5 ml/kg (phospholipid concentration 80 mg/ml); half the dose was instilled into each main bronchus through a feeding catheter (No. 5) and the infant reconnected to the ventilator at the same settings as before treatment.

Table 1. Clinical details of treated infants

Case	Birth weight (g)	Gestational age (weeks)	Sex	Delivery mode	Apgar score	Outcome	Comments
1	920	26	F	Breech	3, 6	Died (age 3 days)	Twin 2. Grade 2 IVH pre-Rx. Autopsy – severe RDS, grade IV IVH, PIE
2	860	24	M	LSCS (Breech)	2, 5	Died (age 12 h)	Immature. Severe RDS, PIE. Bilateral PMTs
3	950	27	M	LSCS (APH)	7, 9	Died (age 21 h)	APH dexameth. CXR – moderate RDS. Grade 2 IVH post Rx. Persistent shunting
4	1080	29	M	LSCS (Placenta praevia)	7, 9	Survived	Grade 2 IVH pre Rx. Right ventricular outflow tract obstruction
5	1115	29	M	SVD	4, 4	Survived (IPPV × 7 days, O_2 × 30 days	Intraventricular cyst, NEC
6	1185	27	M	LSCS (APH, oblique lie)	2, 8	Survived (IPPV × 58 h, O_2 × 7 days)	Hypotension requiring dopamine
7	1230	29	M	LSCS (Twins)	5, 7	Died (age 52 h)	PROM × 36 h. Twin 1. Hypotension after birth. BP 30/16. Severe PIE. Bilateral PMTs
8	1120	26	F	LSCS (Breech, APH)	3, 3	Died (age 11 h)	PROM × 8 weeks. Oligohydramnios. Low set ears. Difficult to ventilate. Severe PIE
9	1335	27	F	SVD	5, 9	Survived (IPPV × 7 days, O_2 × 69 days)	Grade 1 IVH pre Rx. PDA (Indomethacin). BPD

Results

In all infants oxygenation as indicated by transcutaneous oxygen monitor and confirmed by arterial oxygen sampling increased immediately (Table 2). In most infants it was possible to decrease ventilator inspiratory pressures significantly in the first hour following treatment. PIP decreased an average of 5 cm H_2O by age 1 hour, range decrease of 14 to increase of 2 cm H_2O with no adverse affect on ventilation as indicated by PCO_2 values. Most infants experienced a brief decrease in oxygenation during surfactant administration but oxygenation had improved compared to pre-treatment values in all infants by age 5 min (Table 2). In Case 4 the improvement in oxygenation was transient and congenital heart disease was subsequently diagnosed. A transient fall in blood pressure was noted in 7 infants. A significant fall in blood pressure occurred in Case 8 from 49/30–29/24 mm Hg at age 5 min.

Five infants died. Case 1 had a significant IVH present at birth. Cases 2 and 7 developed pulmonary interstitial emphysema (PIE) and fatal pneumothoraces.

Case 3 (Table 2b) became hypoxic at age 3 hours, one hour following surfactant treatment, due to persistent shunting. A tolazoline infusion was begun at age 6 hours with increase in $tcPO_2$ from 0.2–4 kPa. PaO_2 rose briefly

Table 2a. Oxygenation: response following surfactant administration

Case	Pretreatment	5 min	1 h	6 h
1	6.78 (1.0)	8.8 (1.0)	21.12 (1.0)	9.08 (0.8)
2	6.74 (1.0)	36.04 (1.0)	11.05 (0.75)	5.27 (1.0)
3	3.8 (1.0)	7.68 (1.0)	2.09 (1.0)	16.66 (1.0)
4	5.47 (1.0)	7.69 (1.0)	4.16 (1.0)	2.76 (1.0)
5	7.36 (1.0)	–	35.22 (0.9)	7.9 (0.4)
6	8.41 (0.8)	13.3 (0.7)	10.5 (0.55)	9.21 (0.5)
7	5.5 (1.0)	29.06 (1.0)	11.8 (0.65)	7.13 (0.75)
8a	3.21 (1.0)	7.7 (0.1)	9.5 (1.0)	2.82 (1.0)
8b	0.8 (1.0)	6.7 (1.0)	(dead)	
9	6.3 (0.9)	22.86 (0.9)	15.64 (0.6)	5.86 (0.45)

FiO_2 values are given in parentheses.

Table 2b. Data for Case 3 (severe shunting following surfactant treatment)

	Pretreatment	Post treatment				
		5 min	15 min	30 min	1 h	3 h
$TcPO_2$ (kPa)	2.78	15.3	19.9	12.9	4.29	5.64
$TcPCO_2$ (kPa)	10.91	9.9	6.3	2.6	3.79	0.97
PIP (cm H_2O)	34	28		17	20	24
PEEP (cm H_2O)	3	3		1	1	3
BP (mm Hg)	45/25	–		39/28	47/33	60/41

FiO_2 value is 1.0 throughout.

to 16.66 at 9 hours. Persistent hypoxia subsequently recurred although ventilation remained easy (PCO_2 ranging from 1.55–3.12 kPa).

At autopsy Case 8 had pulmonary hypoplasia (Table 1). Cord pH was 7.32. Intubation was required at birth and the lungs were difficult to ventilate. Arterial blood gas at age 70 minutes was pH 6.82, PCO_2 16.19 kpa and PaO_2 3.21 kPa with PIP 40 cm H_2O and rate 76/minute in 100% oxygen. Following surfactant $tcPO_2$ rose from 1.4–7.7 kPa. The fall in PCO_2 was more gradual. At age 4 hours pH was 7.4, PCO_2 6.54 and PaO_2 9.79 kPa on PIP 32 and rate 60/minute in 95% oxygen. PIE developed. The infant was re-treated at 9 hours with resulting improvement in $tcPO_2$ from 0.8–6.7 kPa at 5 minutes. However, resistant hypotension developed.

Neonatal Mortality in VLBW Infants

During the period Jan 1985–Dec 1986 31 of 108 normally formed inborn live VLBW infants died at the Rotunda Hospital, Dublin (Matthews and Clarke 1987). None had received surfactant treatment. Survival was 68% (13 of 19) at 27–28 weeks gestation compared with 97% survival at 29–30 weeks gestation. Forty-eight percent (13 of 27 infants) with birth weight 750–999 g survived compared with 94% (62 of 66) at 1000–1500 g. The survival rate was 89% (42 of 47) in infants delivered by caesarean section, 65% (27 of 41) in vertex deliveries and 40% (8 of 20) in breech deliveries. Thirteen extremely immature infants were not resuscitated. Of the other 18 deaths, autopsy revealed the main factors to be RDS [11], PIE [6], pneumothorax [3], and IVH [6].

Discussion

Many studies have shown the benefit of the use of surfactant in selected groups of infants in controlled trials (Merritt et al. 1986; Gitlin et al. 1987; Raju et al. 1987). Concern has been expressed about changes which may follow the use of surfactant, including an increased incidence of patent ductus arteriosus (Fujiwara et al. 1980) and fears of an increased incidence of air leaks. Three of the deaths in our infants may have been related to changes in lung mechanics following the use of surfactant. Although two of our infants died from pneumothoraces many studies show a decreased incidence of air leaks (Merritt et al. 1986; Gitlin et al. 1987). The development of persistent hypoxia (Case 3) following the administration of surfactant may have been coincidental or may have been related to inadvertent delay in decreasing ventilator pressures resulting in tamponade of the pulmonary circulation. It has been shown that in some infants a decrease in PaO_2 may be associated with increased mean airway pressure (MAP) (Stewart et al. 1981). It is suggested that in some patients who have higher pulmonary artery pressures with a patent ductus arteriosus and/or foramen ovale an increase in airway pressure could lead to a right to left shunt and resultant hypoxia. Alvolar distention could also result in ventilator perfusion mismatch. Improved lung compliance following surfactant treatment may increase the risk of adverse effects from excessive MAP in some infants.

A transient fall in blood pressure could possibly be due to impaired venous return associated with improved lung compliance and excessive transmission of airway pressure to the capillaries or possibly secondary to increased pulmonary blood flow following treatment. A direct effect of a component of the surfactant on systemic circulation is unlikely.

Neonatal pulmonary hypoplasia associated with premature rupture of membranes and oligohydramnios is a well recognised entity (Thiebault et al. 1985). Pulmonary hypoplasia and intrauterine crowding deformities may occur in a spectrum of severity. The course of infants with pulmonary hypoplasia resembles severe RDS with persistent pulmonary hypertension (Thiebault et al. 1985) as occurred in Case 8. Of 76 infants in Thiebault's series the 3 survivors of the 21 with positional deformities as well as those without deformities had clinically normal pulmonary function at age 1 year. The response of Case 8 to surfactant indicates that surfactant deficiency may be an important additional factor superimposed on the underlying alveolar hypoplasia in such infants.

Concern about possible long-term effects of treatment remain, although no short-term adverse immunogenic reactions have been observed (Avery et al. 1986). Infectious agents are another potential problem.

Others have noted a high mortality in "poor responders" (H. Halliday, personal communication) i.e. infants who continue to require $FiO_2 > 0.6$ 4 hours following treatment. Some such infants may be suffering from the effects of asphyxia.

In view of the risk of side effects and the desire to avoid unnecessary treatment, studies in individual hospitals are required to determine which infants are at high risk of dying or suffering significant morbidity from RDS. In our hospital a significant number of deaths in infants with birth weight below 1000 g and gestational age $\leq$ 28 weeks remains due to RDS or its sequelae (Clarke et al. 1987; Matthews and Clarke 1987). In our experience infants born by breech or vertex delivery are at high risk of dying compared to those delivered by caesarean section (Matthews and Clarke 1987). Use of rapid "bedside" techniques prior to birth or at delivery to estimate lung maturity, such as the shake test (Clements et al. 1972), may avoid unnecessary treatment in some infants.

Our experience following surfactant treatment in a small group of VLBW infants with severe RDS suggests the need to treat selected infants before RDS becomes established. However the deaths of 3 infants, possibly related to changes in lung mechanics, emphasizes the need for very close monitoring of surfactant treated infants. Paradoxically, treated infants may be less stable than might be presumed due to the improving blood gases.

Acknowledgements. We thank Doctors Tore Curstedt (Department of Clinical Chemistry, Karolinska Hospital, Stockholm) and Bengt Robertson (Department of Paediatrics, St. Goran's Children's Hospital, Stockholm) for supplying us with Curosurf.

References

Avery ME, Taeusch HW, Floros J (1986) Surfactant replacement. N Engl J Med 315:825–826

Clarke TA, McKenna PJ, Matthews TG (1987) Factors associated with neonatal deaths in very low birth weight infants. Ir J Med Sci 156:211–214

Clements JA, Platzker ACG, Tierney DF et al (1972) Assessment of the risk of the respiratory distress syndrome by a rapid test for surfactant in amniotic fluid. N Engl J Med 286:1077

Fujiwara T, Maeta H, Chida S et al (1980) Artificial surfactant therapy in hyaline membrane disease. Lancet 1:55–59

Gitlin JD, Soll RF, Parad RB et al (1987) Randomized controlled trial of exogenous surfactant for the treatment of hyaline membrane disease. Pediatrics 79:31–37

Matthews TG, Clarke TA (1987) Very low birth weight infants – survival related to obstetric and neonatal variables. Rotunda Hospital Dublin; Clinical Report 1986, pp 88–90

Merritt TA, Hallmann M, Bloom BT et al (1986) Prophylactic treatment of very premature infants with human surfactant. N Engl J Med 315:785–790

Raju TNK, Vidyasgar D, Bhat R (1987) Double-blind controlled trial of single-dose treatment with bovine surfactant in severe hyaline membrane disease. Lancet 1:651–656

Stewart AR, Finer NN, Peters KL (1981) Effects of alterations of inspiratory and expiratory pressure and inspiratory/expiratory ratios on mean airway pressure, blood gases and intracranial pressure. Pediatrics 67:474–481

Thiebault DW, Beatty EC, Hall RT et al (1985) Neonatal pulmonary hypoplasia with premature rupture of fetal membranes and oligolhydramnios. J Pediatr 107:273–277

Surfactant Replacement with SF-RI 1 in Premature Infants with Respiratory Distress Syndrome: A Clinical Pilot Study

L. Gortner, F. Pohlandt, P. Bartmann, B. Disse, and E. Weller

Patients and Methods

In a clinical pilot study, a bovine surfactant (SF-RI 1, Dr. Karl Thomae, FRG) was administered intratracheally to 34 very low birth weight infants with RDS.

The surfactant preparation was given if an $FiO_2 > 0.5$ was necessary to achieve a $paO_2 > 50$ mm Hg and/or if the peak inspiratory pressure (PIP) necessary to achieve normoventilation exceeded the following limits:

PIP > 22 cm H_2O in infants < 750 g birth weight (b. w.)
PIP > 24 cm H_2O in infants 750–1000 g b. w.
PIP > 28 cm H_2O in infants > 1000 g b. w.

In the first 19 patients the dose was 20–30 mg/kg b. w. (group 1), in the other 15 patients 40–50 mg/kg b. w. (group 2). Retreatment with surfactant (identical dose) was permitted, if the aforementioned limits for FiO_2 or PIP after 12–24 hours had been exceeded. The birth weight of our patients was in the range 430–1500 g (median 770 g). The corresponding gestational age ranged from 24–31 weeks (median 27 weeks).

Results

Of 34, 24 (71%) survived (b.w. 430–1500 g), and 9 patients died after variable periods of time (b. w. 570–1010 g). Survival rate in group 1 was 58% (11 of 19 patients), in group 2 however 87% (13 of 15 patients).

Conclusions

A dose of 40–50 mg/kg b. w. SF-RI 1 is more effective in the treatment of RDS in premature infants than 20–30 mg/kg b. w. Clinical data of our patients treated with 40–50 mg/kg b. w. are comparable to those published in other studies with surfactant preparations of natural origin.

Ventilator Settings After Surfactant Replacement with SF-RI 1: Clinical Experiences

L. Gortner, F. Pohlandt, P. Bartmann, B. Disse, and E. Weller

Patients and Methods

Thirty-four very low birth weight infants, whose clinical data were already given (Gortner et al. this book) were treated with a bovine surfactant intratracheally (SF-RI 1, Dr. Karl Thomae GmbH, FRG). Mechanical ventilation was provided by time-cycled, pressure-limited ventilators. Peak inspiratory pressure (PIP) and positive end-expiratory pressure (PEEP) were adjusted in order to keep paO_2 between 40–60 mm Hg and $paCO_2$ between 40–45 mm Hg. Routinely, respiratory rate was 100/min with an inspiratory time of 0.2 and an expiratory time of 0.4 s.

Mean airway pressure (MAP) was calculated according to Zmora and Merritt (1980). Blood gases were measured transcutaneously ($tcpO_2$, $tcpCO_2$), using a single electrochemical sensor (Transend TM Cutanous Gas System, Sensor-Medics). Values were controlled by arterial blood gas analysis (AVL 938) in intervals of 4–6 hours. Values are given as means ± SEM.

Results

The ratio PaO_2/PAO_2 increased from the pre-treatment value of 0.13 ± 0.014 ($n = 21$) to 0.27 ± 0.028 ($n = 19$) after 15 min. The ratio slowly declined to 0.20 ± 0.024 ($n = 20$) within 12 hours.

PIP could not be lowered as rapidly as FiO_2. The pressure decreased over the first 12 hours after treatment from 29.1 ± 1.9 cm H_2O ($n = 21$) to 23.85 ± 1.5 cm H_2O ($n = 20$). MAP was affected in a similar way, decreasing from 9.2 ± 0.59 cm H_2O ($n = 21$) to a minimum of 7.3 ± 0.47 cm H_2O ($n = 20$) after 12 hours.

When using high frequency positive pressure ventilation, changes in pulmonary mechanics after surfactant application – a longer pulmonary time constant – must be taken into account.

Conclusions

The changes in ventilator settings after surfactant replacement in our experience are comparable to those, reported by Hallman et al. (1985) with respect to the arterial/alveolar oxygen-ratio and MAP. When using high frequency positive pressure ventilation a decrease in the respiratory rate (prolongation of expiratory time) is mandatory in order to avoid inadvertent PEEP after surfactant application.

References

Gortner L, Pohlandt F, Bartmann P, Disse B, Weller E (1987) Surfactant replacement with SF-RI 1 in premature infants with RDS: a clinical pilot study. (This book)

Hallman M, Merritt TA, Jarvenpaa A-L, Boynton B, Mannino F, Gluck L, Moore T, Edwards D (1985) Exogenous human surfactant for treatment of severe respiratory distress syndrome: a randomized prospective clinical trial. J Pediatr 106:963–969

Zmora E, Merritt TA (1980) Use of a side-hole endotracheal tube adapter for tracheal aspiration. Am J Dis Child 134:250–254

Human Surfactant from Amniotic Fluid in the Treatment of Respiratory Distress Syndrome*

M. Hallman, M. Pohjavuori, K. Bry, and T. A. Merritt

Introduction

The major issue in surfactant substitution has been the availability of an efficient and safe preparate. At present the quantitative and especially the qualitative requirements of surfactant are unclear. The evaluation is complex because the severity of RDS varies widely. There are a number of possible preparates that function well both in vitro and in vivo in animals. The first and the best known was that used by Fujiwara et al. (1980). However, the suitability is ultimately evaluated on the basis of controlled trials comparing the outcome and the long-term follow-up.

Surfactant deficiency in RDS is characterized by lack of the major phospholipids and proteins. Although there are definite differences in the ontogeny of individual surfactant components, it is well established that surfactant pool size is the major determinant of alveolar instability in RDS. A logical approach to evaluate risks and benefits of surfactant substitution is to use a preparate that closely resembles the complex that is missing from the airways. The present short review summarizes the approach of harvesting natural human surfactant from the amniotic fluid (HA-surfactant) for treatment of RDS. Secondly, we point out that surfactant substitution allows to evaluate our current therapy of RDS, particularly the cardiovascular abnormalities that play a significant role in the pathogenesis of RDS.

Recovery of Amniotic Fluid Surfactant

Quantity Available in Amniotic Fluid

In the human fetus, unlike the experimental animals thus far studied, large quantities of surfactant components are carried by the fetal lung fluid into the amniotic fluid (Hallman 1984). The pool size of surfactant at birth, expressed on the basis of body weight, or lung dry weight, is about one order of magnitude higher than that in the adult. In the human fetus at term, our

* Supported by grants from the Finnish Academy, the Sigrid Juselius Foundation, the Foundation for Pediatric Research in Finland, and National Institutes of Health (Hl-35036).

estimate is 100–200 mg total surfactant/kg, whereas in the adult, it ranges between 10 and 20 mg/kg. These figures are based on the measurement of the extracellular lavagable surfactant pool, and on the assumption that 50 % of total surfactant is present in the intracellular space. The large surfactant pool at term is in part due to the fact that type II cells constitute nearly 100 % of alveolar lining cells, whereas in the adult the corresponding percentage is 25 (Pinkerton et al. 1982). In the adult the alveolar surface area per body weight is higher than in the newborn. The calculated number of type II cells per lung is 3–6 times higher in the adult than in the newborn.

At term, sometimes up to one half of the surfactant storage resides in the transcellular space: the airways, amniotic fluid and the stomach. If we assume that fetal surfactant is entirely and exclusively cleared by fetal swallowing, then

Table 1. Isolation of human surfactant from amniotic fluid

Refrigerate amniotic fluid within 1 h and store at +4° C for no more than 48 h before the first centrifugation
Centrifuge 300 xg for 5 min. At this point the supernatant may be frozen at – 20° C for no more than 4 months
Collect supernatant through sterile 20 μm nylon mesh
Centrifuge supernatant 10,000 xg, 90 min
Collect pellet. It can be frozen for no longer than 4 months, or processed further
Vortex pellet with sterile 0.9 % NaCl until completely homogenous
Prepare gradient for ultracentrifuge using sterile tubes and solutions. Layer gradient with 13 ml of 0.60 M Sucrose-10 mM Tris-Cl-0.1 mM EDTA-20 mM NaCl, pH 7.4 at 0° C; 13 ml of 0.27 M Sucrose-Tris-Cl-EDTA-NaCl; 2 ml of 0.9 % NaCl. Then top off with the preparate suspended in 0.9 % of NaCl (one tube contains material from 50–100 ml of amniotic fluid).
Centrifuge in swinging bucket rotor (SW 27) at 27,000 RPM for 60 min at + 4° C
Remove the turbid band between 0.60 M and 0.27 M Sucrose layers
Filter the liposomes through 20 μm mesh nylon filter. Do not use vacuum
Dilute with excess of water
Centrifuge at 10,000 xg for 120 min
Collect pellet in small amount of water
Remove 10 μl duplicates for determination of phospholipid phosphorus
Take 2 % of material for bacterial and fungal cultures
When bacterial and fungal cultures are found to be negative combine specimens containing together 100 μmol phospholipid
Measure surface activity
Freeze dry the specimen
Expose the freeze dried specimens to strong UV-light for 16 h
The weight of surfactant estimated as follows: μmol phospholipid phosporus / 0.80 = 1 mg surfactant
New batch of surfactant is made when enough processed, sterile, surface active material, to make up at least 8 doses (> 650 mg)
Freeze dried surfactant powder suspended using Vortex maximum speed and sterile plastic Pasteur pipette
Surfactant carrier is 100 mM NaCl, 1.5 mM $CaCl_2$. Final concentration is 25 mg surfactant/1 ml carrier
Surfactant suspension contains no visible particles, and it passes freely 20 μM nylon mesh
Divide the batch into individual doses of known size (currently 80 mg each)
Heat for 30 min at 56° C. During heating, vortex briefly every 120 s

the calculated half-life of the complex is 36 hours (range 15–74 hours, provided that the range of the amniotic fluid volume is 0.4–2 liters and fetal swallowing 0.45 liters/24 hours).

Method of Collection and Isolation

Collection of the amniotic fluid for surfactant therapy is indicated when there is: (a) term pregnancy, (b) cesarean section with intact fetal membranes, and (c) no evidence of maternal or fetal infection or amnionitis. The amniotic fluid is suctioned into a sterile container, *via* a catheter with multiple holes, introduced into the amniotic cavity through a small hole in fetal membranes.

Tables 1, 2 show the isolation and testing of HA-surfactant (Hallman et al. 1983). The surfactant is routinely studied for surface activity, for microbial contamination, for saturated phosphatidylcholine and total phospholipid content. In order to eliminate inpurities and minimize variability in properties, the batch size has been increased to a minimum of 8 doses.

Artificial rupture of fetal membranes to speed up the term delivery is routinely used in many perinatal centers. In theory, this would allow more surfactant to be recovered than during cesarean section alone. However, the risks of microbial contamination are higher.

Viable microbes can be eliminated by lipid extraction. When SP 35 was added to the lipid extract of HA-surfactant (contains SP 18 and SP 4–6), it improved both in vitro and in vivo surface activity as defined by the ability to stabilize the lungs of ventilated immature 27-day-old rabbits. In addition, the inhibitor resistance of this artificial human surfactant was better than that from rabbit lung lavage.

Table 2. Routine testing of human surfactant batch from amniotic fluid

- Surface activity
  - Dilute specimen to 4 mg/ml using the same carrier
  - Apply specimen to pulsating bubble surfactometer chamber
  - Make the bubble with the radius of 0.55 mm, and wait for 15 s while continuously recording the pressure in the chamber
  - After 15 s, make the bubble to pulsate between 0.55 and 0.4 mm radius at frequency of 0.33 Hz
  - Continue pulsation for 5 min while continuously recording the pressure
- Bacterial and fungal cultures
- Herpes and cytomegalovirus tests
- Analysis of total phospholipid and saturated phosphatidylcholine
- Acceptable surfactant
  - Demonstrates surface activity using pulsating bubble surfactometer
    - Surface adsorption to 25 dynes/cm within less than 5 s
    - Minimum surface tension less than 10 dynes/cm within 1 min
    - Minimum surface tension less than 5 dynes/cm within 5 min
  - No detectable bacteria or fungi
  - Negative viral tests
  - Saturated phosphatidylcholine/total phospholipid ratio higher than 0.45

According to current evidence the quality of HA-surfactant can be improved over the present one. Additionally, the organic solvent method of HA-surfactant isolation would potentially allow vaginally derived surfactant to be used for treatment of RDS, increasing the yield of surfactant.

Abnormal Perinatal Adaptation

In preterm infants there are a number of factors that adversely affect the respiratory adaptation (Hallman, and Gluck 1982). Besides surfactant deficiency, the preterm infants have high alveolar permeability, small dimensions of the air spaces, elastic chest cage, immature respiratory muscles and their coordination, lack of interstitial and alveolar connective tissue support, broad intersaccular septum, low oncotic pressure, and inefficient biochemical and immunological defence mechanisms. Furthermore, low capacity of the pulmonary vascular bed, immature cardiac myocytes, and inadequate constrictive response of the ductal ring following increase in oxygen tension, all contribute to exceptionally high tendency towards cardiac failure and pulmonary flooding (Friedman and George 1984).

As proposed by Enhörning and Robertson (1973) a logical approach to treat surfactant deficiency is to administer homologous, natural surfactant. Since amniotic fluid is a major source of surfactant we designed a method for its recovery, in order to study the benefits and risks of homologous natural surfactant.

Human Surfactant Substitution – Spectrum of Clinical Responses

Altogether 112 small preterm infants (gestational age 24.0–29.9 weeks) participated in one of the three randomized clinical trials in University Central Hospital, Helsinki; 61 of them received a total of 119 doses of amniotic fluid surfactant, each 60–105 mg in 3–4 cc of saline (Hallman et al. 1985; Merritt et al. 1986). The 51 control infants received air as a placebo. The lung effluent (amniotic fluid or tracheal aspirate) was evaluated for surfactant phospholipids prior to exogenous surfactant. The patients were eligible for surfactant when the lung effluent was "immature": the L/S ratio was less than 2 and phosphatidylglycerol undetectable at birth.

The following data describe briefly the spectrum of clinical responses following exogenous surfactant in diverse, sometimes desperate clinical conditions. We describe the patients in five categories.

Small Preterm Infants with "Mature" Endogenous Surfactant (Gestational Age 27.2–29.7 Weeks)

Two infants received surfactant at birth, one at the age of two hours. The surfactant-rich liquid was cleared rapidly from the airways. None of the infants had excessive phospholipid recovery during routine suctioning of the airways, or any other evidence of side effects. All three infants were extubated within two days and required no supplemental oxygen beyond the age of one week.

"Immature" Infants Requiring a Single Dose (Gestational Age 26.4–29.9 Weeks)

Altogether 6 of the 18 cases who received surfactant in respiratory failure ("rescue") had an acute increase in arterial PO_2, and had no relapse. Of the 24 infants who were treated within 10 min from birth, 11 required no retreatment and had a mild respiratory course. No infant died or developed BPD.

"Immature" Infants Requiring Retreatment with Subsequent Good Response (Gestational Age 26–29.7 Weeks)

Altogether 10 of the 18 "rescue" treated infants and 11 of the 24 "prophylactically" treated infants required more than one dose. Retreatment was allocated whenever there was a "relapse" (deterioration of the respiratory status following initial surfactant-induced improvement), or whenever the therapeutic effect was deemed to be unsatisfactory ("prophylactic retreatment": inspiratory oxygen requirement remains above 40–50 % and the mean airway pressure above 8–10 cm H_2O). The "prophylactic retreatment" is preferred because only one of these 11 infants demonstrated a true "relapse". There were 2 cases of BPD among the infants who required retreatment. One infant died.

"Immature" Infants with Transient or Small Response to Exogenous Surfactant

Of the small preterm infants three demonstrated a transient increase in arterial PO_2 from 38 ± 12 to 86 ± 15 mm Hg, of less than 1 hour in duration. Two infants died and one developed BPD and quadriplegia. Two infants had underlying severe birth asphyxia and one severe infection. There were a total of 6 placebo-treated "controls" who could not be resuscitated or died within 5 hours after birth of severe respiratory failure. One surfactant-treated and two placebo-treated infants additionally received prostacyclin in an attempt to relieve pulmonary vasospasm. In all cases there was an acute increase in arterial PO_2 from 21 ± 10 to 58 ± 18 mm Hg, but all died within a few hours. According to clinical and autopsy findings these infants had RDS and persistence of pulmonary hypertension.

Two additional preterm infants had severe fluid overload at birth. There was only little acute improvement of respiratory failure after exogenous surfactant (inspiratory O_2 was decreased by 20–30 %, no decrease in mean airway pressure). The respiratory failure was relieved first when the fluid overload was treated. One infant developed BPD.

"Immature" Very Small Preterm Infants (Gestational Age 24–25.9 Weeks)

A total of 16 surfactant-treated infants were studied. Ten received "prophylactic" surfactant, five "rescue" surfactant. One infant had severe fluid overload and was excluded from present data. There were ten placebo-treated infants; 7 of them died during the first neonatal week.

The seven first infants who received surfactant (6 were retreated) required less than 50 % O_2. However, at a mean age of 18 hours (range 8–24 hours), the clinical condition acutely deteriorated. The mean arterial pressure was low (20 ± 3 cm H_2O), and precordium was hyperactive suggesting symptomatic patent ductus arteriosus. The shock was reversed by infusing albumin and red blood cells, or by *decreasing* the positive endexpiratory pressure. One infant survived. Among those who died, four had hyaline membrane disease, three grade 3–4 intraventricular hemorrhage, five patent ductus arteriosus.

In an attempt to prevent the shock-like condition that apparently caused an early death of surfactant-treated infants, the following change in the management was made. During the first two neonatal days, 4 % of albumin and packed red blood cells were given, not only to treat the iatrogenic blood loss but an additional 20–30 ml/kg/day was infused, in an attempt to maintain an adequate perfusion. Since then there have been significantly less problems: seven of the eight infants have survived.

Conclusions

We have supplemented natural human surfactant isolated from amniotic fluid (HA-surfactant) in RDS in order to decrease possible side effects due to immunization and to maximize the efficacy. Quantitatively this is a viable alternative to treat RDS, or even postneonatal infants, children and adults with respiratory distress syndrome. HA-surfactant as a possible source of infection has been considered. There has been no increase in the incidence of infections. Comparative trials or data on immunological safety of the various surfactants are very incomplete or non-existent. The incidence of BPD appears to decrease following HA-surfactant. Similar to other natural surfactants, HA-surfactant acutely decreases the severity of RDS.

The surfactants tested at present may represent first or second generation of a future line of more effective and more safe preparates. It is likely that a gene technological homologous surfactant will eventually replace the preparates currently available.

The surfactant dose has to be adjusted according to clinical needs and factors (especially patent ductus arteriosus and leaky alveolar-capillary barrier allowing proteinous inhibitors into the airways) that deteriorate respiratory function. In the future rapid and accurate surfactant tests may help to determine the suitable dose of surfactant. Although a small dose of surfactant (60–80 mg, *vs* the endogenous pool size 120–180 mg) given to the preterm babies with "mature" surfactant profile had no harmful effects, the potential toxicity of surfactant excess has not yet been ruled out.

The present studies have demonstrated evidence that in some rare cases of RDS the respiratory failure was not due to lack of surfactant, although a surfactant defect was invariably present. These infants had either severe fluid overload, or severe asphyxia and persistent pulmonary hypertension.

A major issue appears to be how to manage the infants receiving exogenous surfactant and how to maximize the synergism with other therapies. In the immature small preterm infants exogenous surfactant both shortened the need of respirator therapy, diminished the life-threatening complications, and improved the quality of life of the survivors.

According to our clinical experience, successful treatment of the most immature infants (at least gestational ages of 24–25 weeks) requires a change in our current practice of fluid management: infants who received 20 ml/kg of blood products, 60–80 ml/kg of intravenous fluids and dopamine, developed a serious cardiac failure during the first day. Since it is known that increase in cardiac preload increases the cardiac output even among the smallest infants, the fluid allowance, especially the blood products, were increased during the first two neonatal days: these infants received 140–170 ml/kg of total fluids, including 30–40 ml/kg of albumin and red blood cells per day. The preliminary results using the new protocol is promising, since the mortality of these extremely immature, surfactant-treated infants has dropped from 86 % to 13 %.

References

Enhörning G, Grossman G, Robertson B (1973) Tracheal deposition of surfactant before the first breath. Am Rev Respir Dis 107:921–926

Friedman WF, George BC (1984) Treatment of congestive heart failure by altering loading conditions of the heart. J Pediatr 106:697–706

Fujiwara T, Chida S, Watabe Y, Maeta H, Morita T, Abe T (1980) Artificial surfactant therapy in hyaline membrane disease. Lancet 1:55–59

Hallman M (1984) Antenatal diagnosis of lung maturity. In: Robertson B, Van Golde LMG, Batenburg JJ (eds) Pulmonary surfactant. Elsevier, Amsterdam, pp 419–448

Hallman M, Gluck L (1982) Respiratory distress syndrome – update 1982. Pediatr Clin North Am 29:1057–1075

Hallman M, Merritt TA, Schneider H, Epstein BL, Mannino F, Edwards DK, Gluck L (1983) Isolation of human surfactant from amniotic fluid and a pilot study of its efficacy in respiratory distress syndrome. Pediatrics 71:473–482

Hallman M, Merritt TA, Järvenpää A-L, Boynton B, Mannino F, Gluck L, Moore T, Edwards D (1985): Exogenous human surfactant for treatment of severe respiratory distress syndrome: a randomized prospective clinical trial. J Pediatr 106:963–969

Merritt TA, Hallman M, Bloom B et al. (1986) Prophylactic treatment of very premature infants with human surfactant. N Engl J Med 315:785–790

Pinkerton KE, Barry BE, O'Neil JJ, Raub JA, Pratt PC, Crapo JD (1982) Morphologic changes in the lung during the lifespan of Fischer 344 Rats. Am J Anat 164:155–174

Clinical Response to Human Surfactant*

T. A. Merritt, M. Hallman, Y. E. Vaucher, C. Berry, and A.-L. Jarvenpaa

Introduction

Human surfactant derived from term amniotic fluid has biochemical and biophysical characteristics nearly identical to surfactant obtained from airway lavage of infants delivered at term and free from respiratory disease [1]. Human surfactant contains, in addition to predominantly dipalmitoylphosphatidylcholine and phosphatidylglycerol, the three surfactant apoproteins (35k, 18k, and 9k) that are essential constituents of surfactant [2]. As described by Hallman et al. the biochemical and surface active properties of human surfactant when instilled into the airways caused the rapid reversal of respiratory failure in preterm infants with documented Respiratory Distress Syndrome (RDS). This reversal of ventilatory failure was characterized by an immediate improvement in PaO_2, thereby permitting a rapid reduction in FiO_2 and a more gradual lowering of ventilatory pressures over the first 72 hours after birth [3]. In contrast to organic solvent extracts of animal lung mince or lavage, we reasoned that human surfactant would involve exposure to species specific proteins, might reduce the possibility of immunologic sensitization to bovine surfactant proteins and be free from organic solvent residues that might prove injurious to human preterm infants.

Because of concerns regarding potential transfer of infectious agents from amniotic fluid derived surfactant to infant recipients, techniques were developed to insure sterility throughout the isolation process and to document the absence of bacterial, fungal, and viral contamination (including HIV) [4] in the reconstituted surfactant used clinically. Rigorous quality control procedures evaluating adsorption and surface tension characteristics were performed on each series of doses made for the subsequent clinical trials. In vitro activity was documented and in vivo activity in initial animal studies and in an ongoing fashion studies confirmed the efficacy of this material [5, 6].

Clinical Methods

Because only 40 %–50 % of infants delivered at less than 30 weeks gestation develop respiratory distress syndrome [7, 8], our initial randomized controlled

* Supported by USPHS-NH-HL-35036, FDA Office of Orphan Drug Products, Division of Research Resources RR00827, the Finnish Academy, and the Sigrid Juselius Foundation.

clinical trial focused on treating infants with documented surfactant deficiency in the highest risk groups for both mortality and morbidity. Using chest radiographic findings consistent with RDS (typically a diffuse pulmonary "white-out" or atelectasis characterized by a diffuse reticulogranular pattern), results of phospholipid analysis of tracheal effluent consistent with surfactant immaturity (low lecithin/sphingomyelin ratio and absence of phosphatidylglycerol), and absence of bacteria on Gram stain of the tracheal aspirate to exclude congenital pneumonia, we elected to treat only infants with documented RDS. Infants less than 10 hours of age, born at 30 weeks or less gestation were eligible to be randomized when they had evidence of RDS and required $FiO_2 \geq 0.6$ to sustain $PaO_2 \geq 50$ torr, and intermittent mandatory ventilation with mean airway pressures ≥ 8 cm H_2O as measured at the distal airway.

Randomization procedures were identical in San Diego and Helsinki after parental permission was obtained. Frozen human surfactant (phospholipid 20 mg/ml) was warmed to 37° C and agitated prior to administration. Surfactant (60 mg minimum or 60 mg/kg birth weight in 3.0 ml of 0.6 % NaCl) was instilled via a small catheter through the endotracheal tube positioned above the carina. As the surfactant was instilled, the infant was gently rocked from side to side and mechanically ventilated for 30 seconds. Thereafter the infant was reconnected to mechanical ventilation and frequent blood gas measurements made. Initially, clinicians were able to reduce the FiO_2 while maintaining the $PaO_2 \geq 50$ torr, and thereafter the peak inspiratory pressures were reduced after the FiO_2 was ≤ 0.4. The control groups received a "dummy" instillation of air into the endotracheal tube and received mechanical ventilation throughout the initial period conforming to the practices of each respective nursery. Because in our pilot study, we encountered a relapse at variable interval after initial improvement in some infants, we elected to retreat infants with surfactant when the F_iO_2 again reached 0.5–0.8 or increases in peak inspiratory pressures ≥ 5 cm H_2O were necessary to maintain $PaO_2 \geq 50$ torr. Soon after entry into the controlled study, each patient was evaluated for evidence of a patent ductus arteriosus (PDA). If evidence of PDA left-to-right shunting was observed, infants received indomethacin 0.2 mg/kg IV initial dose and 0.1 mg/kg I. V. at 12 and 24 hours. If closure was not evident within 48–72 hours after indomethacin treatment, surgical ligation was performed. Morbidities common among very low birth weight infants were tabulated from a chart review and compared among treatment groups.

The objective of our rescue trial was to determine whether restoring surfactant sufficiency would reverse ventilatory failure and improve both short-term morbidities and reduce mortality. Further, if toxic or other adverse effects were not observed, it might in a subsequent trial be efficacious to consider prophylactic surfactant therapy to prevent RDS or to lessen its severity.

Results

Fifty-three infants (25 surfactant treated and 28 controls) were eligible for treatment in our combined trial and compared in a blinded fashion. Surfactant treated infants weighed at birth 987 ± g 182 and control infants 1056 ± 246 g at birth. Eight infants were excluded from final analysis because of pneumonia [3], congenital heart disease [3], pulmonary hypoplasia [1], and transient tachypnea [1]. Neither birth weights nor gestational age were different among the treatment groups, nor were there significant differences in sex predominance, use of prenatal steroids, or tocolytic agents [7].

Ventilatory failure occurred at 3.7 ± 1.1 hours in the infants randomized to receive surfactant and at 4.6 ± 1.8 hours in the control group. Of the 22 infants receiving surfactant, 13 required only a single dose while 9 infants received a second or subsequent dose at 9.4 ± 6.3 hours after the initial treatment. Within 24 hours after birth 77 % of the surfactant treated infants in contrast to 83 % of the control infants manifested evidence of left-to-right PDA shunting and ligation was necessary in approximately 30 % of infants in both groups. We compared the F_iO_2 requirements needed to maintain PaO_2 between 50–70 torr, mean airway pressures, a/AO_2 ratios, and the ventilatory index (F_iO_2 MAP/PaO_2) between the 2 treatment groups through the first 144 hours. Significant differences existed favoring surfactant treatment for FiO_2 and mean airway pressures over the first 72 and 48 hours respectively. The a/AO_2 was significantly improved over the first week and the ventilatory index improved simultaneously. The neonatal morbidities between the two groups showed fewer airleaks (both pneumothorax and pulmonary interstitial emphysema), fewer infants requiring supplemental oxygen past 30 days, and less mortality among the surfactant treated infants at both institutions (Table 1).

Table 1. Mortality and morbidity among infants with RDS in rescue trial compared with controls (reproduced with permission)

	Group		
	Control (n)	*Surfactant (n)*	*P* value[a]
Death or bronchopulmonary dysplasia	14	6	0.019
O_2 0.3 for 30 days	8	2	0.015
Intraventricular hemorrhage	15	10	0.224
Grade 1 or 2	8	3	
Grade 3 or 4	7	7	
Pneumothorax	7	1	0.025
Pulmonary interstitial emphysema	5	0	0.028
Sepsis	3	0	0.125
Retinopathy of prematurity	2	0	0.256
Necrotizing entercolitis	3	1	0.261

[a] Fisher exact test.

Based upon these encouraging results, we undertook a randomized, controlled trial to determine whether human surfactant administered endotracheally at birth to very premature infants (24–30 weeks) would prevent the respiratory distress syndrome or reduce its severity [9]. Thirty-one treated infants (birth weight 938 ± 286 g) were compared in a blinded fashion with 29 control infants (birth weight 964 ± 174 g). Again we documented surfactant deficiency by finding the L/S $\leq$ 2.0 and absence of phosphatidylglycerol in the amniotic or tracheal fluids at birth. The methods and dosage of surfactant administration were similar to that of our rescue trial except that at birth the resuscitation team intubated the infants and a few manual ventilations were performed to document endotracheal tube placement prior to instillation of surfactant or placebo (air). Infants were then brought to the intensive care nursery where routine care was given. Retreatment with surfactant was continued if there was an increase in the $FiO_2 > 0.5$ for satisfactory oxygenation or if there was a need to increase the peak inspiratory pressures by 5.0 cm H_2O. Up to 4 treatments were allowed. Retreatment occurred at 12.2 ± 1.8 hours for the second dose and 20.6 ± 2.9 hours for the third dose when retreatment criteria were met. Only 3 infants received a fourth dose at 40–48 hours after birth. The two treatment groups were compared for outcome based upon the frequency of neonatal morbidity between the groups and the occurrence of bronchopulmonary dysplasia and death. This randomized, controlled, clinical trial showed the frequency of airleaks was significantly reduced but there was no effect upon the occurrence of intraventricular hemorrhage (any grade), infections, necrotizing enterocolitis, pulmonary hemorrhage, or on patent ductus arteriosus (Table 2). Significantly fewer infants died and developed bronchopulmonary dysplasia in the surfactant treated group. Again, infants treated with surfactant had a significant reduction in FiO_2, mean airway pressures, improved a/AO_2 ratios, and a lower ventilatory index over the first week after birth. Of note was the lessening of the duration of neonatal care for surfactant treated groups.

To monitor for evidence of immunologic sensitization among infants treated with surfactant, sequential sera were evaluated for surfactant antisurfactant immune complexes to be reported by Dr. David Strayer (this volume). On days 14 or 15 after birth, complement levels (C_3, C_3A, C_4, CH_{50}, and C_{1q}) were compared between the two treatment groups in the latter study. There were no significant differences. At 6 months serum IgE in 26 or 36 infants treated in San Diego (both surfactant treated and placebo) were < 5 units/ml. Histories suggestive of rash or eczema or reactive airways disease (other than in infants with bronchopulmonary dysplasia) could not be elicited from the parents of infants who were treated with surfactant at 6 months–1 year after birth. Infants were administered a prick-skin test using human surfactant at 1 : 10 and 1 : 100 dilutions, and histamine as a control, and negative skin tests were routinely found.

Infants in both San Diego and Helsinki have been enrolled in neurodevelopmental screening clinics. Even though the developmental testing procedures were similar, but not identical, there is no evidence that surfactant treatment is favoring the survival of infants with neurodevelopmental handicap

Table 2. Neonatal morbidity among infants treated prophylactically with human surfactant for respiratory distress and infants treated with placebo

	Surfactant Group	Control Group
No. of infants	31	29
Pneumothorax	2	7
Pulmonary interstitial emphysema	1	14
Pulmonary hemorrhage	2	3
Patent ductus arteriosus	23	24
No treatment	4	4
Indomethacin	14	10
Indomethacin and ligation	5	10
Necrotizing enterocolitis	2	4
Infections		
None	24	19
For < 7 days	2	2
For ≥ 7 days	5	8
Intraventricular hemorrhage	19	21
None	12	8
Grade I or II	11	11
Grade III or IV	8	10
Ventriculoperitoneal shunt	0	1[a]

There was no significant difference between the groups in any of the variables except pneumothorax ($P < 0.02$) and emphysema ($P < 0.001$).
[a] Grade III.

Table 3. Neurodevelopmental outcome for surfactant-treated ($n = 19$) and control ($n = 9$) groups of San Diego infants

Treatment Group	Motor delay		Mental delay	
	None/mild (%)	Moderate/severe (%)	None/mild (%)	Moderate/severe (%)
Surfactant	74	26	79	21
Control	66	33	66	33

None/mild motor or mental delay represent the mean ± 2 SD (68–128), and moderate/severe represent ≤ 67. Developmental tests (Bayley scales of infant development) were performed at 12–18 months of age corrected for prematurity.

in infants now followed over 2 years. Indeed, in San Diego the overall developmental assessments are improved among infants receiving surfactant compared to placebo treated infants. The results of our previous trials are currently being evaluated in a bi-center randomized clinical trial evaluating concurrently the relative efficacy of prophylactic surfactant therapy to treatment after the onset of ventilatory failure. We believe that randomized, controlled, clinical trials continue to be the best method to evaluate the efficacy of new therapies. To this end our present approach will continue to be the restriction of surfactant therapy to carefully monitored trials with extensive evaluation of survivors for potential evidence of unexpected, but nonetheless possible, adverse effects (Table 3).

Discussion

Surfactant therapy holds a great promise in improving survival and reducing selected morbidities in very low birth weight preterm infants. Insufficient information presently exists for more general use of surfactant outside of controlled clinical trials. Particularly important is the need to assess efficacy and outcome of a variety of surfactant concurrently with simultaneous controls. At present insufficient data permit advocating prophylactic surfactant treatment over more selective usage in a rescue mode. A more complete understanding of the pharmacokinetics and turnover of exogenously administered surfactant is critical for a more rational approach to retreatment. Human surfactant has proven to be efficacious in the treatment of RDS. It is anticipated that second-generation surfactants derived from human surfactant proteins and phospholipids will prove equally efficacious and may present less concerns regarding potential toxicities.

References

1. Hallman M, Merritt TA, Schneider HA, et al. (1983) Isolation of human surfactant from amniotic fluid and a pilot study of its efficacy in respiratory distress syndrome. Pediatrics 71:473
2. Revak S, Merritt TA, Hallman M, Cochrane CG (1988) Use of human low molecular weight (LMW) apoprotein in the reconstitution of surfactant biologics activity. J Clin Invest 81:826-833
3. Hallman M, Merritt TA, Jarvenpaa AL (1985) Exogenous human surfactant for treatment of severe respiratory distress syndrome: a randomized, prospective, clinical trial. J Pediatr 106:963
4. Martin LS, McDougal S, Loskoski SL (1985) Disinfection and inactivation of the human T lymphotropic virus type III/lymphadenopathy-associated virus. J Infect Dis 152:400
5. Schneider H, Hallman M, Benirschke K, et al. (1982) Human surfactant: a therapeutic trial in premature rabbits. J Pediatr 100:619
6. Revak S, Merritt TA, Hallman M, Cochrane CG (1986) Reconstitution of surfactant activity using purified apoprotein and phospholipids measured in vitro and in vitro. Am Rev Respr Dis 134:1258

7. Collaborative Group on Antenatal Steroid Therapy (1981) Effects of antenatal dexamethasone administration on the prevention of respiratory distress syndrome. Am J Obstet Gynecol 141:276
8. Collaborative Group on Antenatal Steroid Therapy (1984) Effects of antenatal dexamethasone administration in the infant: long-term follow-up. J Pediatr 104:259
9. Merritt TA, Hallman M, Bloom BT (1986) Prophylactic treatment of very premature infants with human surfactant. NEJM 315:785

Human Surfactant and Calf Lung Surfactant Extract: Moderation of Respiratory Distress in Preterm Infants by a Single Prophylactic Dose in a Randomized and Controlled Clinical Trial*

B. T. Bloom

Introduction

Neonates with Respiratory Distress Syndrome (RDS) are deficient in pulmonary surfactant (Avery and Mead 1959). A low protein content surface active extract of calf lung lavage (CLSE) has been reported effective in the prevention of RDS and its complications when instilled in the trachea of preterm infants immediately after birth (Enhorning et al. 1985; Kwong et al. 1985; Shapiro et al. 1985). Human surfactant (HS) isolated from amniotic fluid has been reported effective in reducing severity of established RDS (Hallman et al. 1985). When delivered in a prophylactic trial HS was noted to reduce morbidity and mortality (Merritt et al. 1986).

Our original trial was designed to confirm and extend HS treatment observations. We were unable to secure an adequate HS supply for a multidose trial, therefore a single dose (100 mg phospholipid) trial was designed. Our study design was altered to include CLSE when Dr. Topper (Kansas University Medical Center) encouraged our participation in a multicenter study. Our goals were to determine if HS must be delivered in a multidose protocol and address the benefits and bioequivalency of HS and CLSE.

Materials and Methods

Surfactants. HS (IND# 26,308) was extracted from amniotic fluid by the method of Hallman et al. (1983). CLSE (IND# 27,169) was extracted as previously reported (Metcalfe et al. 1980) and provided by ONY Inc. (Buffalo, NY).

Trial Design. Mothers in labor at ≤ 31 weeks gestation were approached to obtain informed consent. Treatment randomization was determined by availability of the surfactants. Upon delivery, infants were intubated by a neonatologist, followed by instillation of either 100 mg of HS, CLSE or air. Subsequently 5–10 positive pressure breaths (peak pressure 30–40 cm H_2O, inspiratory time 0.8–1.5 seconds and 0.4 F_iO_2) were given.

* Funded by grants from the Wesley Foundation, Wesley Medical Research Institutes, The Children's Miracle Network Telethon, and the March of Dimes # 6-511.

Infants whose parents refused surfactant administration [2], who were delivered without informed consent [2] and those delivered when no surfactant was available [26] were utilized for controls. Discussion of the treatment group if known by the staff was discouraged. CLSE/HS assignment was known only to the principal investigator. In spite of these attempts, it appears that blinding caregivers to surfactant administration is virtually impossible.

Statistical Analysis. Mean airway pressures, F_iO_2 and blood gases were measured and the a/A ratio (Gilbert and Keighley 1974) and ventilation efficiency index (Notter et al. 1985) were calculated at 4 hour intervals for 96 hours. No adjustments for missing data were made for those infants who expired. Duration of mechanical ventilation (IMV), nasal CPAP, and supplemental O_2 were also recorded. These parameters were analysed by Mann-Whitney U with z corrected for ties in two group data sets. Infant demographics, maternal data and outcome variables (bronchopulmonary dysplasia defined as $FiO_2 > 0.21$ at 28 days) were compared by Chi-square and Chi-square with continuity correction where appropriate.

This protocol was approved by the Investigational Review Board of HCA Wesley. This analysis was undertaken to justify continuation of this single HS dose protocol. Infants with congenital malformations which affect lung development, premature rupture of fetal membranes longer than 3 weeks, and those with congenital infections were excluded from this analysis.

Results

This report constitutes an interval analysis of 78 consecutive inborn infants, including 30 control, 30 CLSE treated, 10 HS treated infants and 8 treated infants who were excluded from analysis. The reasons for exclusion were: 22–23 weeks gestational age [1], congenital Herpes [1], enteroviral sepsis [1], Group B Streptococcal sepsis [2], hydrops fetalis [1] and oligohydramnios deformation sequence with hypoplastic lungs [2]. The infant with enteroviral sepsis/pneumonia and one of the infants with hypoplastic lungs survived.

A listing of the analyzed variables are included in Table 1. There was no significant difference between the groups in birthweight, gestational age, male: female ratio, race or Apgar scores. No difference in the occurrence of necrotizing enterocolitis, apnea or retinopathy of prematurity was noted. A reduction in occurrence of IVH in the treated groups was noted (overall 17.1 %, CLSE 6.6 %), yet this did not reach statistical significance. No difference was noted in PDA as documented by contrast echocardiography in the first 48 hours, yet intervention (Indocin or Ligation) was reduced in the CLSE treated group (CLSE v. HS $p = 0.01$ / CLSE v. controls $p = 0.01$).

A reduction in the duration of IMV, $FiO_2 \geq 0.3$ and supplemental oxygen, as well as a reduction in occurrence of pneumothorax, bronchopulmonary dysplasia and death was noted in both HS and CLSE groups when compared to controls.

Table 1. Infant, maternal and outcome data (mean ± SD) from a randomized controlled study of treatment with human surfactant (HS) or calf lung surfactant extract (CLSE) in an inborn population of infants delivered at ≤ 31 weeks' gestation. P value determined by chi-square with continuity correction

	Treatment group			P value		
	Control	HS	CLSE	HS vs Control	CLSE vs Control	CLSE vs HS
Infant characteristics						
Birthweight (g)	1085 ± 297	1159 ± 324	1066 ± 254	NS	NS	NS
OB gestational age (weeks)	27.5 ± 2.4	27.6 ± 2.1	27.7 ± 2.0	NS	NS	NS
Dubowitz (weeks)	28.9 ± 2.3	28.1 ± 2.3	28.1 ± 2.0	NS	NS	NS
Small-for-dates	4	0	3	NS	NS	NS
Males	16	6	16	NS	NS	NS
Singleton	24	9	25	NS	NS	NS
Race (white/black/other)	23/5/2	8/1/1	27/2/1	NS	NS	NS
1-Min APGAR	4.7 ± 2.4	5.4 ± 1.8	5.3 ± 2.2	NS	NS	NS
5-Min APGAR	6.9 ± 1.7	7.4 ± 1.9	6.9 ± 1.9	NS	NS	NS
Maternal characteristics						
Antenatal Steroids (h)	0.7 ± 4.0	3.6 ± 11.4	2.4 ± 13.1	NS	NS	NS
Beta agonists (h)	36 ± 112	51 ± 123	26 ± 93	NS	NS	NS
Vaginal delivery	4	2	5	NS	NS	NS
Presentation (vertex/breech/other)	17/10/3	6/3/1	22/5/3	NS	NS	NS
Anesthetic (epidural/other/gen.)	20/2/8	6/0/4	23/3/4	NS	NS	NS
Pregnancy-induced hypertension	4	0	7	NS	NS	NS
Chronic hypertension	1	0	3	NS	NS	NS
Oligo-/Polyhydramnios	1/1	0/0	1/1	NS	NS	NS
Labor	22	6	21	NS	NS	NS
Intrauterine asphyxia	1	1	1	NS	NS	NS
Fetal distress	13	4	10	NS	NS	NS
Abruption	7	4	6	NS	NS	NS
Chorioamnionitis	3	2	6	NS	NS	NS
Vaginal bleeding						
Minimal or none	25	7	27			
Acute large	3	1	2			
Chronic	1	2	1			
Outcome data						
Pulmonary hemorrhage	2	0	1	NS	NS	NS
Necrotizing enterocolitis	5	1	4	NS	NS	NS
Retinopathy (none/I/II/III)	12/6/1/2	4/2/1/0	18/8/1/1	NS	NS	NS
RLF (survivors)	1	1	1	NS	NS	NS
IVH (no IVH/Gr I–II/Gr. III–IV	22/4/4	7/1/1	28/1/1	NS	NS	NS
Apnea	25	7	28	NS	NS	NS
Nasal CPAP (days)	9 ± 11	4 ± 6	9 ± 13	NS	NS	NS
PDA (none/PDA no tx/treated)	8/4/18	4/0/6	13/9/8	NS	0.031	0.067
IMV (days)	23 ± 34	12 ± 21	9 ± 15	NS	0.005	NS
$FiO_2 > 0.21$ (days)	42 ± 57	22 ± 41	12 ± 24	NS	0.005	NS
$F_iO_2 \geq 0.3$ (days)	23 ± 42	8 ± 19	6 ± 14	0.1	0.001	NS
Home O_2	2	1	0	NS	NS	NS
Pneumothorax	11	1	1	NS	0.004	NS
Pulmonary interstitial emphysema	15	5	2	NS	0.001	0.008
Bronchopulmonary dysplasia	13	2	4	NS	0.022	NS
Deaths	8	2	1	BS	0.030	NS

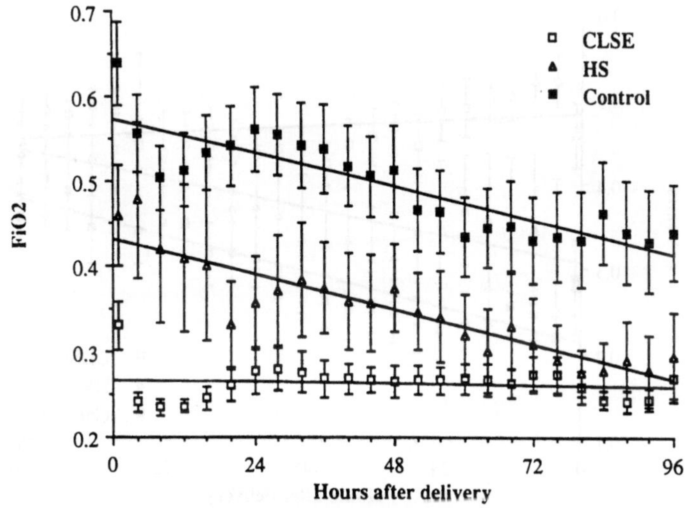

Fig. 1. Mean and standard error of the fractional inspired oxygen concentration *(F_iO_2)* for infants treated with human surfactant *(HS)*, calf lung surfactant extract *(CLSE)* or conventional therapy *(Control)* over the first 96 h after delivery. Linear regression equations noted are HS y = 0.43 – 0.0017x, CLSE y = 0.27 – 0.0001x, Controls y = 0.57 – 0.0017x

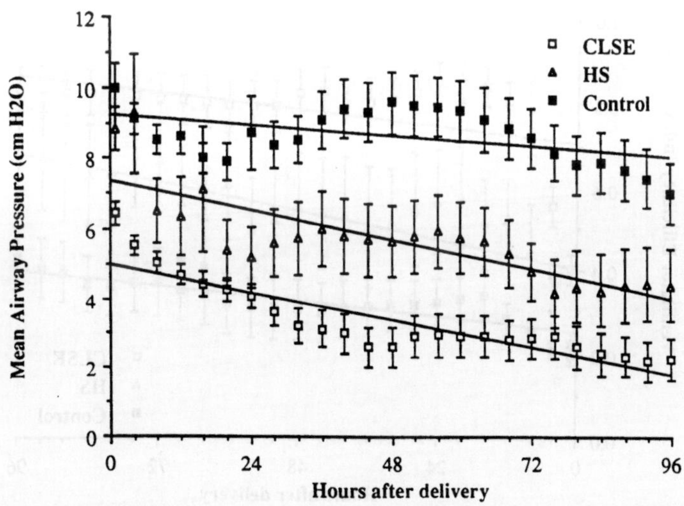

Fig. 2. Mean and standard error of the mean airway pressure for infants treated with human surfactant *(HS),* calf lung surfactant extract *(CLSE)* or conventional therapy *(Control)* over the first 96 h after delivery. Linear regression equations noted are HS y = 7.37 – 0.035x, CLSE y = 4.98 – 0.033x, Controls y = 9.22 – 0.012x

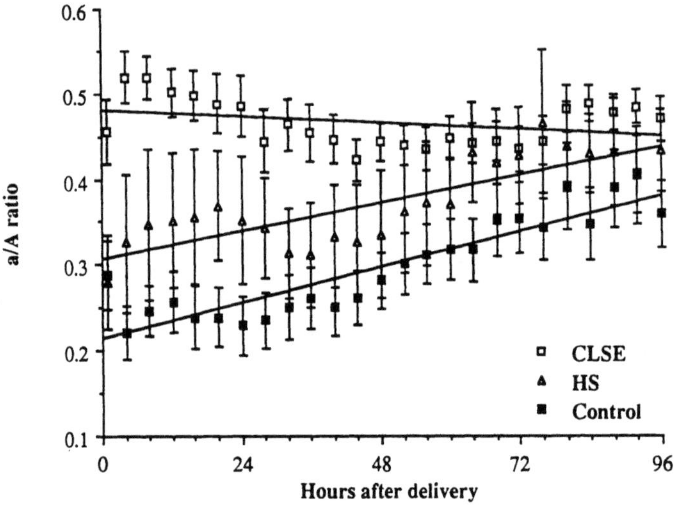

Fig. 3. Mean and standard error of the a/A ratio for infants treated with human surfactant *(HS)*, calf lung surfactant extract *(CLSE)* or conventional therapy *(Control)* over the first 96 h after delivery. Linear regression equations noted are HS y = 0.31 + 0.0014x, CLSE y = 0.48 - 0.003x, Controls y = 0.22 + 0.0017x

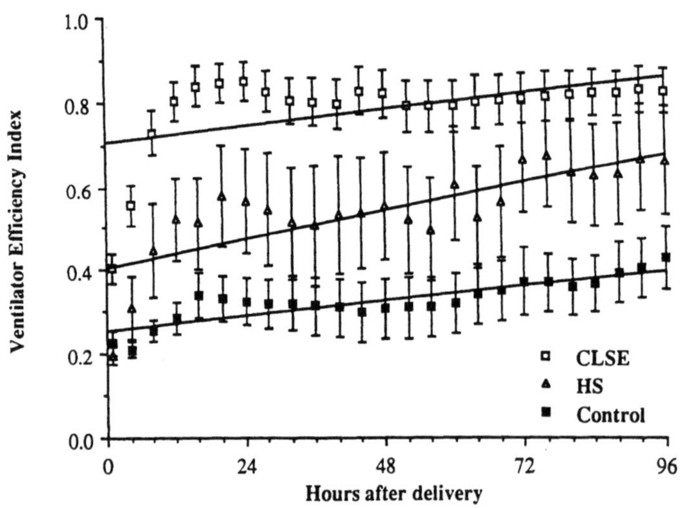

Fig. 4. Mean and standard error of the ventilator efficiency index for infants treated with human surfactant *(HS)*, calf lung surfactant extract *(CLSE)* or conventional therapy *(Control)* over the first 96 h after delivery. Linear regression equations noted are HS y = 0.41 + 0.0028x, CLSE y = 0.26 + 0.0014x, Controls y = 0.71 + 0.0016x

These reductions and a lower incidence of PIE reached statistical significance for the CLSE group.

Reductions in measured support parameters (reported as mean ± standard error) were noted throughout the first 96 hours for both HS and CLSE treated infants (Figs. 1–4). CLSE treated infants required less support during this 96 hour period than did the HS treated group. Differences between FiO_2, mean airway pressures, ventilation efficiency index and a/AO_2 ratio for CLSE, HS and control infants reached significance at the time intervals noted in Table 2.

Eight control infants died from RDS [4], BPD [2], peritonitis [1] and pulmonary hemorrhage [1]. Two HS infants died from RDS, one had developed PIE. One CLSE infant died from candida sepsis.

Table 2. Significance of differences notes in Figs. 1–4, as determined by Mann-Whitney U, in support levels (mean airway pressure and F_iO_2 and lung function [a/A ratio and ventilator efficiency index (VEI)] over the first 96 h after delivery in patients treated with human surfactant (HS) or calf lung surfactant extract (CLSE)

	FiO_2			VEI			a/A ratio			Mean airway pressure		
Time (h)	HS vs Control	CLSE vs Control	CLSE vs HS	HS vs Control	CLSE vs Control	CLSE vs HS	HS vs Control	CLSE vs Control	CLSE vs HS	HS vs Control	CLSE vs Control	CLSE vs HS
1	NS	0.001	0.05	NS	0.001	0.002	NS	0.01	0.05	NS	0.001	0.002
4	NS	0.001	0.01	NS	0.001	0.01	NS	0.001	0.05	NS	0.001	0.002
8	NS	0.001	0.02	NS	0.001	0.05	NS	0.001	NS	NS	0.001	0.01
12	NS	0.001	0.05	0.05	0.001	0.02	NS	0.001	NS	0.05	0.001	0.02
16	NS	0.001	NS	NS	0.001	0.005	NS	0.001	NS	NS	0.001	0.02
20	0.05	0.001	NS	0.05	0.001	0.05	NS	0.001	NS	0.05	0.001	0.05
24	NS	0.001	0.05	0.05	0.001	0.02	NS	0.001	NS	0.05	0.001	0.1
28	NS	0.001	NS	NS	0.001	0.05	0.05	0.001	NS	0.05	0.001	0.05
32	NS	0.001	0.02	NS	0.001	0.05	NS	0.001	0.05	NS	0.001	0.05
36	NS	0.001	0.05	NS	0.001	0.05	NS	0.001	0.05	NS	0.001	0.05
40	NS	0.001	NS	NS	0.001	NS	NS	0.001	NS	0.05	0.001	0.05
44	NS	0.001	NS	NS	0.001	0.05	NS	0.001	NS	0.05	0.001	0.05
48	NS	0.001	0.05	0.05	0.001	0.05	NS	0.002	NS	NS	0.001	0.05
52	NS	0.005	NS	NS	0.001	0.05	NS	0.01	NS	NS	0.001	0.05
56	NS	0.002	NS	NS	0.001	NS	NS	0.02	NS	NS	0.001	0.05
60	NS	0.01	NS	NS	0.001	NS	NS	0.005	NS	NS	0.001	0.05
64	NS	0.005	NS	NS	0.001	NS	NS	0.02	NS	NS	0.001	0.05
68	NS	0.02	NS	NS	0.001	NS	NS	0.05	NS	NS	0.001	0.05
72	NS	0.05	NS	0.05	0.001	NS	NS	NS	NS	0.05	0.001	NS
76	NS	0.02	NS	0.05	0.001	NS	NS	NS	NS	0.05	0.001	NS
80	NS	0.01	NS	0.05	0.001	NS	NS	NS	NS	NS	0.001	NS
84	NS	0.005	NS	NS	0.001	NS	NS	0.02	NS	NS	0.001	NS
88	NS	0.005	NS	NS	0.001	NS	NS	NS	NS	NS	0.001	NS
92	NS	0.01	NS	NS	0.001	NS	NS	NS	NS	NS	0.001	NS
96	NS	0.01	NS	NS	0.001	NS	NS	0.02	NS	NS	0.001	NS

Conclusion

Administration of either HS or CLSE was not related to an increase in morbidity. CLSE and HS both reduced the amount of respiratory intervention required for support of these very preterm infants. CLSE in our population appeared more effective, however, statistical significance was reached only at specific interval measurements. There appears to be an advantage of CLSE in relationship to the occurrence of significant PDA.

Continuation of this study will permit further evaluation of HS, while the efficacy of CLSE is documented in concurrence with previously published studies. These preliminary data suggest a benefit, although at this point not of statistical significance, of HS in a single dose prophylactic protocol for infants $\leq$ 31 weeks over conventional therapy. In addition, a subjective observation that a single dose of HS does not appear to be as effective as the multidose protocol, indirectly supports development of multidose CLSE trials. Our study also emphasizes the need for further large comparison studies of available surfactants and dosage protocols.

Acknowledgements. Drs. Nelson, Dirksen, Roan, Reed and Johnson, the Nursing and Respiratory Therapy staffs at HCA Wesley for their participation in the care given these infants. The NICU ward clerks for their efforts in monitoring maternal admissions. Our obstetric colleagues and their residents for fluid collection and support. Paula Reinhardt RN, Alice Bloom and Tom Rose RRT, for assistance in data collection. Maureen Byrne-Clark for her editorial assistance. Drs. Merritt, Topper, Claflin, Egan and Kwong for their input and support.

References

Avery ME, Mead J (1959) Surface properties in relation to atelectasis and hyaline membrane disease. Am J Dis Child 97:517–523

Enhorning G, Shennan A, Possmayer F et al. (1985) Prevention of neonatal respiratory distress syndrome by tracheal instillation of surfactant: a randomized clinical trial. Pediatrics 76:145–153

Gilbert R, Keighley JF (1974) The arterial/alveolar oxygen tension ratio. An index of gas exchange applicable to varying inspired oxygen concentrations. Am Rev Respir Dis 109:142

Hallman M, Merritt TA, Schneider H (1983) Isolation of human surfactant from amniotic fluid and a pilot study of its efficacy in respiratory distress syndrome. Pediatrics 71:473–482

Hallman M, Merritt TA, Jarvenpaa A-L et al. (1985) Exogenous human surfactant for treatment of severe respiratory distress syndrome: A randomized prospective clinical trial. J Pediatr 106:963–969

Kwong MS, Egan EA, Notter RH, Shapiro DL (1985) Double-blind clinical trial of calf lung surfactant extract for the prevention of hyaline membrane disease in extremely premature infants. Pediatrics 76:585–592

Merritt TA, Hallman M, Bloom BT et al. (1986) Prophylactic treatment of very preterm infants with Human Surfactant. N Engl J Med 315:785–790

Metcalfe IL, Enhorning G, Possmayer F (1980) Pulmonary surfactant-associated proteins: their role in the expression of surface activity. J Appl Physiol 49:34

Notter RH, Egan EA, Kwong MS et al. (1985) Lung surfactant replacement in premature lambs with extracted lipids from bovine lung lavage: Effects of dose, dispersion technique and gestational age. Pediatr Res 19:569–577

Shapiro DL, Notter RH, Morin FC III et al. (1985) Double-blind randomized trial of a calf lung surfactant extract administered at birth to very premature infants for prevention of respiratory distress syndrome. Pediatrics 76:593–599

Artificial Surfactant: Prophylaxis for Respiratory Distress Syndrome

C. J. Morley

In 1977, following the pioneering work of Enhorning and Robertson [1] and our own experiments with natural surfactant in premature rabbits [2, 3] and lambs we realised that concentrated pure natural surfactant should be a very useful addition to the treatment of very premature babies with respiratory distress syndrome (RDS). However, we believed that it would not be possible to obtain surfactant from animal lungs, check its purity, chemical composition and biophysical properties and also guarantee its sterility for use in routine treatment of very premature babies. It seemed that any surfactant derived from animal or human sources could be a source of infection and foreign proteins.

At that time Dr. Bangham and I felt that it should be possible to imitate the biophysical properties of natural surfactant with a sterile mixture of phospholipids which could be widely used for the treatment of babies if shown to be effective. When working on this mixture we read of the work of Hallman [4] who showed that babies with severe RDS lacked phosphatidylglycerol (PG) in their surfactant. Experiments with various phospholipid mixtures confirmed that a simple mixture could be made to imitate the surface properties of natural surfactant if it contained 30 % PG and 70 % dipalmitoylphosphatidylcholine (DPPC) in a crystalline state [5]. DPPC is a very straight stable molecule which is capable of floating on an air/water interface and when present in excess in a liquid state spreading over the surface it is able to raise the equilibrium surface pressure from zero to about 45 mN/m. When a monolayer is compressed it exhibits the almost unique physical property of maintaining rigidity and sustaining a very high surface pressure. This is probably the property of pulmonary surfactant which maintains alveolar patency during expiration. The surfactant monolayer rich in DPPC solidifies under compression and thereby splints the alveoli open. Unfortunately, pure DPPC alone could not be used as an exogenous surfactant because it is not surface active at body temperature. This is because it is solid below its transition temperature of 41° C. For DPPC to spread at body temperatures a liquifying molecule is required. Unsaturated PG has this effect and the combination of crystalline PG and DPPC in a ratio of 7 : 3 w/w when used at 37° C spreads rapidly over an air/water interface, yet sustains high surface pressures on compression. It appeared that this might be a useful exogenous surfactant for the treatment of surfactant deficiency states. This surfactant has subsequently been called Artificial Lung Expanding Compound (ALEC) [6].

It is well known that phospholipids in water form various structures such as myelin forms and liposomes and that free individual molecules rarely exist [7].

In this state they are not readily available to form a surface monolayer. The mechanism by which surfactant phospholipids reach the surface and form a monolayer has not been clearly demonstrated. We showed that ALEC in a dry crystalline form placed on the surface spread a monolayer very rapidly [2, 5]. Initially ALEC was used in a clinical trial, as a powder, blown into the lungs of very premature babies at birth through an endotracheal tube which was being used for resuscitation [8]. Subsequently, we realised that if the ALEC crystals were mixed with cold water below their transition temperature of about 25° C they did not form liposomes but maintained their crystalline properties. This meant that ALEC could be delivered more easily to babies' lungs as a crystalline suspension [9].

Studies on the effect of ALEC as a powder on the lung/thorax compliance of 27 day fetal rabbits were undertaken in Stockholm with Bengt Robertson, Gertie Grossmann and Burkhard Lachmann [2]. The results showed that, even though it was rather difficult to get the powder down the tracheostomy tube with an internal diameter of less than 1 mm, at 30 min the compliance (ml/cm H_2O/kg) was significantly higher in the natural surfactant treated animals (0.49, range 0.06–1.24) and the ALEC treated animals (0.28, range 0.05–1.14) than the controls (0.07, range 0.01–0.32). However, by 60 min both natural surfactant treated (0.29, range 0.05–1.33) and ALEC treated (0.27, range 0.02–1.08) had similar values which were significantly higher than the controls (0.07, range 0.01–0.25). The problem with these experiments was that the dose of ALEC entering the lungs was variable and in some animals the powder may have blocked the tracheostomy tube. The effect of ALEC treatment has been shown in subsequent clinical studies to have a slower effect on lung expansion than natural surfactant, in some cases taking up to 6 hours before an effect could be seen followed by slow improvement. It is interesting that an effect was shown in these very short term rabbit experiments.

On the basis of this information and the fact that ALEC contained only pure phospholipids which were deficient in premature babies' lungs we were given permission in 1979 to proceed to the preliminary clinical trial. The plan was to treat babies of less than 35 weeks gestation as near to the first breath as possible with ALEC powder. This could only confidently be delivered into the lungs if the babies were intubated. However, we did not feel that it was ethical to intubate babies specifically for treatment with ALEC, even if they were at risk of RDS. In consequence, babies were treated only if they required intubation for rescuscitation. One 25 mg dose of ALEC was blown down the endotracheal tube from a resuscitation bag modified to hold a gelatin capsule containing the ALEC which could then be pierced during inflation so that the powder was blown into the lungs. To ensure that the surfactant was given consistently it was only given by Dr. Anne Greenough or myself. We were not able to attend the deliveries of all premature babies and so the babies we missed who were intubated became the controls. The babies were rescuscitated by the duty paediatric team.

The preliminary results of this trial were published in 1981 [8]. However babies continued to be enrolled until January 1982 by which time there were 130 babies in the trial [9, 10]. One baby with an autopsy diagnosis of Potter's syndrome was excluded leaving 129 for analysis. There were 54 ALEC treated babies and 75 controls. There was no statistical difference between the groups in gestational age (mean ± SD: controls 30.1 ± 2.6 weeks; ALEC 29.7 ± 2.2 weeks), birthweight (controls 1398 ± 468 g; ALEC treated 1359 ± 404 g), male sex (controls 68%, ALEC treated 54%), Caesarian section delivery (controls 61%, ALEC treated 66%), antenatal steroids (controls 1%, ALEC treated 5%), or babies less than 30 weeks gestation (controls 45%, ALEC treated 52%).

The incidence of complications in the two groups were patent ductus arteriosus (controls 15%; ALEC treated 20% p = NS), pneumothorax (controls 28%; ALEC treated 11% $p < 0.05$), number of surviving babies who were in oxygen for more than 10 days (controls 28%; ALEC treated 19% p = NS), periventricular haemorrhages (controls 24%; ALEC treated 15% p = NS) and deaths (controls 4%; ALEC treated 20% $p < 0.01$).

In babies of less than 30 weeks gestation who had the most complications (controls = 34 and ALEC treated = 28), the main effects of ALEC was a reduction in the incidence of pneumothoraces (controls 61%; ALEC treated 9% $p < 0.001$) and deaths (controls 41%; ALEC treated 7% $p < 0.001$). The mean age at death for the treated babies was 80.5 hours and the controls 21.0 hours. It was possible that the increased number of survivors in the ALEC group might have had a more prolonged illness. In consequence it was interesting to note that the proportion in oxygen for more than 20 days was not increased.

This preliminary trial revealed two main things. Firstly, and most important, the ALEC treatment did not appear to cause any harm and secondly it produced beneficial effects which warranted further investigation. The non-randomised nature of this trial meant that the results were not readily acceptable although there was no obvious bias or choice in the babies receiving ALEC. Because of the beneficial effects and the problems with the trial design, a second trial was undertaken in collaboration with the neonatal unit in Nottingham and with the guidance of Dr Sheila Gore at the MRC Biostatistics unit [9, 10]. This second trial had several differences from the preliminary study:

1. The babies were all treated with ALEC as a cold crystalline suspension. We changed from using ALEC powder to suspension because the powder was difficult to deliver easily to very premature babies in sufficient quantity and we realised that the powder could be kept in a crystalline state if the saline temperature was below the transition temperature of the ALEC.
2. All babies were treated at birth because we saw in the first trial that many of the babies who were not intubated, and were therefore treated at birth with ALEC, developed severe RDS.
3. The surfactant was placed into the pharynx at delivery to try to ensure that it was inhaled with the first breath following the studies and reasoning of Enhorning [11] who showed that premature rabbits inhaled a pharyngeal dose of surfactant and suggested that its distribution in the lung would be more even if it travelled throughout the lung with the meniscus at the first breath.

4. Up to 4 doses were given over the first 24 hours to maximise the effect because we had seen from the first trial that some babies appeared to improve and then deteriorated after just one dose. The dose was increased to approximately 50 mg and subsequently to 100 mg per dose, again to maximise the treatment.

Thus, the protocol for this trial was as follows:

1. Babies were born in a collaborating hospital with a member of the team at the delivery between 23–34 weeks gestation inclusive. This was the "best fit" gestation prior to delivery (i.e. LMP, scans etc.).
2. Babies were individually randomized from envelopes immediately before delivery.
3. Surfactant suspension was prepared just before delivery, with equipment kept at 4° C, by hand shaking ALEC with 1 ml of saline in the vial and aspirating it gently into a syringe to prevent foam formation. It was kept cold to maintain its crystalline state.
4. The control substance was 1 ml of cold saline. This was chosen because it is the volume the hospitals use during routine endotracheal suction.
5. As near to the first breath as possible, the baby received either ALEC or saline into the pharynx so that it might be inhaled. If the baby was intubated for resuscitation, a second dose was instilled through the endotrachael tube. If intubated at 1 and 24 hours, third and fourth doses were given.
6. The treatment given was not disclosed to the nurses or doctors caring for the baby.
7. All clinical decisions were taken by the duty paediatric teams.
8. Exclusions were retrospective and the babies declared ineligible only if stillborn, or if they had gross congenital malformation.
9. The trial was conducted with the permission of the Ethics Committee at each hospital.

A total of 341 babies entered the trial. Of the 14 ineligible babies, 7 were fresh stillbirths and 7 had malformations. There were 8 clinical cases of lung hypoplasia. Because of the difficulty of differentiating this from severe RDS we deferred to Jonathan Wigglesworth (Professor Paediatric Pathology, Hammersmith Hospital London). Four were definite cases and are included in the total of ineligible babies. The other 4 are included in the trial analyses. Three babies could not be resuscitated but as they were randomised and treated they are included in the analysis. The trial eventually contained 327 eligible babies, 163 ALEC treated and 164 controls. There were no statistically significant differences between the groups for antenatal factors which might influence subsequent lung disease.

The babies with the most severe neonatal problems are those under 30 weeks gestation. The trial enrolled 136 babies eligible for analysis. Five were ineligible because they were stillborn or had major malformations. Sixty nine were surfactant treated and 67 controls. The two groups are well matched for birthweight (mean ± SD; controls 1047 ± 270 g, ALEC treated 1092 ± 229 g), gestational age (controls and ALEC treated both 27.4 ± 1.6 weeks),

L : S ratio (both controls and ALEC treated 1.7 ± 0.6), sex (males: controls 61%, ALEC treated 57%), delivery (caesarian section: controls 61%, ALEC treated 57%) and steroids (controls 9% and ALEC treated 10%).

There was a reduced mortality of 9% in the ALEC treated group compared with 14% in the controls for the whole trial and 19% in the ALEC treated babies less than 30 weeks compared with 34% in the controls ($p < 0.02$). Of the 24 control deaths, 21 had obvious RDS and related complications (88%), 1 had possible lung hypoplasia (4%), 1 was not resuscitatable (4%) and one was septicaemic at birth (4%). Of the 14 ALEC treated babies who died, 7 had complications associated with RDS (50%), 2 possible lung hypoplasia (14%), 1 was non viable at 23 weeks gestation (7%), 1 was unresuscitatable (7%) and 4 had infections (29%). The deaths from RDS and associated complications were significantly decreased in the ALEC treated group ($p < 0.05$).

The incidence of periventricular haemorrhages (PVH) was reduced in the ALEC treated babies to 8% compared with 18% in the controls for the whole trial ($p < 0.05$) and 19% in the ALEC babies of less than 30 weeks compared with 40% of the controls ($p < 0.01$). In Cambridge, PVH were diagnosed by ultrasound scan performed weekly by a group unaware of the babies' trial status. ALEC treatment was associated with a significant reduction (controls 10% and ALEC treated 2% $p < 0.05$). PVH was associated with approximately two thirds of the deaths in both groups.

The incidence of pneumothoraces was 18% of the ALEC treated babies and 20% of the controls (p = NS). In the babies of less than 30 weeks gestation 28% of the ALEC treated and 36% of the controls developed a pneumothorax (p = NS).

There was a slight reduction, compared with the controls, in the proportion of surviving babies treated with ALEC who were in prolonged oxygen (> 28 days). This was significant only in the babies of less than 30 weeks gestation (controls 35% and ALEC treated 20% $p = 0.05$). This effect is important because the reduced mortality in the ALEC group might have been expected to result in more surviving very premature babies requiring prolonged oxygen therapy.

The identical overall 12% incidence of patent ductus arteriosus (PDA) (diagnosed clinically) in both treatment groups raises no concern about PDA being an adverse effect of ALEC prophylaxis.

Measurements of respiratory compliance (ml/cm H_2O) were made at 1, 6, 24, 48 hours and 7 days. There was no significant difference in the whole trial but there were important effects in the babies less than 30 weeks gestation, there was little effect at 1 hour (controls 0.54 ± 0.05 and ALEC treated 0.64 ± 0.07 p = NS), but significant improvements in the surfactant treated group at 6 hours (controls 0.62 ± 0.08 and ALEC treated 1.07 ± 0.18 $p < 0.01$), and 24 hours (controls 0.62 ± 0.05 and ALEC treated 1.01 ± 0.11 $p < 0.001$). At 48 hours (controls 0.71 ± 0.10 and ALEC treated 0.84 ± 0.11 p = NS) and 7 days (controls 0.96 ± 0.17 and ALEC treated 1.19 ± 0.13 p = NS) the compliance was higher in the ALEC treated group but the difference is not significant. The compliance of the ALEC treated babies rose rapidly in the first 6 hours to a plateau whereas the controls improved slowly over the first few days.

Examination of the inspired oxygen concentrations over the first 4 days showed that the ALEC treated group required less oxygen, but from 12 hours the ALEC treated babies had significantly lower inspired oxygen requirements than the controls:

	Controls	ALEC-treated	
1 h	58 ± 3	51 ± 2	$p = 0.06$
6 h	53 ± 3	45 ± 3	$p = 0.06$
12 h	52 ± 3	42 ± 3	$p = 0.01$
18 h	50 ± 3	41 ± 3	$p = 0.02$
24 h	52 ± 3	41 ± 3	$p < 0.01$
36 h	50 ± 3	40 ± 3	$p = 0.02$
48 h	46 ± 3	37 ± 3	$p < 0.05$
72 h	41 ± 3	30 ± 2	$p < 0.01$
96 h	36 ± 3	28 ± 2	$p = 0.02$

Despite these results this second trial, with 341 babies from 23 to 34 weeks gestation, only had a 50 : 50 chance of detecting a halving of mortality because most deaths occur below 30 weeks gestation and at the second informal interim analysis (May 1984) the trial only had randomized over 90 babies of 25–29 weeks gestation. The ALEC treated group had a reduced mortality from 38% (18/47) to 21% (10/47). On the basis of this it was decided that this trial should be extended to a multicentre study to define the real effect of ALEC on mortality in very premature babies [12].

The multicentre trial was limited to babies of 25–29 weeks gestation because they have a high incidence of RDS and a reasonable chance of survival. It was designed to assess the prophylactic effect of ALEC on mortality. Secondary targets were the effect of ALEC on respiratory therapy, the severity of RDS and complications of prematurity.

The collaborators' prior opinion about the possible effect of ALEC on mortality in this group was a reduction from 36%–27%. The trial was therefore designed to identify this target. Three hundred eligible babies gave a conditional power of 74% (i.e. including approximately 100 babies of 25–29 weeks gestation from the Cambridge/Nottingham trial).

The trial was conducted by 10 neonatal units each randomizing approximately 20 eligible babies and their results pooled with all the babies of 25–29 weeks gestation from the Cambridge wing of the Cambridge/Nottingham Trial. The protocol was discussed at the collaborators' meeting and each centre was given a manual detailing the trial definitions and protocol.

The protocol was identical to that of the Cambridge/Nottingham trial except for exclusion of non-resuscitated babies and the different gestational ages. The trial was conducted with the permission of the Ethics Committee at each hospital.

The randomization ratio was 1 : 1 by permuted blocks of variable length within centre. There were two gestational age strata: 25 and 26 weeks; and 27, 28 and 29 weeks to prevent imbalance in this factor. Randomization occurred just before birth. Twenty babies were randomized but later found to be ineligible. Fifteen controls were ineligible: 3 resuscitation not attempted, 4 stillborn and 8 had major malformations and 5 ALEC treated babies were ineligible: 3 stillborn and 2 had major malformations. Apart from stillbirths, the ineligible babies were not evenly distributed in the two groups and exclusion of the babies where resuscitation was not attempted or babies who were malformed weighs against ALEC.

The analysis compares babies in the groups to which they were randomized. Four were incorrectly treated (a pair of twins was muddled, one baby was given saline instead of ALEC and one had the wrong code in the envelope).

The Cambridge/non-Cambridge strata was included only for interest and is not intended as an important part of this randomized trial. The effect of the four levels of stratification i.e., 25–26 weeks Cambridge; 25–26 weeks non-Cambridge; 27–29 weeks Cambridge; 27–29 weeks non-Cambridge is reflected in the Mantel-Haenszel analysis of the mortality and other complications from which X^2 values and confidence intervals are derived.

No adjustment of nominal significance levels has been made because of the 2-stage trial design, the second stage being an extension and not a predefined stopping point, but so minded readers can adopt $X^2 = 4.74$ instead of $X^2 = 3.84$ as the critical value for significance at the 5% level.

Three hundred and twenty eight babies were randomized, of whom 20 were ineligible. There was a reasonable balance between groups for factors which could influence the incidence of complications, except for an excess of multiple births in the control group which is allowed for in the regression analyses.

The mortality while in the neonatal unit was reduced from 30% to 19% for the ALEC treated babies, a one third reduction, the 95% confidence interval for the difference being from 2 to 21 extra survivors ($p < 0.01$). The effect is evident in all strata: Cambridge (controls 29%, ALEC 16%) and non-Cambridge centres (controls 30%, ALEC 21%) and by gestation: 25–26 (controls 50%, ALEC 44%); 27–29 (controls 24%, ALEC 9%).

ALEC treatment almost halved neonatal mortality (first 28 days) from 27%–14% ($p < 0.002$) the 95% confidence interval for the difference being from 4–22 additional survivors. This effect was shown by regression analysis to be equivalent to one extra week of gestation and was evident in Cambridge (controls 26%, ALEC 11%) and non-Cambridge centres (controls 27%, ALEC 17%) and by gestation: 25–26 (controls 47%, ALEC 30%); 27–29 (controls 21%, ALEC 9%).

As more control than ALEC treated babies died, comparison of other outcomes was biased against ALEC.

When estimating the effect on cerebral haemorrhages the analysis concentrates on parenchymal haemorrhages because these are highly correlated with subsequent problems. These were detected by ultrasound or postmortem. Cranial ultrasound was not standardised between centres. Therefore, haemorrhages were recorded as: 1 = no bleeding; 2 = bleeding but not parenchymal; 3 = parenchymal brain haemorrhage (PBH). All babies were included except for 5 ALEC treated and 6 control babies who were not scanned. The incidence of PBH was reduced by one third from 24% to 16% ($X^2 = 3.43$ $p = 0.06$), with 95% confidence interval for the difference from 1 extra to 17 fewer PBH. There is no significant heterogeneity amongst strata ($X^2_{(3)} = 4.38$) where the incidence was Cambridge (Controls 30%, ALEC 14%) and non-Cambridge (controls 18%, ALEC 18%), 25–26 weeks (Controls 32%, ALEC 33%), 27–29 weeks (Control 21%, ALEC 10%). The incidence of haemorrhages without parenchymal involvement was 19% in the two groups.

RDS was graded before the trial by the number of hours the babies received more than 30% oxygen in the first ten days: No RDS = < 24 hours; Mild = 24–47 hours; Moderate = 48–120 hours; Severe = > 120 hours. Deaths in the first 5 days were graded severe if the baby died while receiving > 30% oxygen. Ventilation was not used in the grading because 80% of very premature babies are ventilated from birth even though some do not have RDS. An inspired oxygen level of 30% was chosen as the discriminator because many premature babies require low concentrations of oxygen to treat problems unrelated to RDS.

The incidence of different grades of RDS was: 36% of ALEC treated babies had no RDS compared with 25% of the controls, and 45% of the controls had severe RDS compared with 33% of ALEC treated babies. Trend analysis regressing the ALEC proportions in each RDS grade on ordered severity for the four grades was consistent with more ALEC babies in the less severe RDS grades (X^2 for trend = 4.84 $p < 0.03$).

Regression analysis for neonatal mortality, hours of ventilation, hours in more than 30% oxygen and hours in oxygen during the first 10 days for 10 day survivors compare ALEC with the effect of being female and of one extra week of gestation, adjustment being made also for centre and multiple births, for which treatment groups are unbalanced. Sex and one week of gestation are included for comparison because of their known effect on outcome. Hours in more than 30% inspired oxygen were included as a better discriminator of the severity of the respiratory disease than hours in oxygen because many very premature babies are in low levels of oxygen for many weeks.

The first 10 days were chosen as the time when ALEC given at birth might have most effect on the need for ventilation and oxygenation. Longer collection of data was impracticable. Using multiple regression analysis for the effect on respiratory support: babies had on average, 19 hours less in > 30% oxygen ($p < 0.05$), 20 hours less ventilation ($p < 0.05$) and 17 hours less oxygen ($p < 0.14$) than the controls. These effects were similar to the reduction produced by one extra week of gestational age.

The incidence of pneumothorax or severe interstitial emphysema was 26% of ALEC treated babies and 30% of the controls ($X^2 = 0.45$ p = NS).

Patent ductus arteriosus (PDA) was diagnosed from a classical murmur, bounding pulses, plethoric lung fields and the need for appropriate treatment. The ALEC treated group had an incidence of 26% and the controls 23% ($X^2 = 0.30$ p = NS).

Chronic lung disease was defined when babies required oxygen therapy after day 10, i.e. at 241 hours. In the 10 day survivors, there is a slightly reduced incidence of chronic lung disease in ALEC treated group (36%) compared with controls (43%) (X^2 = 2.58 p = NS). The 95% confidence interval is from an increase of 3 to a reduction of 20 cases.

Necrotising enterocolitis (NEC) was diagnosed when babies had abdominal distension, bloody stools and no other aetiology. The incidence of NEC was 9% of ALEC treated babies compared with 12% of controls, 95% confidence intervals being from 4 more to 10 less cases per 100 ALEC treated babies (X^2 = 0.90 p = NS).

In conclusion; this protein free artificial surfactant (ALEC), given at birth to unselected babies under 30 weeks gestation reduces mortality, the incidence of intraventricular haemorrhage, the severity of respiratory disease and the need for prolonged oxygen therapy and was apparently harmless. Artificial surfactant with this formulation could be a safe effective addition to the therapy of premature babies at risk of respiratory distress syndrome and its associated complications.

Acknowledgements. This work was supported by the Medical Research Council and the University of Cambridge Baby Research Fund. It would not have been possible without the hard work and support of very many colleagues in Cambridge and other British neonatal units.

References

1. Enhorning G, Robertson B (1973) Lung expansion in the premature rabbit fetus after tracheal deposition of surfactant. Pediatrics 50:58–66
2. Morley CJ, Bangham AD, Johnson P, Thorburn GD, Jenkin G (1978) Physical and physiological properties of dry lung surfactant. Nature 271:162–163
3. Morley C, Robertson B, Lachmann B, Nilsson R, Bangham A, Grossmann G, Miller N (1980) Artificial surfactant and natural surfactant. Comparative study of the effects on premature rabbit lungs. Arch Dis Child 50:758–765
4. Hallman M, Feldman BH, Kirkpatrick E, Gluck L (1977) Absence of phosphatidylglycerol (PG) in respiratory distress syndrome in the newborn: study of the minor surfactant phospholipids in newborns. Pediatr Res 11:714–720
5. Bangham AD, Morley CJ, Phillips MC (1979) The physical properties of an effective lung surfactant. Biochim Biophys Acta 573:552–556
6. Bangham AD, Miller NGA, Davies RJ, Greenough A, Morley CJ (1984) Introductory remarks about Artificial Lung Expanding Compounds (ALEC) Colloid and Surfaces 10:337–347
7. Bangham AD, De Grier J, Greville GD (1967) Osmotic properties and water permeability of phospholipid liquid crystals. Chem Phys Lipids 1:225–246
8. Morley CJ, Bangham AD, Miller N, Davis JA (1981) Dry artificial surfactant and its effect on very premature babies. Lancet 1:64–68
9. Morley CJ (1986) The Cambridge experience of artificial surfactant. In: Strang L (ed) Proceedings of the international symposium of the physiology and pathophysiology of the fetal and neonatal lung. MTP, Lancaster
10. Morley CJ, Greenough A, Gore SM, Miller NGA, Bangham AD, Pool J, Davis JA, Vyas H (1988) Randomized trial of artificial surfactant (ALEC) given at birth to babies from 23–34 weeks gestation. Early Human Development 17:41–54

11. Enhorning G, Grossmann G, Robertson B (1973) Pharyngeal deposition of surfactant in the premature rabbit fetus. Biol Neonate 22:126–132
12. Ten Centre Study Group (1987) Ten centre trial of artificial surfactant (artificial lung expanding compound) in very premature babies. BMJ 294:991–996

Rapid Response of Oxygenation, Slow Response of Compliance After Liposomal Phospholipid Substitution in Respiratory Distress Syndrome*

M. Obladen, P. Stevens, and E. Kattner

Introduction

Why Artificial Surfactant? The efficacy of natural surfactant both to cure and to prevent respiratory distress syndrome (RDS) of the neonate has been established [2–5, 8, 14–16, 23, 25, 26]. However, some doubts remain concerning the safety of this treatment from immunological and virological considerations. In addition, the possibility to prepare surfactant substitutes from human amniotic fluid [8, 15, 16] is limited due to shortage of material and risk of infection. If the use of artificial surfactants is considered beyond neonatal RDS, for instance in the treatment of adult respiratory distress syndrome, large amounts of surface active material will be required, which can hardly be obtained from natural sources.

Protein-free surfactants have been studied extensively in Britain both in dry form [17, 18, 30] and in saline soluted liposomes [7, 19, 29], but the results were inconsistent. Whereas Morley [19] found an increased survival and reduced incidence of parenchymal brain hemorrhage in infants treated with artificial surfactant, Wilkinson [30] and Halliday [7] could not show beneficial effects and Milner [17] found no useful change in static total compliance or in blood gases 20 and 40 minutes after the administration of artificial surfactant.

In Vitro and Animal Studies. During the past 7 years, we have developed a technique to prepare surface active large unilamellar vesicles from dipalmitoyl phosphatidylcholine and phosphatidylglycerol (DPPC-PG 8 : 2).

This protein-free material has been studied in vitro [20], in surfactant deprived adult rats [22], and in preterm rabbits of 27 days gestation [21]. Despite identical in vitro activity, the large phospholipid liposomes had less influence on oxygenation and restitution of compliance in the animal models than natural surfactant preparations.

Aim of Study. The aim of our study was to compare the substitution effect of an artificial surfactant preparation on oxygenation with that of lung mechanics in order to obtain information as to whether liposomal surfactant action is based on its surface active properties.

* Supported by Deutsche Forschungsgemeinschaft, Ob 43/5-1.

Methods

Study Design and Entry Criteria. With approval of Bochum University's Ethical Committee, a pilot study was performed restricted to 10 infants with severe RDS.

Infants were admitted to the protocol when the following entry criteria were fulfilled:

1. Birthweight less than 1500 g
2. Gestational less than 30 weeks
3. Age less than 6 hours
4. Grade 2 or more RDS present on X-ray
5. Respiratory failure requiring artificial ventilation
6. Phosphatidylglycerol absent in tracheal aspirate
7. Parental written informed consent
8. Prognosis regarded poor by independent consultant

A single dose of 50 mg/kg of surface active large unilamellar vesicles containing 80 % dipalmitoylphosphatidylcholine and 20 % egg phosphatidylglycerol was instilled into the tracheal tube.

The freeze-dried material was obtained from Nattermann (Köln, FRG) and was stored at –18° C until use.

Wilhelmy Balance and Electron Microscopy. After warming to room temperature, the samples were dissolved in a phosphate buffered electrolyte solution by gentle shaking and vortexing for 1 minute. No ultrasound was applied. All samples were studied by Wilhelmy balance and by electronmicroscopy [20] as previously described.

All samples were highly surface active in the Wilhelmy balance, adsorbing to a film in less than one second and giving a maximal surface tension of 57 and a minimal surface tension of 0–2 mN/m at compression to 20 % area. The stability index of the samples ranged from 1.96–2.00.

Pneumotachography. Pneumotachographic measurements of tidal volume (VT) and compliance of the respiratory system (Crs) was performed using the technique described above [12].

Blood Gases. Using intermittent arterial puncture and continuous transcutaneous recording (Transoxode Draeger, Lübeck, FRG), blood gases were measured. Oxygenation was evaluated by calculating transcutaneous/alveolar PO_2 ratio. Differences during the course were evaluated with the Wilcoxon signed rank test.

X-ray Studies. Before, 6 and 24 h after the administration of artificial surfactant, X-ray studies were performed and were evaluated according to Giedion [6].

Tracheal Aspirate Studies. Tracheal suctioning was performed as clinically indicated. Before surfactant substitution and after 1 and 4 days, the lung profile was determined from tracheal suctionings using two-dimensional thin layer chromatography as described by Kulovich [13] and modified by Stevens [27].

Results

Patients and Clinical Data. Table 1 shows the initial clinical data and Table 2 the course of the 10 infants studied: 7 infants survived, 1 each died from RDS (day 1), from intraventricular hemorrhage (day 2) and from bronchopulmonary dysplasia (day 42). Of the survivors, the mean duration of mechanical ventilation was 34 days (range 13–42) and bronchopulmonary dysplasia was present three times. No ductus arteriosus with significant left to right shunting was observed.

Mean initial FiO_2 requirement was 0.71 (0.40–1.0), mean initial compliance was 0.26 ml/cm H_2O/kg (range 0.14–0.38). Within 6 hours after substitution, a marked increase in transcutaneous/alveolar PO_2 ratio was found ($p < 0.01$). This increase, however, did not parallel the increase in compliance, which occurred later.

Dissociated Response. As indicated in Table 2, different types of response could be distinguished: Rapid response in both oxygenation and compliance, dissociated response with a rapid improvement in oxygenation and a delayed increase in compliance, and no significant change after supplementation.

Table 3 shows the change in peak inspiratory pressure, compliance of the respiratory system, oxygen requirement and transcutaneous/alveolar PO_2 ratio after supplementation. Mean initial FiO_2 requirement was 0.71 (0.40–1.0), mean initial compliance was 0.25 ml/cm/H_2O/kg (range 0.14–0.38). Already 30 minutes after substitution, a marked increase in transcutaneous/alveolar PO_2

Table 1. Clinical data and pulmonary status of ten preterm infants with severe RDS before liposomal phospholipid substitution

Patient no.	Sex	Birth weight	Gestation (weeks)	Antenatal steroids	L/S	PI/PS	RDS grade	Initial FiO_2
1	F	1010	28	Yes	1.9	0.6	II	0.46
2	M	820	27	Yes	2.1	1.1	III	1.0
3	M	810	27	No	1.8	0.0	III	1.0
4	F	1110	29	No	0.0	0.0	III	0.85
5	F	1030	29	No	0.0	0.0	III	0.80
6	M	1250	29	Yes	3.4	0.0	II	0.95
7	F	820	28	No	11.8	4.3	III	0.40
8	M	1260	28	No	1.6	0.3	III	0.50
9	F	110	28	No	0.0	0.0	II	0.63
10	M	970	26	No	2.6	0.5	II	0.50

L/S, lecithin/sphingomyelin ratio; PI/PS, phosphatidylinositol/phosphatidylserine ratio.

Table 2. Type of response and clinical course in the ten infants shown in Table 1, after single tracheal instillation of 50 mg/kg surface active liposomes

Patient no.	Age at substitution	Response	Δ FiO_2	Δ CRS	Relapse	Ventilator (days)	Intraventricular hemorrhage	Outcome
1	2 h 15′	Rapid	–0.16	+0.11	24 h	39	–	Survived after pneumothorax, grade I BPD
2	1 h 30′	Rapid	–0.75	+0.52	16 h	(3)	Grade IV	Died at 3 days from IVH
3	2 h 30′	No	–0.20	+0.01	–	(1)	–	Died at 14 h from RDS and pneumothorax
4	4 h 50′	Dissociated	+ 0.15	+ 0.02	No	20	–	Survived. Operated on for jejunal obstruction
5	5 h 30′	Dissociated	–0.20	+0.02	No	34	–	Survived with grade II BPD
6	2 h 50′	No	–0.10	–0.01	–	42	Grade II	Survived with grade II BPD
7	4 h	Rapid	–0.15	+0.19	20 h	35	Grade III	Survived with ventricular dilation
8	3 h 40′	Dissociated	–0.15	± 0	No	13	–	Good
9	3 h 10′	Dissociated	–0.08	–0.01	18 h	30	Grade III	Survived with ventricular dilation
10	2 h 30′	Rapid	–0.26	+0.03	10 h	(42)	Grade III	Died at day 42 from BPD

Δ FiO_2, change in the need for oxygen 1.5 h after substitution; Δ Crs, change in compliance.

Table 3. Results after tracheal substitution of surface active phospholipid liposomes in ten preterm infants severe RDS: peak inspiratory pressure (P_{insp}), dynamic compliance of the respiratory system (C_{RS}), need for oxygen (FiO_2) and transcutaneous/alveolar PO_2 ratio (Tc/A-PO_2) before and after liposomal substitution

Time	P_{insp} (cm H_2O)	C_{RS} (ml cm H_2O^{-1} kg^{-1})	FiO_2	Tc/A-PO_2
Before substitution	22.3 ± 4.0	0.25 ± 0.05	70.9 ± 23.7	0.18 ± 0.10
After substitution				
0.5 h	22.5 ± 3.8	0.25 ± 0.05	57.5 ± 19.9*	0.30 ± 0.16**
1.5 h	22.6 ± 3.8	0.28 ± 0.07	51.9 ± 25.7**	0.31 ± 0.17**
4 h	22.0 ± 4.7	0.32 ± 0.09	51.6 ± 27.1**	0.33 ± 0.20**
12 h	21.6 ± 6.5	0.34 ± 0.17*	53.0 ± 26.1*	0.24 ± 0.15
24 h	19.9 ± 5.2*	0.30 ± 0.17	47.8 ± 22.2	0.27 ± 0.13

Maximal oxygenation response is present after 1.5 h, and significant compliance response is present after 12 h. Values are means ± SD, Wilcoxon signed rank test.
* $p < 0.05$; ** $p < 0.01$.

ratio was found ($p < 0.01$). This increase, however, did not parallel the increase in compliance, which occurred later in the majority of cases. A significant increase in compliance was found after 12 hours ($p < 0.05$). In 8 of the infants, a marked but usually not immediate improvement in oxygenation occurred within 1.5 hours after liposome administration. The slow rise in compliance observed after 4 hours in 7 infants usually did not parallel the rise of the $tcPO_2/A\ PO_2$ ratio. The dissociated response observed in the majority of the cases is illustrated in Fig. 1.

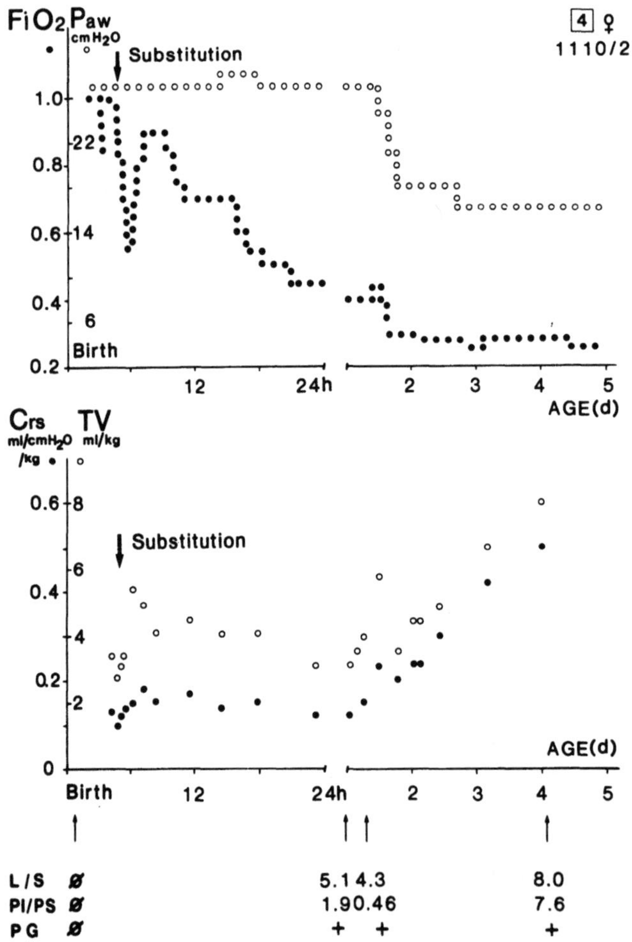

Fig. 1. Dissociated response after tracheal administration of surface active phospholipid liposomes (infant 4). *Top* Inspired oxygen *(dots)* and peak inspiratory pressure *(circles)*. *Middle* Compliance of the respiratory system *(dots)* and specific tidal volume *(circles)*. *Bottom* Phospholipid composition of tracheal aspirate. *L/S*, lecithin/sphingomyelin ratio; *PI/PS*, phosphatidylinositol/phosphatidylserine ratio; *PG*, phosphatidylglycerol. Note improvement in oxygenation within 1.5 h and increase in compliance 24 h after substitution

Rapid Responder. An immediate response in both oxygenation and compliance (the type of reaction seen in the animal models) occurred only in 4 of the infants. Figure 2 shows this type of response. The "rapid responders", even though showing typical clinical RDS, were relatively mature in their phospholipid lung pattern and all had phosphotidylinositol in their tracheal aspirate. In the rapid responding infant whose course is shown in Fig. 2, grade IV intraventricular hemorrhage occurred at 32 hours of age.

Relapse. Figure 2 also shows relapse, which we observed in 5 of the 8 responders after a mean interval of 18 hours.

A typical X-ray course after substitution is shown in Fig. 3 from the infant whose course is shown in Fig. 2: At 1 hour of age, the infant had grade II RDS

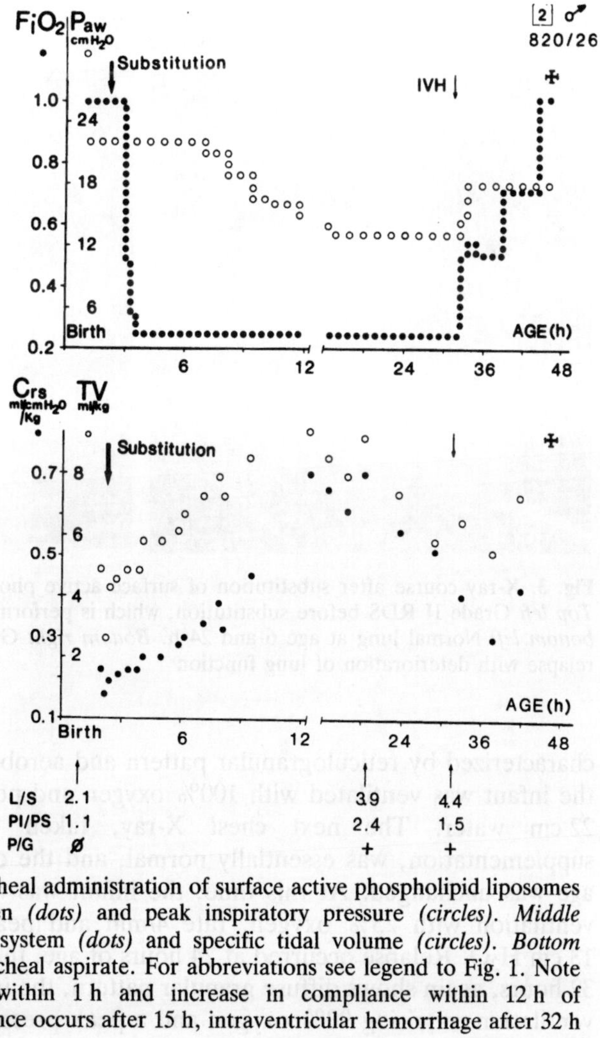

Fig. 2. Rapid response after tracheal administration of surface active phospholipid liposomes (infant 2). *Top* Inspired oxygen *(dots)* and peak inspiratory pressure *(circles)*. *Middle* Compliance of the respiratory system *(dots)* and specific tidal volume *(circles)*. *Bottom* Phospholipid composition of tracheal aspirate. For abbreviations see legend to Fig. 1. Note improvement in oxygenation within 1 h and increase in compliance within 12 h of substitution. Relapse of compliance occurs after 15 h, intraventricular hemorrhage after 32 h

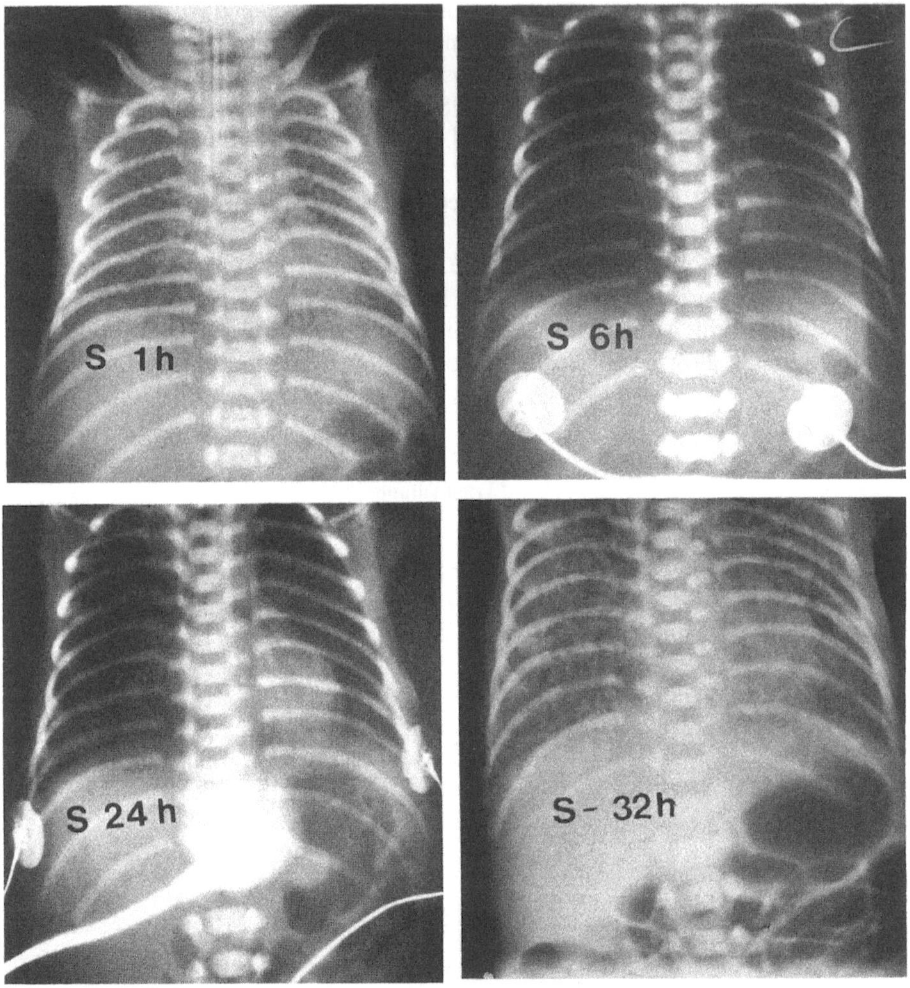

Fig. 3. X-ray course after substitution of surface active phospholipid liposomes (infant 2). *Top left* Grade II RDS before substitution, which is performed at age 1.5 h. *Top right* and *bottom left* Normal lung at age 6 and 24 h. *Bottom right* Granular pattern at 32 h of age, relapse with deterioration of lung function

characterized by reticulogranular pattern and aerobronchogram. At this time, the infant was ventilated with 100% oxygen and peak inspiratory pressure of 22 cm water. The next chest X-ray, taken 4 hours after liposome supplementation, was essentially normal, and the chest X-ray at 24 hours of age was unchanged. At this time, the infant was on intermittent mandatory ventilation with 25% oxygen, rate 4/min and peak inspiratory pressure of 13 cm H_2O. Relapse occurred at 24 hours of age, the last chest X-ray, taken at 32 hours, again shows diffuse granular pattern, the infant was back on artificial ventilation requiring 90% oxygen and a peak pressure of 20 cm of water.

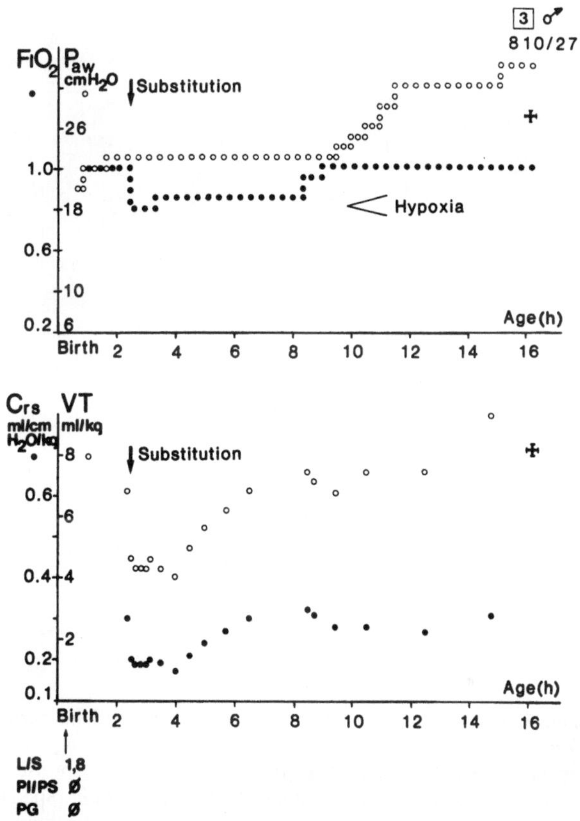

Fig. 4. No response after administration of surface active phospholipid liposomes (infant 3). *Top* Inspired oxygen *(dots)* and peak inspiratory pressure *(circles)*. *Middle* Compliance of the respiratory system *(dots)* and specific tidal volume *(circles)*. *Bottom* Phospholipid composition in tracheal aspirate. For abbreviations see legend to Fig. 1. Only minimal improvement in both oxygenation and compliance occurred in this infant within 3 h after substitution. The infant died from progressive respiratory failure and final pneumothorax

Non-Responder. One of the infants (Fig. 4, infant No. 3) showed no response to liposome administration and died from progressive respiratory failure at the age of 16 hours despite ventilation with 100% oxygen and very high peak airway pressure. In this infant, the morphology of the liposome preparation differed markedly from that seen in the other infants: Fig. 5. shows the electronmicrographs of the typical preparation containing exclusively large unilamellar liposomes. The other sample, which was given to the non-responder, has smaller vesicles and shows a pattern of regular aggregation. Both samples were prepared from the same batch, displayed the same surface tension lowering abilities in the Wilhelmy balance (maximal surface tension 37, minimal surface tension 2.0 mN/m, Stability index 2.0) and we are not aware of any difference in the type of dissolution of the sample.

Tracheal Aspirate Studies. Tracheal aspirates were studied to investigate whether phosphatidylglycerol administration interferes with the preterm infants' de novo phosphatidylinositol synthesis. Phosphatidylglycerol, formerly absent in all infants, was present in all 8 tracheal aspirates sampled 24 hours after substitution. At this time, lecithin/sphingomyelin ratio had increased from 2.52–4.18 and phosphatidylinositol/phosphatidylserine ratio from 0.68 to

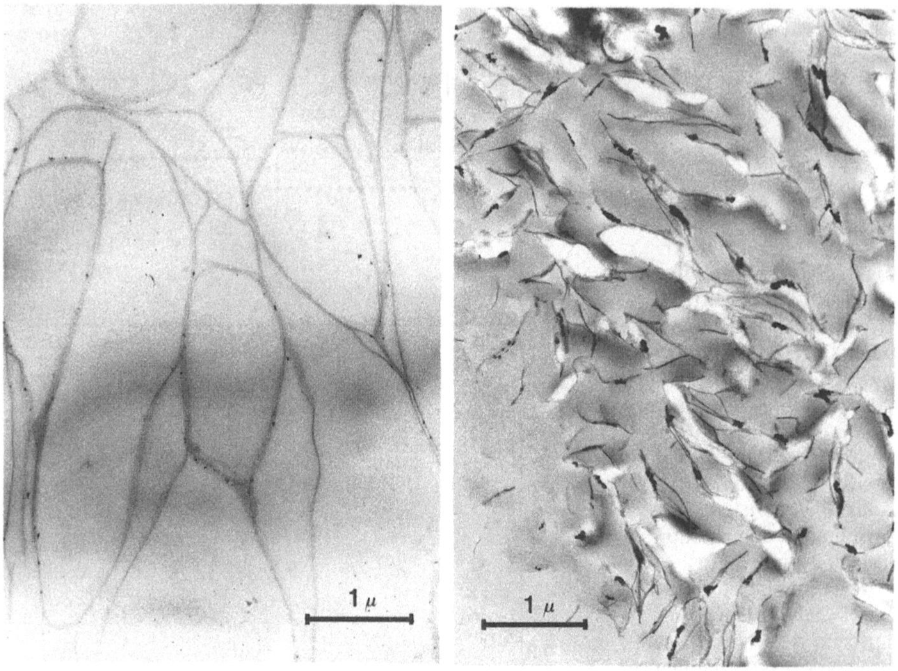

Fig. 5. Electron micrographs obtained from artificial surfactant samples after fixation with osmiumtetroxide and glutaraldehyde, thin sectioning, and staining with lead citrate. *Left* Specimen used for infant 1 (resulting in rapid response of oxygenation and compliance) shows exclusively large unilamellar vesicles. *Right* Specimen used for infant 3 (without significant response of oxygenation or compliance) shows smaller unilamellar vesicles and some undiluted crystals. Both samples were prepared under identical conditions and from the same batch, displayed identical and rapid film adsorption, and were able to lower the surface tension to 0 mN/m in the Wilhelmy balance

1.03. On the fourth day after substitution, phosphatidylglycerol was present in the lung effluent of 3 of the 9 infants living at this time, lecithin/sphingomyelin ratio was 4.54 and phosphatidylinositol/phosphatidylserine ratio was 3.71.

Discussion

Different Response, Different Disease? The aim of our study was to correlate oxygenation response with changes in compliance after supplementation of surface active liposomes. The observed dissociation of both responses offers a variety of explanations and speculations:

1. Surface active phospholipid liposomes without protein may not act as a surfactant at all but may provide substrate for endogenous surfactant biosynthesis, explaining the increase in compliance after 12 hours. Whereas the functions of the different phospholipids in the surfactant system are quite clear, some uncertainty exists concerning the functions of surfactant

apoproteins. The major 35 KD glycoprotein may contribute to the formation of an alveolar surfactant reserve by supporting the formation of tubular myelin, may regulate recycling and surfactant homeostasis [31], but probably is not essential for film adsorption. An accumulation of glycolipid in the alveolar space has been observed in adult respiratory distress syndrome [24]. The minor ethanol/ether soluble 8 KD lipoprotein augments liposome uptake by isolated granular pneumocytes [1] and seems to be important for surfactant function by enhancing rapid film adsorption [28].

2. Ultrastructural heterogenity of the liposomes may be responsible for different in vivo activity. In a previous in vitro study [20] we have shown that slight variations in the type of preparation, in the temperature during preparation, and in the use of ultrasound may greatly influence vesicle size, the speed of film adsorption, and the surface activity of binary mixed liposomes.
3. In a rescue study with full-blown RDS, the time of administration of exogenous surfactant may have been too late to expect an optimal response, because extensive hyaline membranes could block the site of surfactant action despite sufficient surface activity of the preparation. Relapse was found in several trials and led to repeated substitution doses [8, 25]. Especially when the administration of surfactant was performed late during the course of RDS, the benefits of the surfactant instillation were not sustained and subsequent deterioration led to prolonged oxygen therapy [25]. In prophylactic studies, where an exogenous surfactant is administered before the first breath [2, 16], the beneficial effect of a single dose of 75–80 mg/kg seemed to be sustained.
4. Respiratory distress syndrome of the neonate may not be a homogenous disease. James et al. [11] could show that surfactant deficiency is not specific for hyaline membrane disease, as similar phospholipid abnormalities were observed in babies with congenital pneumonia and transient tachypnoea of the newborn. An "adult" type of RDS, characterized by large amounts of surfactant inhibitor [10] in the alveoli may not respond to a single dose of 50 mg/kg of surface active liposomes.

Comparison with Other Trials. To date, no studies comparing natural to artificial surfactants have been published. Natural surfactant seems to have an immediate action [4, 15, 23], whereas in protein-free artificial surfactant the response seems to be delayed [17, 19]. Our data show that after the substitution of protein-free liposomes there is little response in lung compliance for several hours and that improved oxygenation did not result in all infants. Probably, apoproteins will be required for optimal function of exogenous surfactant; especially the 8 KD lipoprotein has been shown to enhance film adsorption from the subphase [28]. The delayed response in compliance observed by us and other investigators may suggest that the material did not act as a surfactant at all, but as a substrate donor for endogenous surfactant production. Recently, Hallman could show that oral administration of inositol led to an increased surfactant synthesis and to improvement of RDS [9].

Relapse was found in several trials and led to repeated substitution doses [8, 25]. Especially when the administration of surfactant was performed late during the course of RDS, the benefits of the surfactant instillation were not sustained and subsequent deterioration led to prolonged oxygen therapy [25]. In prophylactic studies, where exogenous surfactant is administered before the first breath [2, 16], the beneficial effect of a single dose of 75–80 mg/kg seemed to be sustained.

Conclusions. We have observed improved oxygenation and increased compliance after a single dose of protein-free surface active liposomes in RDS. The effect, however, varied greatly in different infants and usually was not sustained. Most infants required prolonged artificial ventilation. Moreover, the effects on oxygenation and compliance were not simultaneous. The results with our preparation did not encourage us to proceed to a larger controlled trial. But we still see a need for the development of an artificial surfactant which can be manufactured easily and in sufficient amounts. Further research should be performed with incorporation of specific apoproteins into surface active liposomes. Trials comparing artificial with natural surfactants should be planned. Direct measurement of lung mechanics may be a useful tool to recognize whether an exogenous preparation acts as a surfactant at all. If the effect is due to lowering of the surface tension at the alveolar wall, an immediate increase in compliance has to be expected after surfactant administration.

Acknowledgements. The authors thank Mrs. Martina Flörecke for technical assistance in lung function studies, Dr. Heinz Schwarz, Max-Planck-Institut for Biology, Tübingen, for performing the electronmicroscopy, and to Dipl.-Math. Ursula Friedlieb for statistical evaluation.

References

1. Claypool WD, Wang DL, Chander A, Fisher AB (1984) An ethanol/ether soluble apoprotein from rat lung surfactant augments liposome uptake by isolated granular pneumocytes. J Clin Invest 74:677–684
2. Enhorning G, Shennan A, Possmayer F, Dunn M, Chen CP, Milligan J (1985) Prevention of neonatal respiratory distress syndrome by tracheal instillation of surfactant: a randomized clinical trial. Pediatrics 76:145–153
3. Enhorning G (1987) Surfactant supplementation: Toronto trial. In: Walters DV, Strong LB, Geubelle F (eds) Physiology of the fetal and neonatal lung. MTP Press, Lancaster, pp 275–285
4. Fujiwara T, Chida S, Watabe Y, Maeta H, Morita T, Abe T (1980) Artificial surfactant therapy in hyaline membrane disease. Lancet:55–59
5. Gitlin JD, Soll RF, Parad RB, Horbar JD, Feldman HA, Lucey JF, Taeusch HW (1987) Randomized controlled trial of exogenous surfactant for the treatment of hyaline membrane disease. Pediatrics 79:31–37
6. Giedion A, Haefliger A, Dangel P (1973) Acute pulmonary X-ray changes in hyaline membrane disease treated with artificial ventilation and positive end-expiratory pressure. Pediatr Radiol 1:145

7. Halliday H, McReid MC, Meban C, McClure G, Lappin TRJ, Thomas PS (1981) Controlled trial of artificial surfactant to prevent respiratory distress syndrome. Lancet 1:476–478
8. Hallman M, Merritt TA, Schneider H, Epstein BL, Mannino F, Edwards DK, Gluck L (1983) Isolation of human surfactant from amniotic fluid and a pilot study of its efficacy in respiratory distress syndrome. Pediatrics 71:473–482
9. Hallman M, Jarvenpaa AL, Pohjavuori M, Bromberger P (1986) Double blind randomized trial on inositol. Supplementation in respiratory distress syndrome. Pediatr Res 20(4):411A
10. Ikegami M, Jobe A, Jacobs H, Lam R (1984) A protein from airways of premature lambs that inhibits surfactant function. J Appl Physiol 57(4):1134–1142
11. James DK, Chiswick ML, Harkes A, Williams M, Hallworth J (1984) Non-specificity of surfactant deficiency in neonatal respiratory disorders. Br Med J 288:1635–1638
12. Kattner E, Kosack K, Obladen M (1987) Pneumotachographic measurements of lung function in preterm infants with RDS (this book)
13. Kulovich MV, Hallman MB, Gluck L (1979) The lung profile. I. Normal pregnancy. Am J Obstet Gynecol 135:57
14. Kwong MS, Egan EA, Notter RH, Shapiro DL (1985) Double-blind clinical trial of calf lung surfactant. Extract for the prevention of hyaline membrane disease in extremely premature infants. Pediatrics 76:585–592
15. Merritt TA, Hallman M, Holcomb K, Strayer D, Bloom B, Revak S, Cochrane CG (1986) Human surfactant treatment of severe respiratory distress syndrome: pulmonary effluent indicators of lung inflammation. J Pediatr 5:741–748
16. Merritt TA et al. (1986) Prophylactic treatment of very premature infants with human surfactant. N Engl J Med 315:785–790
17. Milner AD, Vyas H, Hopkin IE (1983) Effects of artificial surfactant on lung function and blood gases in idiopathic respiratory distress syndrome. Arch Dis Child:458–460
18. Morley CJ, Miller N, Bangham AD, Davis JA (1981) Dry artificial lung surfactant and its effect on very premature babies. Lancet 1:64–68
19. Morley CJ (1987) The Cambridge experience of artificial surfactant. In: Walters DV, Strong LB, Geubelle G (eds) Physiology of the fetal and neonatal lung. MTP Press, Lancester, pp 255–274
20. Obladen, Popp D, Schöll CH, Schwarz H, Jähnig F (1983) Studies on lung surfactant replacement in respiratory distress syndrome. Rapid film formation from binary mixed liposomes. Biochim Biophys Acta 735:215–224
21. Obladen M, Kampmann W, Zimmermann I, Lachmann B (1985) Artificial surfactant in preterm rabbits with and without respiratory distress syndrome: difference of in vitro and in vivo activities. Eur J Pediatr 144:195–199
22. Obladen M (1985) Liposomale Phospholipid-Substitution und Lungenfunktion bei surfactant-deprivierten Ratten. Klin Pediatr 197:152–154
23. Raju TNK, Bhat R, McCulloch KM, Maeta H, Vidyasagar D, Sobel D, Anderson M, Levy PS (1987) Double-blind controlled trial of single-dose treatment with bovine surfactant in severe hyaline membrane disease. Lancet 1:651–655
24. Rauvala H, Hallman M (1984) Glycolipid accumulation in bronchoalveolar space in adult respiratory distress syndrome. Prog Lipid Res 25:1257–1262
25. Smyth JA, Metcalfe IL, Duffty P, Possmayer F, Bryan MH, Enhorning G (1983) Hyaline membrane disease treated with bovine surfactant. Pediatrics 71:913–917
26. Shapiro DL, Notter RH, Morin III FC, Deluga KS, Golub LM, Sinkin RA, Weiss KI, Cox CH (1985) Double-blind, randomized trial of a calf lung surfactant extract administered at birth to very premature infants for prevention of respiratory distress syndrome. Pediatrics 76(4):593–599
27. Stevens P, Thiemann U, Obladen M (1984) Computer-corrected neonatal lung profile. Prog Respir Res 18:230–234
28. Takahashi A, Fujiwara T (1986) Low molecular weight proteolipid in lung surfactant: its role in surfactant function. Pediatr Res 20(4):443A
29. Ten Centre Study Group (1987) Ten centre trial of artificial surfactant (artificial lung expanding compound) in very premature babies. Br Med J 294:991–996

30. Wilkinson A, Jenkins PA, Jeffrey JA (1985) Two controlled trials of dry artificial surfactant: early effects and later outcome in babies with surfactant deficiency. Lancet 2:287–291
31. Wright JR, Wager RE, Hawgood S, Dobbs L, Clements JA (1986) Surfactant apoprotein M_r = 26,000–36,000 enhances uptake of liposomes by type II cells. J Biol Chem 262(6):2888–2894

Impact of Surfactant Therapy on Bronchopulmonary Dysplasia*

T. A. Merritt, M. Hallman, and D. Edwards

Introduction: Influence of Surfactant Therapy on Acute Lung Injury

Bronchopulmonary dysplasia (BPD) continues as an important form of chronic lung disease in infancy with occurrence rates from 5–68% among infants receiving assisted ventilation after birth [1–5]. Since BPD may develop from nearly all forms of neonatal acute lung injury (respiratory distress syndrome, persistent pulmonary hypertension, pneumonia, ventilator dependent apnea) common pathophysiologic mechanisms in these different forms of lung injury that ultimately lead to BPD can be deduced. The role of inflammatory cells and their secretion or release of protease [6], leukotrienes [7], and glycoplipids [8] that alter pulmonary function or damage lung epithelial cells and the connective tissue matrix of the lung leading to pulmonary edema, airway hyper-reactivity, and pulmonary hypertension, and consistent with a characteristic clinical course in infants with BPD have been described.

Infants with respiratory distress syndrome resulting from a deficiency or inadequate endogenous pool size of pulmonary surfactant are at greatest relative risk for developing bronchopulmonary dysplasia [9]. Immaturity of the connective tissue of the pulmonary acini, the limited quantity and immature composition of surfactant (both phospholipid and apoprotein), the requirement for intermittent mandatory ventilation with barotrauma and requirement for supplemental oxygen for prolonged intervals contribute to pathogenesis of lung injury. The association of pulmonary airleaks, especially interstitial emphysema, with later BPD suggests that treatments that lower the frequency of these morbidities may also have an impact on the frequency of BPD [10, 11].

Since the primary characteristics of RDS pathogenesis is the expression of surfactant deficiency with end-expiratory alveolar collapse, it should then be possible to prevent, or at least moderate, the development of RDS and thus the lung injury which contributes to PBD, by instilling into the airways a concentrated surfactant at birth, or soon after the clinical manifestations of RDS develop. Surfactant administered to the airways must be able to reach the distal airways in a relative homogenous fashion, and adsorb onto the air-liquid interface as the fetal lung liquid is absorbed. The stabilizing monolayer film of

* Supported by USPHS, NIH-HL 35036, FDA Office of Orphan Drugs, Division of Research Resources RR 00827, the Finnish Academy; and the Sigrid Juselius Foundation.

surfactant once formed should permit varying surface tensions within the acini, increase alveolar stability on expiration, provide for more uniform alveolar expansion during inspiration, decreased transvascular filtration of edema fluid and proteinaceous inhibitors, and thus decrease the work of breathing while allowing endogenous surfactant synthesis and secretion to develop after birth.

Enhorning and coworkers [12] first demonstrated the anticipated improvement in pulmonary function by restoring surfactant in a series of fetal animals studies using pharyngeal and/or tracheal deposition of the natural surfactant (approximately 30 mg/ml) prior to lung aeration. Surfactant instillation resulted in increased compliance, improved histological, as well as radiological expansion, and an increased survival rate [13]. These short-term studies demonstrated dramatic resolution of hypoaerated "white-out" on chest radiographs characteristics of RDS, but were unfortunately not carried out until the animal survived the newborn period or had the opportunity to develop changes characteristic of chronic lung disease or resolution of lung disease. When the same surfactant preparation was tested on preterm rhesus monkeys (130–132 day gestation) there was also improved gas exchange and survival compared to controls [14]. Adams and coworkers [13] found that treating preterm lambs (120 days gestation) with tracheal instillation of natural surfactant in a dose of 50–170 mg total lipid/kg body weight prevented the early features of RDS. Jobe et al. [15] gave natural sheep surfactant to 120 day fetal lambs either at birth or at about 20 minutes after birth when respiratory failure was established. Used in a "preventative" or preventilatory approach or to rescue an animal in respiratory failure, surfactant treatment dramatically increased PaO_2 from approximately 50 torr to 270 ± 35 torr in 100% O_2 [16]. Although the lambs were ventilated using infant ventilators, they deteriorated within the eight hours as demonstrated by increase in pCO_2, decrease in PaO_2, and in lung-thorax compliance. Nonetheless, natural surfactant deposition improved gas exchange sufficiently to enable a substantial lowering of F_iO_2. Egan and associates [17] found similar improvements in fetal sheep given a natural lipid extract of cow lung lavage while animals treated with synthetic phospholipid mixtures actually fared worse than controls. Surfactant deficient fetal animals not only have decreased arterial-alveolar oxygen tension ratios, hypercarbia, and acidosis, but increased solute permeability (primarily plasma proteins) into the epithelial air spaces and the pulmonary interstitium and then through epithelial "tight junctions" into the alveolar air space. Over distension of the lung by mechanical ventilation in surfactant deficiency has also been shown not only to increase the protein solute "leak" into the airspace but to cause extensive bronchiolar epithelial desquamation with necrosis and protein transudation into the terminal bronchials within minutes of positive pressure ventilation. As plasma proteins "leak" into the alveolus they can aggregate the surfactant or inhibit its biophysical function of lowering surface-tension by poorly understood mechanisms [18]. Exogenous surfactant instillation is reported by Jobe et al. to decrease protein flux into the alveoli in ventilated preterm lambs [19], and to reduce the epithelial disruption in rabbits as described by Nilsson and coworkers [20]. Maeta and coworkers ventilated premature baboons for 9–10 days with and without surfactant [21]. From these studies they concluded that surfactant therapy not only had acute effects of improved oxygenation and lung mechanics but also prevented residual lung disease.

Clinical Methodology

In order to anticipate the beneficial clinical responses of surfactant instillation to preterm infants with RDS based on animal studies, one should anticipate: (a) improvement in PaO_2 sufficient to permit a reduction in supplemental oxygen to near room air levels; (b) lowered $PaCO_2$; (c) reduced mean airway pressures (and therefore a probable reduction in extraalveolar airleaks); (d) reduction in transudation of proteins into the alveoli; and (e) a reduction in both the duration of assisted ventilation and the need for supplemental oxygen, with lower mean airway pressures (CPAP) or time until extubation. If present concepts regarding barotrauma oxygen exposure and oxidant damage to the lung, as well as a decreased occurrence of airleaks (pneumothorax, pneumomediastinum, interstitial air) are correct, a reduction in the number of infants developing BPD should accompany the use of an effective surfactant. Alternatively, it could be argued that the increased number of early survivors following surfactant therapy would increase the number of infants requiring assisted ventilation for either milder RDS, pulmonary edema associated with PDA, or apnea. These early survivors would increase the number of susceptible infants with immature antioxidant defenses, inadequate nutrition, or those with prolonged intubation that might increase the prevalence of chronic lung disease.

Clinically indicated chest radiographs were reviewed by pediatric radiologists at the University of California, San Diego and Children's Hospital, Helsinki without knowledge of treatment assignment. The severity of respiratory distress was graded according to the system of Edwards, and radiologic diagnosis and grading of BPD conformed to that of Toce et al. [24].

Results

Human surfactant given either at birth or after ventilatory failure (at approximately 4 hours after birth) associated with RDS documented by immature phospholipid profile ($L/S \leq 2.0$ and absence of phosphatidylglycerol) significantly reduced the frequency of pneumothorax, pulmonary interstitial emphysema, requirement for supplemental oxygen ≥ 0.3 at 30 days, and radiographic bronchopulmonary dysplasia (Table 1) [22, 23]. Among infants treated with surfactant after ventilatory failure, there was a very significant difference between the two groups on radiographs obtained after surfactant administration ($p < 0.00005$) (Fig. 1). This difference remained significant in subsequent films ($p < 0.05$). Although in the rescue treatment trial, mortality was not significantly reduced by multidose surfactant treatment (i.e., administration of at least 70 mg for infants ≤ 1.0 kg or 70 mg/kg for infants > 1.0 kg) there was a significantly reduced mortality when surfactant was used as a prophylactic treatment (with subsequent doses). Using the grading scale of

Table 1. Outcome of the rescue and prophylactic trials using human surfactant

	Rescue trial			Prophylactic trial		
	Surfactant	Control	P value	Surfactant	Control	P value
No. of patients	22	23		31	29	
Birth weight (g)	1055	987		938	964	
Gestational age (weeks)	27.2	27.0		26.7	26.8	
Pneumothorax *(n)*	1	7	$p < 0.025$	2	7	$p < 0.02$
Pulmonary interstitial emphysema	0	5	$p < 0.028$	1	14	$p < 0.001$
$O_2 > 0.3$ for 30 days *(n)*	2	8	$p < 0.015$	6	11	$p < 0.01$
Bronchopulmonary dysplasia *(n)*	3	9	$p < 0.02$	5	9	$p < 0.003$
Death *(n)*	3	5	NS	5	15	$p < 0.001$

Toce et al. [24]. 8/53 surfactant treated infants had mild BPD and only two had moderate BPD, contrasted to 18/52 in the control group (p < 0.01). Of the latter 18 infants, 9 had severe disease of whom 2 infants died at 7 and 13 months from cor pulmonale and ventilatory failure, while 9 infants had moderate BPD in the control group. Although 8 infants in the surfactant treatment protocol developed radiographic evidence of BPD, only a single infant required supplemental oxygen beyond 3 months and no deaths associated

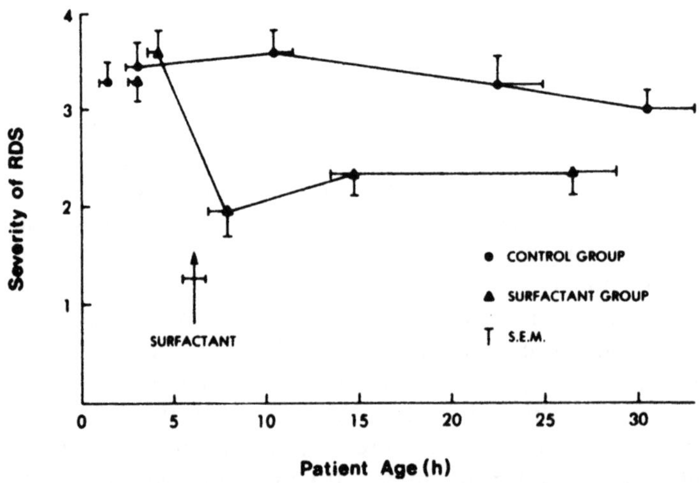

Fig. 1. Scores for the severity of RDS in infants treated with human surfactant and in controls. A higher score indicates more severe RDS as seen on radiographs. Values were not significantly different between groups before surfactant administration, but were afterwards

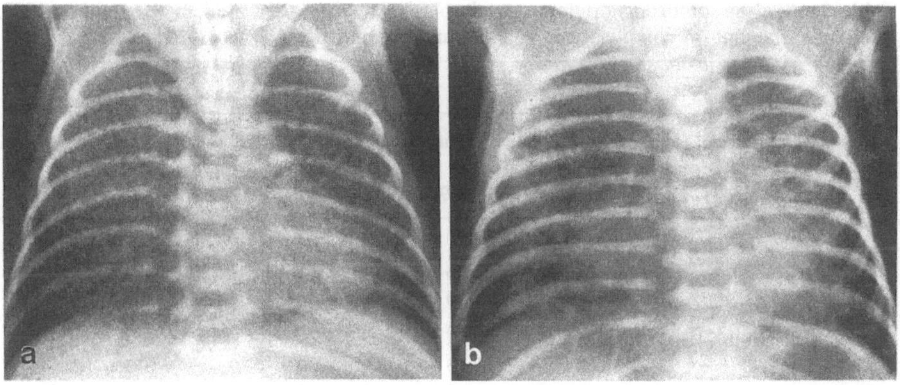

Fig. 2a, b. Transient interstitial process in some infants receiving human surfactant. (Reproduced with permission)

with BPD occurred in these infants. These observations suggest that human surfactant therapy used either at birth or after development of severe RDS, and when given in multiple doses for evidence of relapse, significantly reduced both the frequency and the radiographic severity of BPD. Similarly the clinical course of BPD was substantially milder in infants receiving surfactant. Radiographic evidence of surfactant associated "interstitial infiltrates" occurring soon after instillation of surfactant and apparent radiographically for several days should not be confused with the insidious development of BPD after the third week of life (Fig. 2) with characteristic radiographic features.

Discussion

Clinical trials of exogenous surfactants given at birth or after established RDS have had variable effects on the occurrence of BPD. Using calf lung surfactant extract. Enhorning et al. reported 22/39 (56%) of infants treated at birth developed BPD compared to 22/33 (66% p = NS) of control infants [25]. While Shapiro and coworkers reported a lessening of RDS "severity", the rate of BPD among their control and treatment groups were similar, and Kwong et al. reported that 6/14 CLSE treated infants in contrast to 9/13 placebo treated infants developed BPD (p = NS) [26]. These studies provided for a single dose treatment at birth and lung maturity status (i.e., probability of developing RDS was not determined). In a double-blind trial of TA Surfactant, Raju and colleagues [27] treated infants with a single dose of 100 mg/kg after very low birth weight infants developed severe RDS (about 4–5 hours after birth). Although oxygenation and ventilation was improved soon after treatment, and pulmonary interstitial emphysema and pneumothorax frequencies were significantly reduced, 9/15 surfactant treated versus 4/7 placebo treated infants developed BPD (p = NS), although the authors commented on a milder form of BPD in the surfactant treated group (Table 2). Using artificial lung

Table 2. Analysis of bronchopulmonary dysplasia in trials published by November 1987: effect of surfactant on complications of RDS and prematurity

	Trial							Total	% of total infants
	Hallman et al. [8]	Enhorning et al. [25]	Kwong et al. [26]	Shapiro et al. [29]	Merritt et al. [22]	Gitlin et al. [30]	Raju et al. [27]		
No. of infants									
S	22	39	14	16	31	18	17	157	
C	23	33	13	16	29	23	13	150	
Bronchopulmonary dysplasia									
S	2*	22	6	5	5*	4	9	53	33.7
C	8	22	9	5	9	7	4	64	40.7
Deaths									
S	3	2	1	3	5**	3	2	19	12.1
C	6	7	2	5	15	6	6	41	31.3

S, surfactant-treated group; C, controls.
* P = NS; ** $P < 0.02$

expanding compound (ALEC), British workers reported that neither pulmonary airleaks nor requirement for supplemental oxygen more than 10 days were different between ALEC or placebo treated infants, in spite of a significant improvement in mortality ($p < 0.002$) [28]. BPD was not commented upon in their multicenter trial.

Our experience using human surfactant (often with 2–3 doses in half the infants) has been more optimistic, with significant reduction in BPD confirmed using two trial designs in two separate intensive care nurseries. While human surfactant has been shown not to cause an inflammatory response in the lungs of premature infants, clearly some infants either respond insufficiently to surfactant replacement to prevent lung injury, or are at risk for BPD because of other requirements for ventilatory support and supplemental oxygen during the first weeks after birth, or have a genetic predisposition towards developing this chronic lung disease.

Clearly the goal of surfactant replacement must be to diminish the need for factors known to contribute to the development of BPD in infants with RDS. While calf lung extracts of surfactant and Surfactant TA used as a single dose to either treat infants at birth or to rescue infants at risk for RDS have reduced airleaks and improved survival, BPD has not been significantly reduced amongst survivors treated with these heterologous surfactant in some, but not all, studies. These studies suggest that multi-dose or species specific surfactant is necessary to sufficiently lower requirement for assisted ventilation and oxygen therapy to reduce the occurrence of BPD. As with any new therapeutic modality, extensive testing remains a pre-requisite prior to advocating wider use. It remains critical for future clinical trials to include the occurrence of radiographic BPD and its grade as a critical outcome measure of efficacy, if new therapies in neonatology are to have their real value critically evaluated.

References

1. Bancalari E, Abdenous GE, Feller R et al (1979) Bronchopulmonary dysplasia: clinical presentation. J Pediatr 95:810
2. Northway WH, Rosan RC, Porter DY (1967) Pulmonary disease following respiratory therapy for hyaline membrane disease. NEJM 276:357
3. Northway WH Jr (1979) Observations on bronchopulmonary dysplasia. J Pediatr 95:815
4. Mayes L, Perxett E, Stahlman MT (1983) Severe bronchopulmonary dysplasia: a retrospective review. Acta Paediatr Scand 72:225
5. Rosan HC, Hyaline membrane disease and a related spectrum of neonatal pneumophatheis: The relationship of normal to diseased neonatal lung. In: Rosenbery HS, Balande RP (eds) Perspectives in pediatric pathology, vol 2. Yearbook Medical, Chicago, pp 35
6. Merritt TA, Cochrane CG, Holcomb K et al (1983) Elastase and alpha-1-proteinase inhibitor activity in tracheal aspirates during respiratory distress syndrome. Role of inflammation in the pathogenesis of bronchopulmonary dysplasia. J Clin Invest 108:741
7. Stenmark KR, Eyzaguirre M, Remigio L et al (1985) Recovery of platelet activating factor and leukotrienes from infants with severe bronchopulmonary dysplasia: clinical improvement with cromolyn therapy. Am Rev Respir Dis 131:A236
8. Hallman M, Pitkanen O, Rauvala H, Merritt TA (1987) Glycolipid accumulation in lung effluent in bronchopulmonary dysplasia. Pediatr Res 21:454A
9. Edwards DK, Dyer WM, Northway WH (1977) Twelve years experience with bronchopulmonary dysplasia. Pediatrics 59:839
10. Moylan FMB, Walker AM, Kramer SS et al (1978) Alveolar rupture as an independent predictor of bronchopulmonary dysplasia. Crit Care Med 6:10–13
11. Moylan FMB, Walker AM, Kramer SS et al (1978) The relationship of bronchopulmonary dysplasia to the occurrence of alveolar rupture during positive pressure ventilation. Crit Care Med 6:140–142
12. Enhorning G, Robertson BA (1972) Lung expansion in the premature rabbit fetus after tracheal disposition of surfactant. Pediatrics 50:58
13. Adams F, Towers B, Osher AB, Ikegami M, Fujiwara T, Nozaki M (1978) Effects of tracheal instillation of natural surfactant in premature lambs. I. Clinica autopsy findings. Pediatr Res 12:841
14. Cutz E, Enhorning G, Robertson B, Sherwood W, Hill D (1978) Hyaline membrane disease: effect of surfactant prophylaxis on lung morphology in premature primates. Am J Pathol 78:581
15. Jobe A, Ikegami M, Jones S, Conway D (1983) Permeability of premature lambs lungs to protein and the effect of surfactant on that permeability. J Appl Physiol 55:169
16. Jobe A, Ikegami M, Glatz T, Yoshida Y, Diakomanoliz E, Padbury J (1981) Duration and characteristics of treatment of premature lambs with natural surfactant. J Clin Invest 61:370
17. Egan E, Dillon W, Zorn SA (1984) Fetal lung liquid absorption and alveolar epithelial solute permeability in surfactant deficient, breathing fetal lambs. Pediatr Res 18:566
18. Ikegami M, Jobe A, Jacobs H et al (1984) A protein from airways of premature lambs that inhibits surfactant function. J Appl Physiol 57:1134
19. Jobe A, Ikegami M, Jacobs H et al (1984) Surfactant and pulmonary blood flow distributions following treatment of premature lambs with natural surfactant. J Clin Invest 73:848
20. Nilsson R, Grossman G, Robertson B (1980) Pathogenesis of neonatal lung lesions induced by artificial ventilation; evidence against the role of barotauma. Respiration 40:218
21. Maeta H, Vidyasagar D, Raju T et al (1986) Pathologic findings in surfactant treated and control animals after prolonged ventilation. Pediatr Res 20:435A
22. Hallman M, Merritt TA, Jarvenpaa AL et al (1985) Exogenous human surfactant for treatment of severe respiratory distress syndrome. A randomized prospective clinical trial. J Pediatr 106:963

23. Merritt TA, Hallman M, Bloom BT et al (1986) Prophylactic treatment of very premature infants with human surfactant. NEJM 315:785
24. Toce SS, Farrell PM, Leavitt LA et al (1984) Clinical and roentgenographic scoring systems for assessing bronchopulmonary dysplasia. Am J Dis Child 138:581
25. Enhorning G, Shennan A, Possmayer F et al (1985) Prevention of neonatal respiratory distress syndrome by tracheal instillation of surfactant. A randomized clinical trial. Pediatrics 76:145
26. Kwong MS, Egan E, Notter R et al (1985) A double-blind clinical trial of calf lung lipid for the prevention of hyaline membrane disease in extremely premature infants. Pediatrics 76:585
27. Raju TNK, Vidyasagar D, Bhat B et al (1987) Double-blind controlled trial of single-dose treatment with bovine surfactant in severe hyaline membrane disease. Lancet 1:651
28. Morley CJ, Bangham AD, Miller N et al (1981) Dry artificial lung surfactant and its effect on very premature babies. Lancet 1:64
29. Shapiro DL, Notter RH, Morin FC et al (1985) A double-blind, randomized trial of a calf lung surfactant extract adminstered at birth to very premature infants for prevention of the respiratory distress syndrome. Pediatrics 76:593-599
30. Gitlin JD, Soll RF, Parad RB et al (1987) Randomized controlled trial of exogenous surfactant for the treatment of hyaline membrane disease. Pediatrics 79:31-37

Alveolar Atelectasis as a Mechanism for Infantile Apnoea: A Possible Role for Exogenous Surfactant Therapy?

D. P. Southall, M. Samuels, and D. G. Talbert

Introduction

Traditionally the majority of abnormal infantile apnoea, including apnoea or prematurity, is considered to be the result of a primary disorder of respiratory control. In this paper we review evidence to suggest that the primary problem lies within the lungs and is best explained by an underlying defect in the function of lung surfactant.

Apnoea due to a Defect in Lung Surfactant

As a consequence of investigating 45 infants and young children with cyanotic episodes it became evident that the majority were most adequately described by a condition termed prolonged expiratory apnoea [1, 2]. This form of apnoea, which is responsible for apnoea of prematurity, cyanotic episodes following pertussis, respiratory syncytial virus and other respiratory tract infections, and cyanotic "breath-holding", is characterised by the extremely rapid onset and progression of arterial hypoxaemia. Typically central cyanosis is apparent within 5–10 seconds of the onset of each episode and arterial pO_2 levels fall to below 20 mm Hg within 20 seconds of the onset. At around 25–30 seconds the infant or child loses consciousness due to cerebral hypoxaemia and this is reflected by the development of slow waves and low voltage on the EEG. This rapid development of arterial hypoxaemia is not seen in infants who have acute airway obstruction. Thus in two patients who were being totally obstructed by maternal smothering, hypoxaemia of sufficient severity to produce loss of consciousness and EEG changes did not develop until 70–80 seconds from the onset [3]. The degree of body movements accompanying both prolonged expiratory apnoea and acute airway obstruction are similar suggesting that O_2 consumption was probably increased but to a similar degree in both instances. Therefore the acute hypoxaemia in prolonged expiratory apnoea probably resulted from a massive right to left shunt. On first investigating prolonged expiratory apnoea we suspected that the shunt was occurring within the heart, probably through the foramen ovale. However, contrast echocardiography during cyanotic episodes ruled out an intracardiac right to left shunt and by a process of elimination we considered that the rapid hypoxaemia was due to the sudden development of a major mismatch between ventilation and perfusion within the lungs.

In attempts to find out why there was a mismatch between ventilation and perfusion many other investigations were performed during cyanotic episodes. One possible cause, upper airway obstruction, was excluded by the observation that episodes continued despite bypass of the upper airway using an endotracheal catheter or tracheostomy. Interestingly however, microlaryngoscopy during episodes showed adduction of the vocal cords.

A second possibility that active expiratory efforts, during a period of absent inspiratory airflow, might force the lungs down in volume producing atelectasis was supported to an extent by the frequent presence of such expiratory efforts during some cyanotic episodes. Moreover, many episodes were associated with crying, a phenomenon that includes active expiratory efforts against a partially closed airway. However, EMG recordings from expiratory muscles during cyanotic episodes that were not preceded by crying or an attempt to cry, showed that for the first 10–12 seconds of the cyanotic episodes there was no evidence of increased expiratory muscle activity.

Recordings from oesophageal balloon catheters, DC coupled respiratory inductance or jacket plethysmographs and thermistors or expired CO_2 monitors showed no effective inspiratory efforts during cyanotic episodes but a large effective inspiratory effort (a gasp) at the termination. There was some evidence of a progressive fall in end-expiratory lung volume during cyanotic episodes but large body movements reduced confidence in this interpretation of the data.

Ventilation scans during cyanotic episodes showed a rapid loss of a large proportion of the carrier gas at the onset of the cyanotic episode. Absence of inspiration of carrier gas was then observed for periods up to 60 seconds. DTPA scans suggested evidence of an increase in alveolar permeability immediately following the onset of the cyanotic episodes.

Between major, clinically obvious, cyanotic episodes infants younger than 6 months behaved differently from older infants and young children. In the older group of patients, long-term physiological recordings of respiratory behaviour and arterial blood oxygen saturation measurements were within normal limits for most of the time. Nevertheless, these older infants and young children, typically presenting with cyanotic breath-holding did manifest patterns of breathing between their clinically apparent cyanotic episodes which were not seen in age-matched controls. They showed prolonged apnoeic pauses during sleep; often exceeding 40 seconds in duration. They demonstrated episodes of abnormal hypoxaemia during crying or disturbances of posture during sleep. They had long episodes of abnormal expiratory braking (grunting) often accompanied by hypoxaemia; a pattern that was particularly evident in the period immediately following a major cyanotic episode. Also apparent during the recovery phase were large negative intrathoracic pressure changes accompanied by prolonged inspiratory flows of large volume (supergasps) and changes in dynamic compliance (as shown by the relationship between inspiratory volumes and oesophageal pressures).

In the group of younger patients, long-term recordings of respiratory behaviour were abnormal for much of the time between their major cyanotic events. Frequent episodes of abnormal arterial oxygen saturation occurred which were particularly evident during active sleep. These episodes of abnormal hypoxaemia occurred in association with 3 patterns of breathing. Firstly, as in the older patients there were prolonged (> 20 second) pauses in inspiratory efforts. Secondly and most frequently there were episodes of hypoxaemia associated with continued breathing efforts which were ineffective in that they were not accompanied by inspiratory airflow. During some such episodes there was evidence to suggest that the upper airway was open (cardiac impulses were seen on the airflow signal). The final pattern, which occurred infrequently compared with the second pattern described above, consisted of hypoxaemia accompanied by continued inspiratory efforts and continued inspiratory airflows.

After consideration of all the above evidence, albeit circumstantial, we have put together the hypothesis outlined in Fig. 1. In our opinion our data are best explained by a defect in the lung surfactant system [4]. Such a defect would produce normal patterns of breathing for most of the time provided that ribcage compliance was not increased (a situation that should exist in the older infant and young child). In the presence of a compliant ribcage, as in the young, particularly preterm infant, alveolar atelectasis might occur during states that are associated with a reduction in ribcage compliance, for instance active sleep, or a fall in functional residual volume as with partial upper airway obstruction (for example from an upper respiratory tract infection). In older patients cyanotic episodes might be expected following crying which is known to reduce lung compliance in normal infants [5], thus placing the lung with a defective surfactant into a cascading and unavoidable alveolar atelectasis.

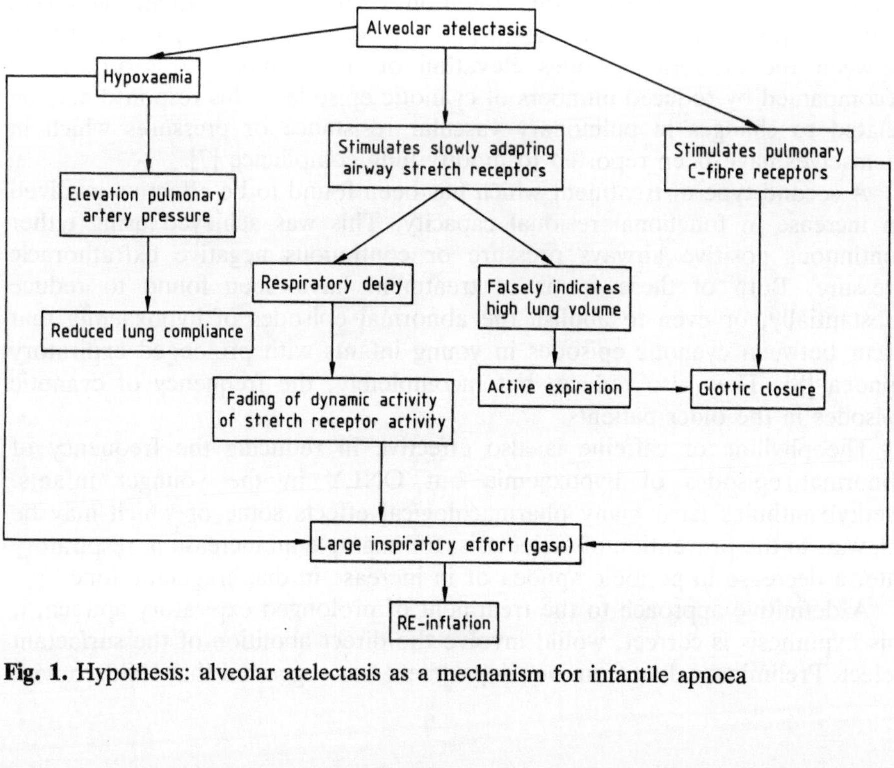

Fig. 1. Hypothesis: alveolar atelectasis as a mechanism for infantile apnoea

Atelectasis would adequately explain the severe arterial hypoxaemia and the fact that cyanotic episodes continue despite a bypass of the upper airway. It would also explain the second and third patterns of hypoxaemia seen between cyanotic episodes in the younger infants and the supergasps, and variable compliance identified immediately following the cyanotic episodes. It does not however explain the prolonged inhibition of inspiratory efforts, the adduction of the vocal cords or the expiratory braking. To explain these accompanying findings we hypothesise that the alveolar atelectasis modifies lung receptor reflexes. In a computer model [4] involving a lung stretch receptor reflex input we were able to show that atelectasis, by inappropriately stretching open large airways [6] (as in the air bronchogram seen in the respiratory distress syndrome) relayed false information (that is maximum lung inflation) to the brain stem respiratory control system thus eliciting an inhibition of inspiratory drive. Stimulation of these lung stretch receptors might also explain the active expiratory efforts against a partially closed larynx (expiratory braking or grunting as also seen in the respiratory distress syndrome). Stimulation of C fibre receptors by atelectasis might also be important in explaining the vocal cord adduction and also the coughing that is sometimes seen at the onset of a cyanotic episode.

Recovery from cyanotic episodes is usually heralded by a gasp and followed by a series of super-gasps. This gasp response is lifesaving in this situation. It involves the large inspiratory pressures required to re-expand the collapsed lungs and is brought on by severe hypoxaemia.

Further evidence that alveolar atelectasis is the mechanism for prolonged expiratory apnoea has been provided by our attempts at treating the abnormal hypoxaemia that accompanies this condition. Firstly in some patients who have frequent cyanotic episodes and abnormal hypoxaemia for long periods of time between the cyanotic episodes elevation of the inspired O_2 saturation is accompanied by reduced numbers of cyanotic episodes. This response may be related to changes in pulmonary vascular resistance or pressures which in themselves have been reported to modify lung compliance [7].

A second type of treatment which has been found to be effective involved an increase in functional residual capacity. This was achieved using either continuous positive airways pressure or continuous negative extrathoracic pressure. Both of these forms of treatment have been found to reduce substantially, or even to abolish the abnormal episodes of hypoxaemia that occur between cyanotic episodes in young infants with prolonged expiratory apnoea [8]. They also reduce, but incompletely, the frequency of cyanotic episodes in the older patients.

Theophylline or caffeine is also effective in reducing the frequency of abnormal episodes of hypoxaemia but ONLY in the younger infants. Methylxanthines have many pharmacological effects some of which may be relevant to the prevention of atelectasis; for example an increase in respiratory rate, a decrease in periodic apnoea or in increase in diaphragmatic tone.

A definitive approach to the treatment of prolonged expiratory apnoea, if this hypothesis is correct, would involve the direct abolition of the surfactant defect. Preliminary data from our department has suggested that children with prolonged expiratory apnoea have reduced proportions of phosphatidylcholine in their tracheal aspirates compared with controls [2] thus implicating a direct abnormality of their surfactant rather than the presence of surfactant inhibitory proteins. If this is correct then exogenous lung surfactant might be effective. Controlled studies to evaluate this possibility are now underway.

Summary

Investigations of 45 infants and young children with recurrent cyanotic episodes have identified respiratory patterns and changes in arterial O_2 saturation which are most adequately explained by the sudden onset of alveolar atelectasis with ensuing disturbances of lung receptors.

The characteristic feature identified during cyanotic episodes is the sudden development and rapid progression of severe arterial hypoxaemia (far below mixed venous levels) with equally rapid recovery. Respiratory efforts may be inhibited for prolonged periods or may continue but without inspiratory airflow. Expiratory efforts and vocal cord adduction may occur. Episodes are usually terminated by large inspiratory efforts associated with a considerable inflow of air into the lungs after the inspiratory effort has ceased. Cyanotic episodes continue despite bypassing of the upper airway. Between episodes there is abnormal expiratory braking and biochemical abnormalities of lung surfactant.

The concept that a primary defect in lung surfactant may be responsible for this syndrome is supported by evidence relating to the development of the lung during early childhood. This has shown that because of the proliferation of small radius structures (alveoli) there is a predisposition to atelectasis from surface tension forces. Theoretical considerations of the effects of a defective surfactant on lung volume/thoracic pressure relationships show that below certain critical lung volumes compliance becomes briefly infinite and subsequently negative with increasingly large negative intrathoracic pressures required to prevent atelectasis. Studies of the structure of the supporting fabric of the lungs have also demonstrated that in early childhood this fabric is potentially less able to support the alveoli and that surfactant is relatively more important than in the older child or adult. These studies of the supporting structure of the lung also show that alveolar atelectasis will lead to dilation of the larger airways within which are the stretch receptors. Activation of these receptors by atelectasis will effectively produce an inappropriate signal that lung volume is too high thus stimulating the brainstem control system to reduce lung volume. This it will try to do by inhibiting further inspiration and initiating active expiratory efforts thus enhancing the disturbance of ventilation.

The construction of a computer model based on a brainstem control system, with inspiratory and expiratory muscles, an upper airway, ventilation in lungs containing a normal or a defective surfactant with feedback from lung receptors and a gas consuming body shows that the clinical pathophysiology seen during cyanotic episodes can be exactly reproduced.

Prolonged expiratory apnoea: sudden atelectasis apnoea braking syndrome (SAABS) appears to be a particularly dangerous cause of severe hypoxaemia in infants and young children. It is proposed that this mechanism most adequately explains apnoea of prematurity, cyanotic episodes induced by pertussis and other respiratory tract infections, and cyanotic breath-holding. There is increasing evidence to suggest that it could be a major cause of sudden and unexpected deaths during infancy and early childhood. Exogenous lung surfactant may be of value in treating this condition.

References

1. Southall DP, Talbert DG, Johnson P, Morley C, Salmons S, Miller J, Helms PJ (1985) Prolonged expiratory apnoea – a disorder resulting in episodes of severe arterial hypoxaemia in infants and young children. Lancet 2:571–577
2. Southall DP, Talbert DG, Johnson P, Morley C, Salmons S, Miller J, Helms PJ (1985) Prolonged expiratory apnoea and hypoxaemia. Lancet 2:1125–1126
3. Southall DP, Stebbens VA, Rees SV, Lang MH, Warner JO, Shinebourne EA (1987) Apnoeic episodes induced by smothering – two cases identified by covert video surveillance. BMJ 294:1637–1641
4. Southall DP, Talbert DG (1987) Sudden atelectasis apnoea braking syndrome (SAABS) In: Hollinger MA (ed) Current topics in pulmonary pharmacology. Elsevier, New York, pp 210–281
5. Javorka K, Tomori Z, Zavarska L (1982) Mechanics of breathing during sneezing and crying in premature newborns. Eur J Respir Dis 63:442–448
6. Nieman GF, Brendenburg CE, Clark WR, West NE (1981) Alveolar function following surfactant deactivation. J Appl Physiol 51:895–904
7. Bancalari E, Jesse MJ, Gelband H, Garcia O (1977) Lung mechanics in congenital heart disease with increased and decreased pulmonary blood flow. J Pediatr 90:192–195
8. Southall DP (1987) Preliminary experience in the use of continuous negative extrathoracic pressure in the treatment of apnoea of infancy and early childhood. Proceedings of Neonatal Soc Meeting, Juli 1987

Changes in Pulmonary and Cerebral Blood Flow After Surfactant Treatment for Severe Respiratory Distress Syndrome

B. McCord, H. L. Halliday, G. McClure, and MMcC. Reid

Introduction

Neonatal respiratory distress syndrome remains a major cause of death and disability in babies. Death is often associated with development of acute complications such as pneumothorax and intraventricular haemorrhage (IVH). Long term disability may result from the sequelae of IVH or bronchopulmonary dysplasia.

As part of a collaborative European Multicentre Study of surfactant replacement [1], we have shown that treated babies have a lower incidence of pneumothorax and IVH [2].

In order to clarify how exogenous surfactant acts we measured pulmonary blood flow in a group of 20 consecutively enrolled babies in the Multicentre Study. In addition we measured cerebral blood flow velocities to determine why treated babies should have a lower incidence of IVH.

Patients

Over a 12 months period pulmonary and cerebral blood flow measurements were made in 20 consecutively enrolled babies from the Belfast arm of the Multicentre Trial [2]. Babies entering the study had severe RDS requiring mechanical ventilation with an FiO_2 greater than 0.6 by 15 hours of age [1, 2]. Babies with severe birth asphyxia (Apgar score less than 5 at 5 minutes), severe IVH (grade III–IV) [3], congenital anomaly or group B Streptococcal infection were excluded.

Methods

Using an ATL Mk V Duplex scanner, measurements of pulmonary blood flow (Qp) and cerebral blood flow velocity (CBFV) were made non-invasively with the Doppler ultrasound probe placed on the praecordium and the anterior fontanelle respectively. Pulmonary blood flow (Qp) is difficult to measure directly but can be estimated as pulmonary venous return from measurement of flow at the aortic valve, if there are no intracardiac shunts [4]. Such intracardiac shunts were excluded prior to study by careful 2D and pulsed Doppler echocardiographic examination. Left ventricular output (Qs) was measured as flow across the tricuspid valve. Increased left to right shunting across a patent ductus arteriosus will be reflected by an increase in Qp and Qp/Qs ratio.

CBFV was measured in two intracranial vessels, the pericallosal and the internal carotid arteries. CBFV was calculated as the mean of the peak flow velocity (MPFV) averaged over 10–15 cycles [5]. These Doppler measurements were made immediately prior to and after surfactant replacement at intervals of 15 minutes, 1, 4, 6, 24 hours and on days 2, 3, 5, 7 and 10 thereafter. The presence of IVH was determined by daily cerebral ultrasound scans performed at the time of each Doppler study. The diagnosis of pneumothorax was confirmed radiologically. Patent ductus arteriosus (PDA) was diagnosed by the presence of a murmur and increased left heart dimensions using M-mode echocardiography [6]. Blood pressure was measured using a non-invasive oscillometric method.

Differences between groups were assessed using two-tailed Mann-Whitney U and Fisher's exact tests where appropriate.

Results

There were 12 treated and 8 control babies. Their clinical characteristics are shown in Table 1 and there were no significant differences between the groups for any of these variables. Table 2 compares the outcome of the two groups. Intraventricular haemorrhages of Grades I and II at enrolment were more common in treated babies although this was not statistically significant. No baby had a pneumothorax prior to entry in the study. After study entry, development of a new IVH or extension of an existing one occurred significantly more in control babies ($P = 0.02$). In addition, pneumothorax

Table 1. Clinical characteristics of study groups

	Group	
	Treated	Control
No. of patients	12	8
Birthweight[a] (g)	1335 (312)	1401 (258)
Gestation[a] (weeks)	28.4 (1.1)	28.8 (1.9)
Males *(n)*	5 (42 %)	4 (50 %)
Apgar scores[b]		
1 min	3 (2–5)	3 (2–7)
5 min	7 (5–9)	7 (5–9)
Age at enrolment[b] (h)	5 (2–13)	7 (2–15)

[a] Mean (SD).
[b] Median (range).

Table 2. Complications of treatment and survival

	Group		
	Treated	Control	*P* value
No. of patients	12	8	
IVH at entry *(n)*	4 (33 %)	1 (12 %)	NS
IVH after entry or extension during study *(n)*	3 (25 %)	7 (87 %)	0.02
Pneumothorax *(n)*	2 (17 %)	6 (75 %)	0.03
Bilateral	0	4 (50 %)	0.03
Pneumothorax and IVH *(n)*	2 (17 %)	6 (75 %)	0.03
Patent ductus arteriosus *(n)*	7 (58 %)	4 (50 %)	NS
Mean arterial blood pressure (mm Hg)			
Before entry[a]	43 (10)	46 (9)	NS
24 h after[a]	47 (6)	47 (13)	NS

[a] Mean (SD).

and particularly bilateral pneumothoraces were more common in control babies (P = 0.03). Pneumothorax and IVH tended to occur together and we noticed a temporal relationship with IVH occurring shortly after the development of pneumothorax. The incidence of PDA and serial Qp/Qs ratios were similar in each group. Blood pressure levels before and after enrolment were also similar in both groups (Table 2). Survival rate was higher in the treated babies although this was not statistically significant.

Figure 1 compares the pulmonary blood flow (Qp) for treated and control infants. There was a small but statistically insignificant rise in pulmonary blood flow by 15 min in treated babies. This was not sustained so that in the 6 hours

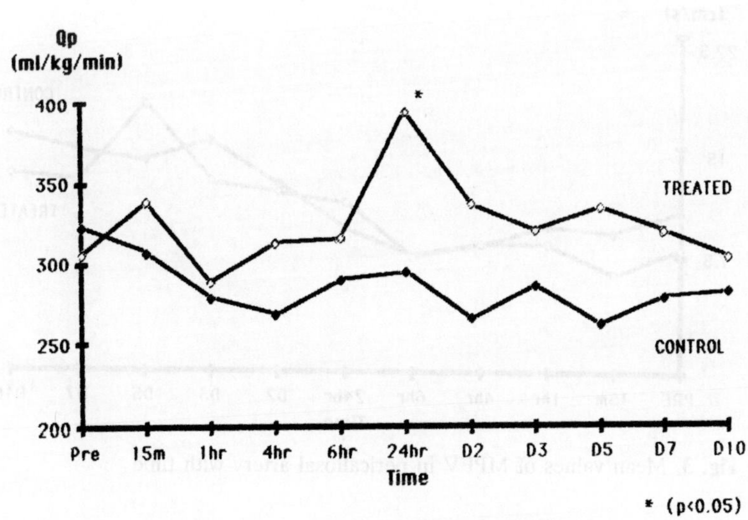

Fig. 1. Mean values of pulmonary blood flow *(Qp)* with time after enrolment

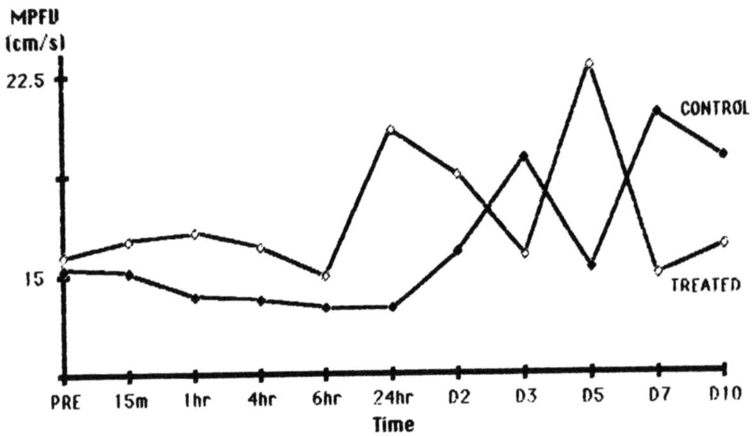

Fig. 2. Mean values of MPFV in internal carotid artery with time

after treatment there were no differences in pulmonary blood flow between the groups. By 24 h pulmonary blood flow was significantly higher in treated babies ($P < 0.05$) but this was not due to a higher incidence of PDA (Table 2).

Cerebral blood flow velocities in the internal carotid artery (Fig. 2) and the pericallosal artery (Fig. 3) were compared for treated and control infants. There were no statistically significant differences in CBFV in either vessel at any time during the study. CBFV in the pericallosal artery was low in both groups and only increased after 24 hours of age.

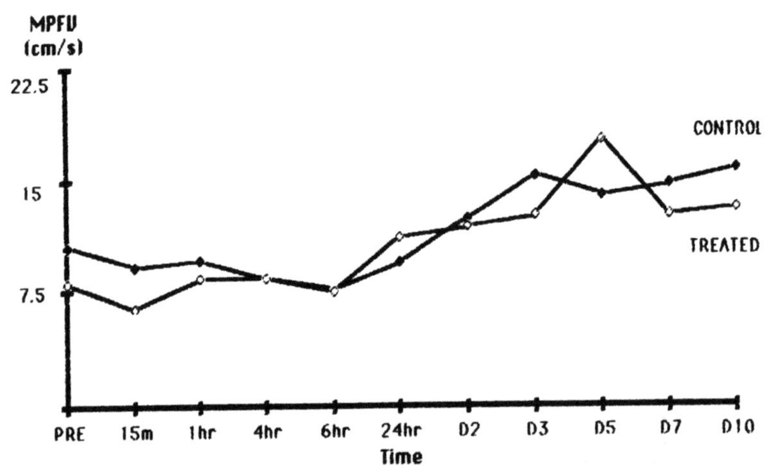

Fig. 3. Mean values of MPFV in pericallosal artery with time

Discussion

We have previously shown that exogenously administered porcine surfactant increases arterial blood oxygen tensions within 15 min [1, 2]. There is an associated reduction in the incidence of intraventricular haemorrhage and pneumothorax [2].

The results of this study showed no increase in pulmonary blood flow until 24 hours following administration of surfactant. It is unlikely that the improved arterial blood oxygen tensions following surfactant treatment are due to changes in pulmonary blood flow. We think that the probable mechanism of action of surfactant is its ability to improve lung compliance and thus facilitate alveolar ventilation. In our previous study we were able to lower mechanical ventilator pressures by 24 hours [2]. The improved compliance together with a reduced ventilator pressure are the probable explanations for the reduced incidence of pneumothorax in our treated babies.

We found a consistent temporal relationship between the development of pneumothorax and intraventricular haemorrhage. There were no significant differences in cerebral blood flow velocities between treated and control babies. However, all our babies showed evidence of low velocities for the first 24 hours of life suggesting maximal cerebral vasodilation [7]. We feel that this lack of vasomotor tone in the cerebral vessels renders infants with severe RDS prone to intraventricular haemorrhage should an acute insult such as pneumothorax occur. Mean arterial blood pressure was similar in both groups of babies before and after treatment for serial measurements. Low mean arterial pressure has been implicated as a cause of severe IVH [8].

Summary

Pulmonary and cerebral blood flow were measured in 20 infants enrolled in a randomised control trial of natural surfactant. Treated infants had a significantly lower incidence of intraventricular haemorrhage and pneumothorax. Pulmonary blood flow did not increase in treated infants until 24 hours after surfactant therapy. There were no significant differences in cerebral blood flow velocities in either the internal carotid artery or the pericallosal artery between treated and control infants. We conclude that surfactant improves arterial oxygenation by increasing pulmonary compliance rather than pulmonary blood flow, and that intraventricular haemorrhage occurs following acute stresses such as pneumothorax in babies who have reduced cerebral vasomotor tone.

Acknowledgements. We thank Lynda Thompson for typing the manuscript and Drs. T. Curstedt and B. Robertson for producing the porcine surfactant in Stockholm. Drs. McCord and Halliday are in receipt of a grant from The Perinatal Trust Fund of Northern Ireland.

References

1. Collaborative European Multicentre Group (1987) An international randomized clinical trial of surfactant replacement therapy in severe neonatal respiratory distress syndrome. Eur J Respir Dis (In Press)
2. McCord B, Curstedt T, Halliday HL, McClure G, Reid MMcC, Robertson B (1987) Surfactant treatment reduces the incidence of intraventricular haemorrhage in severe respiratory distress syndrome. Arch Dis Child (In Press)
3. Papile L-A, Burstein J, Burstein R, Koffler H (1978) Incidence of evolution of subependymal and intraventricular haemorrhage: a study of infants with birth weights less than 1500 gm. J Pediatr 92:529–534
4. Alverson DS, Eldridge M, Dillon T, Kabet SM, Berman W (1982) Non-invasive pulsed Doppler determination of cardiac output in neonates and children. J Pediatr 101:46–50
5. Philip AGS (1985) Symposium on non-invasive neonatal diagnosis. Clin Perinatol 12:181–182
6. Halliday HL, Hirata T, Brady JP (1979) Echocardiographic findings of large patent ductus arteriosus in the very low birthweight infant before and after treatment with indomethacin. Arch Dis Child 54:744–749
7. Drayton MR, Skidmore R (1987) Vasoactivity of the major intracranial arteries in newborn infants. Arch Dis Child 62:236–240
8. Miall-Allen VM, DE Vries LS, Whitelaw AGL (1987) Mean arterial blood pressure and neonatal cerebral lesions. Arch Dis Child 62:1068–1069

Part IV
Surfactant Replacement in ARDS and Other Lung Diseases

The Future for Surfactant Therapy of the Adult Respiratory Distress Syndrome

R. G. Spragg, P. Richman, N. Gilliard, and T. A. Merritt

Few clinical syndromes are as frustrating to the critical care specialist as the adult respiratory distress syndrome (ARDS). Its pathogenesis is frequently obscure. Current therapy is entirely supportive. The clinical course is unpredictable. Interventional strategies, often initially promising, have proven disappointing. Finally, the high mortality rate has not been demonstrably improved despite several decades of effort.

Given this disheartening picture, it is not surprising that those caring for critically ill adults have viewed with both admiration and envy the progress which has been made by neonatologists caring for infants with the respiratory distress syndrome (RDS) using exogenous surfactants. In addition, it has become clear that multiple parallels exist between that syndrome and ARDS. In addition, progress has been made in defining the abnormalities of the surfactant system which are seen in the patient with ARDS. As data regarding the abnormalities of surfactant have accumulated, it has been reasonable to begin cautious clinical trials of surfactant replacement in patients with ARDS.

To understand the rationale for surfactant therapy of ARDS, the problems inherent in studying the benefits of surfactant therapy, and the future of focused investigations in this area, it is necessary to be familiar with the syndrome of adult respiratory distress and the history of therapies which have been directed toward it.

Current Understanding of ARDS

The modern concepts of ARDS were introduced by Ashbaugh et al. with their description of 12 patients with the syndrome [1]. The definition of ARDS as a high-permeability (non-hydrostatic) lung edema became clinically applicable with widespread introduction of the pulmonary artery catheter. This device allows catagorization with relative certainty of patients with low (insufficient to cause pulmonary edema) and high (sufficient to cause pulmonary edema) pulmonary vascular pressures. In the absence of knowledge of pulmonary vascular pressures, it is impossible to assign patients correctly to such catagories for careful clinical studies [2].

A second technologic advance which has promoted study and understanding of ARDS has been use of the fiberoptic bronchoscope. This device has allowed collection of bronchoalveolar lavage fluid for analysis of cells and soluble constituents. It has made possible observations of elevated protein content of secretions from the lower airway in patients with ARDS – observations which confirm the presence of a high permeability edema.

A current operational definition of ARDS agreed upon by many investigators requires: (a) presence of pan-lobar infiltrates on the chest radiograph, (b) a pulmonary artery wedge pressure of ≤ 18 torr, (c) $Pa0_2/F_I0_2$ < 150, and (d) development of the pathologic state over a short (< 7 days) period.

ARDS develops in a wide variety of clinical settings; however, it is most commonly seen in association with sepsis, pancreatitis, multiple organ failure, polytrauma, and burns [3].

When lung tissue from patients who meet the clinical criteria for ARDS is examined, it shows rather typical changes which evolve over one to two weeks. Initial abnormalities include: polymorphonuclear leukocyte accumulation and degranulation; edema of endothelial cells, with swelling of cell organelles; interstitial edema; destruction of Type I epithelial cells; formation of hyalin membranes; and flooding of the alveolar space with protein-rich edema fluid and cell debris. Subsequent changes include marked proliferation of Type II epithelial cells (believed by many to then differentiate to Type I cells) and development of a dense fibrotic reaction centering on the alveolar duct [4–6]. Unfortunately the histopathologic changes in the lungs of patients with ARDS are entirely non-specific, and provide little information on the pathogenetic mechanisms responsible for the tissue injury.

Initial clinical observations of patients with ARDS also show few distinguishing features. As a group, patients satisfying the criteria listed above have diminished lung compliance, marked restriction of lung volumes, and profound hypoxemia due largely to shunting of blood past unventilated lung units.

The similarity of physiologic and histologic changes in patients who develop ARDS following one of a variety of insults has, in the past, prompted investigators to treat the group of patients as homogeneous. Greater experience has now shown clear-cut differences in the natural history of ARDS occurring following certain insults. For example, patients developing ARDS following near-drowning, heroin ingestion, or blood transfusion complicated by a leukoagglutinin reaction may satisfy criteria for the syndrome for only 1 to 2 days. In contrast, patients developing ARDS secondary to sepsis, pancreatitis, or inhalation of toxic substances may fulfill criteria for the presence of the syndrome for many weeks.

It is hoped, but not yet proven, that biochemical differences will be discovered which will also differentiate subgroups of patients with ARDS. For this reason, investigators are often concerned to provide clear diagnostic criteria for catagorization of patients with ARDS.

Only a limited number of studies have investigated biochemical abnormalities present in the patient with ARDS. These have provided, as one general conclusion, that ARDS is an acute inflammatory event. An increased number and fraction of polymorphonuclear leukocytes is seen in lavage fluid from patients with ARDS, and products of these cells are recoverable in lavage fluid. Those products include collagenase and neutrophil elastase (HNE) [7–9].

In addition, the activated neutrophil produces, through function of its surface membrane NADPH oxidase, the superoxide radical. Products derived from superoxide are capable of oxidizing reactive protein groups, causing lipid peroxidation, and breaking DNA single strands [10–12]. Because of their short lifespan and highly reactive nature, most toxic oxygen species are difficult to measure in biologic samples. However, several investigators have demonstrated the presence of H_2O_2 in exhaled breath condensate of patients with ARDS, either during the syndrome or just prior to its appearance [13, 14].

Alpha-1-proteinase inhibitor (A1PI), the native inhibitor of neutrophil elastase, is frequently found in the oxidized form in lavage fluid from patients with ARDS [10]. This observation is consistent with release of oxygen radicals within the lungs of these patients, for plasma A1PI is not oxidized. In addition, oxidation of A1PI renders the molecule incapable of inactivating HNE [15]. As a result, active HNE and (inactive) A1PI have been recovered in lavage fluids from patients with ARDS. In addition, a similar "protease imbalance" is found in patients with pneumonia [16], and in infants with RDS [17].

Initial observations by Ashbaugh et al. on post mortem tissue from two patients with ARDS demonstrated that fluid recovered from minced lungs had a abnormally high surface tension in vitro. Subsequent reports also involved analysis of samples collected post mortem [18, 19]. Although these studies demonstrated evidence of deficiency of lung surfactant function, that observation may be due to terminal or post mortem changes and thus not represent mechanisms contributing to respiratory failure.

Studies by Hallman et al. of both the function and chemical composition of surfactant recovered from bronchoalveolar lavage of patients with ARDS showed biochemical abnormalities, with a decreased fractional composition of phosphatidyl choline and phosphatidyl glycerol, a depression of the lecithin/sphingomyelin ratio, and also marked loss of surface activity. Biochemical abnormalities of lung surfactant were present early in respiratory failure, and tended to normalize during recovery [20].

Loss of surfactant function may be due to several factors (Table 1). Analysis of minimum surface tension (γ min) and time to reach an equilibrium surface tension (TEST) are abnormally elevated in surfactant purified by Folch extraction of lavage fluid obtained from patients with ARDS. This result might be explained by either a primary abnormality of surfactant synthesis, involving lipid or protein constituents; an abnormality of assembly of these components; chemical alteration of the components; or the presence of a lipid inhibitor of surfactant function. Although Type II pneumocytes are not lost from the

Table 1. Possible causes of loss of surfactant function in patients with ARDS

Presence of functional inhibitors
Abnormal synthesis of lipid and/or protein
Abnormal assembly of surfactant components
Oxidative and/or proteolytic alteration of surfactant components

acutely injured lung, their function in production and assembly of lung surfactant may be impaired, and a material of abnormal composition may be produced.

It is clear that lung surfactant function may be impaired in the lungs of patients with ARDS. Although γ min and TEST are abnormal in surfactant purified by Folch extraction from lavage fluid of ARDS patients, those values are markedly more abnormal when equimolar amounts of phospholipid are examined in the original lavage fluid. Multiple species which may be present in lung epithelial lining fluid (ELF) have been shown to inhibit surfactant function. These include, glycolipid [21], albumin [22], hemoglobin [23], fibrin monomer [24], and bilirubin [25]. Clearly, understanding the interactions between ELF inhibitors of surfactant function and both native and exogenous surfactant will have marked implications for replacement therapy.

In addition to abnormalities of lung surfactant production and the presence of inhibitors of surfactant function, attack by proteases present in ELF on surfactant apoproteins may decrease the function of lung surfactant. In particular, the low molecular weight lung surfactant apoproteins appear to promote rapid adsorption in mixtures of phospholipids, and loss of function of those proteins might be predicted to diminish lung surfactant function. HNE is frequently present in ELF and has been shown to be capable of hydrolizing the 35K MW apoprotein [26], suggesting one mechanism whereby surfactant from patients with ARDS might be altered.

In addition, the evidence that oxygen radicals are present in the lungs of patients with ARDS suggests that oxidation of lung surfactant apoproteins or peroxidation of lung surfactant lipids might occur, with deleterious effects on surfactant function.

Whatever the mechanism whereby lung surfactant function is lost, it is tempting to think that this loss may contribute to the pathophysiology of ARDS. Loss of surfactant function would be predicted to promote alveolar collapse, leading to loss of lung volume, shunt, and hypoxemia – major features of the pathophysiology of ARDS. In addition, it has been suggested that diminished surface tension within the alveoli protects against the formation of alveolar edema, and that loss of surfactant function might therefore promote edema formation [27, 28]. Although multiple factors undoubtedly participate in the genesis and maintenance of edema in the lung of patients with ARDS, loss of surfactant function may be a contributing event.

In addition to its biophysical properties, lung surfactant may be capable of modulating both lymphocyte and macrophage function. Natural killer cell function, as well as lymphocyte proliferation in response to mitogens or antigens is decreased by lung surfactant [29, 30], while opsonization and microbial killing by rat macrophages has been reported to be enhanced [31]. On the other hand, human lung surfactant appears neither to be toxic to H. influenzae or S. pneumoniae nor to alter alveolar macrophage uptake or killing of bacteria in vitro [32]. The extent to which changes in the lung surfactant system during ARDS modulate macrophage and lymphocyte function is unknown. As infection is an important late cause of mortality for patients with ARDS, there is excellent reason to encourage study of the non-biophysical aspects of surfactant function in patients with ARDS.

Therapy

The current therapy of the patient with ARDS is entirely supportive. Attention is directed to correction of any underlying condition, for example sepsis, which might have been responsible for development of the syndrome. Meticulous care is taken to avoid development of gastrointestinal bleeding, renal failure, thrombocytopenia, or barotrauma. Gas exchange is supported with supplemental oxygen and mechanical ventilation. Use of positive end expiratory pressure and various modes of mechanical ventilation allow preservation of hemoglobin saturation in most patients. The mortality from ARDS has remained approximately 65% for the last two decades, suggesting little advance in therapy. However, it is possible that increasing severity of illness has been exactly offset by increasing skill in therapy over that period of time.

Several pharmacologic interventions have been studied in prospective double-blind studies. Neither adrenal corticosteroids nor PGE_1 have had apparent benefit on the mortality from ARDS [33–45].

Therapy of ARDS directed toward re-establishing normal lung surfactant function would be rational, and might be expected to result in improved lung expansion, with decreased shunt and increased compliance. Benefits of these changes would include: enhanced blood oxygenation at lower fractions of inspired oxygen; avoidance of toxic levels of oxygen; and use of lower ventilating pressures with reduced barotrauma.

Alterations in lung surfactant function might be achieved by inducing synthesis and/or release of endogenous surfactant, or by removing inhibitors from the alveolar milieu. Numerous factors which may stimulate surfactant secretion have been reviewed recently by Wright and Clements. These include adrenergic agonists, cholinergic agonists, calcium ionophore A23187, cholera toxin, phorbol myristate acetate, and mechanical inflation of the lung. Studies of the mechanisms by which these factors act suggest that both cyclic AMP-dependent protein kinase and protein kinase C are involved in the regulation of surfactant secretion [36]. Unfortunately, efforts to enhance production of native surfactant in experimental adult animals have not been promising and treatment of ARDS by stimulation of endogenous surfactant production and release does not appear to be possible in the near future.

Administration of exogenous surfactant may be done in one of several ways, and the optimal method to maximize efficient distribution is unknown. Some have advocated aerosol delivery, but a major disadvantage of this form of administration is the long time required to administer an amount which is likely to be required. If a jet nebulizer will deliver 0.8 mg DPPC/L air, if airflow is assumed to be 10L/min, and if retention is 32% (the value for inhaled aerosolized protein) (unpublished data), then it will take 23 hours to deliver a dose of 50 mg/kg to a 70 kg adult.

Direct instillation of exogenous surfactant has the advantage of being rapid, and if done correctly may result in homogenous delivery of the instilled material to the lung periphery.

This method of surfactant replacement has been used by Lachmann to treat several terminally ill patients who had ARDS [37], and by us to treat 3 adults [38]. Lachman found that intratracheal instillation of surfactant suspended in a volume equal to the anatomic dead-space was well tolerated. No improvement in gas exchange was seen until a cumulative dose of $>$ 200 mg/kg was given.

In our preliminary report with Drs. Robertson and Curstedt, we found that intrabronchial administration of 50 mg/kg porcine lung surfactant in divided doses to each bronchus was well tolerated and caused a modest improvement in gas exchange in all patients [38]. This improvement was transient in two patients, and sustained in the third. In addition, we found reduced activity of inhibitor of surfactant function in BAL after surfactant replacement.

Future of Surfactant Therapy of ARDS

Investigations which will shed light on the usefulness of surfactant therapy of ARDS should proceed on several levels. First, continuing studies of the structure of lung surfactant may eventually disclose alterations which are responsible for the functional derangements seen in patients with ARDS. These may include alterations of primary, secondary, or tertiary structures. Second, investigations concerned with the mechanism(s) by which lung surfactant inhibitors act may suggest therapeutic approaches which prevent the interaction of inhibitor with surfactant, leaving surfactant function intact. Thirdly, investigations of the role lung surfactant plays in modulating lymphocyte and macrophage function in the lung may provide new endpoints by which to judge the adequacy of lung surfactant function in the adult.

Finally, the utility of surfactant replacement therapy deserves continuing investigation. Several aspects of replacement therapy deserve attention. The amount of exogenous surfactant required to restore lung surfactant function is not known. The results of neonatal studies, animal studies, and theoretical calculations suggest that a dose between 50 and 200 mg/kg may be appropriate. The optimal method of delivering this dose is not known, and variables to be investigated include use of aerosol, volume of instilled surfactant suspension, and utility of maneuvers aimed at maximizing the homogeneity of distribution.

Investigations of surfactant replacement therapy will be enhanced by development of clinical techniques to measure surfactant pool size and turnover. When these techniques are available, it will be possible to make rational decisions regarding the aim of therapy – whether to restore a diminished surfactant pool size to normal or to expand the pool size to overcome the effect of inhibitors. This information, coupled with clinical observations, will help determine the frequency with which exogenous surfactant must be given.

A separate issue is the optimal time in the course of ARDS at which to give replacement therapy. Answers to this question will require knowledge of abnormalities of the lung surfactant system during the time that ARDS is developing and after the syndrome has become established. Unfortunately, longitudinal studies of patients who are at risk for ARDS, and which follow patients on to the development of overt high-permeability lung edema, have proven highly frustrating. Heterogenity of patient populations, difficulty in prospectively identifying an "at risk" population which develops ARDS with an acceptable frequency, and performance of clinical studies on critically ill patients all contribute to this frustration.

Currently, neonatologists require that surfactant preparations for replacement therapy have the expected biochemical composition, excellent in vitro function, and excellent function in vivo in animal models. The pulmonologist studying the efficacy of surfactant therapy of ARDS may have the luxury of also requiring his surfactant to have excellent function in the treatment of RDS. Indeed, neonatologists have led the way in understanding both the biology and pathobiology of lung surfactant. It is likely that they will determine if there are significant functional differences between the available surfactant preparations. However excellence of function and resistance to inhibition of function may prove to be separable, and in that case the optimal surfactant preparation for a lung rich in inhibitor may prove to be different from the preparation which is optimal for the immature lung.

A final consideration which may affect the future of surfactant replacement therapy is the availability of adequate supplies. Currently the lungs of 25 pigs are required to prepare a single 4 g dose of surfactant – a dose which may eventually be shown to be one-fourth the optimal dose. The problems of producing sufficient purified surfactant for investigations are not trivial, and are no less trivial when considering the possibility of using a completely synthetic product containing genetically engineered apoprotein.

Despite these considerations, surfactant trials in patients with ARDS should proceed. Although interventional trials in ARDS have a bleak history, we view the promise of clinical use of surfactant in the adult population with cautious optimism.

Acknowledgement. This work is supported in part by the U.S. Army Medical Research and Development Command under Contract No. DAMD17-88-C-8020. Opinions, interpretations, conclusions, and recommendations are those of the author and are not necessarily endorsed by the U.S. Army. The investigators have adhered to the policies regarding the protection of human subjects as prescribed by 45 CFR 46 (Protection of Human Subjects).

References

1. Ashbaugh DG, Bigelow DB, Petty TL (1967) Acute respiratory distress in adults. Lancet II:319–323
2. Unger KM, Shibel EM, Moser KM (1975) Detection of left ventricular failure in adult respiratory distress syndrome. Chest 67:8–13
3. Hopewell PC, Murray JF (1982) Adult respiratory distress syndrome. In: Moser KM, Spragg RG (eds) In respiratory emergencies. CV Mosby, St Louis, pp 94–122
4. Schlag G, Voigt WH, Redl H, Glatzl A (1980) Vergleichende Morphologie des posttraumatischen Lungenversagens. Anasth Intensivther Notfallmed 15:315–339

5. Bachofen M, Weibel ER (1982) Structural alterations of lung parenchyma in the adult respiratory distress syndrome. Clin Chest Med 3:35–56
6. Pratt PC, Vollmer RT, Shelburne JD, Crapo JD (1979) Pulmonary morphology in a multihospital collaborative extracorporeal membrane oxygenator project. Am J Pathol 95:191–214
7. Christner P, Fein A, Goldberg S, Lippman M, Abrams W, Weinbaum G (1985) Collagenase in the lower respiratory tract of patients with adult respiratory distress syndrome. Am Rev Respir Dis 131:690–695
8. McGuire WW, Spragg RG, Cohen AB, Cochrane CG (1980) Studies on the pathogenesis of the adult respiratory distress syndrome I. J Clin Invest 69:543–553
9. Lee CT, Fein AM, Lippman ML, Holtzman H, Kimbel P, Weinbaum G (1981) Elastolytic activity in pulmonary lavage fluid from patients with adult respiratory distress syndrome. N Engl J Med 304:192–196
10. Cochrane CG, Spragg RG, Revak S (1983) Studies of the pathogenesis of the adult respiratory distress syndrome: evidence of oxidant activity in bronchoalveolar lavage fluid. J Clin Invest 71:754–761
11. Kellog EW, Fridovich I (1977) Liposome oxidation and erythrocyte lysis by enzymatically generated syperoxide and hydrogen peroxide. J Biol Chem 752:6721–6728
12. Schraufstatter IU, Hinshaw DB, Spragg RG, Cochrane CG (1986) Oxidant injury of cells: DNA strand breaks activate poly-ADP-ribose polymerase and leads to depletion of nicotinamide adenine dinucleotide. J Clin Invest 77:1312–1320
13. Baldwin SR, Simon RH, Grum CM, Ketai LH, Boxer LA, Devall LJ (1986) Oxidative activity in expired breath of patients with adult respiratory distress syndrome. Lancet I:11–14
14. Sznajder JI, Fraiman A, Crawford G, Hall J, Schmidt G, Sanders W, Wood LDH (1987) Increased oxidant activity in expiratory breath of patients with hypoxemic respiratory failure. Am Rev Respir Dis 135:A264
15. Janoff A, George-Nascimento C, Rosenberg S (1986) A genetically engineered, mutant human alpha-1-proteinase inhibitor is more resistant than the normal inhibitor to oxidative inactivation by chemicals, enzymes, cells, and cigarette smoke. Am Rev Respir Dis 133:353–356
16. Abrams WR, Fein AM, Kucich U, Kueppers F, Yamiada H, Kozmorycz T, Morgan L, Lippmann M, Goldberg SK, Weinbaum G (1984) Proteinase inhibitory function in inflammatory lung disease. Am Rev Respir Dis 129:735–741
17. Merritt TA, Cochrane CG, Holcomb K, Bohl B, Hallman M, Strayer D, Edwards DK III, Gluck L (1983) Elastase and alpha-1-proteinase inhibitor activity in tracheal aspirates during respiratory distress syndrome. Role of inflammation in the pathogenesis of bronchopulmonary dysplasia. J Clin Invest 72:656–666
18. Petty TL, Reiss OK, Paul GW, Silvers GW, Elkins ND (1977) Characteristics of pulmonary surfactant in adult respiratory distress syndrome associated with trauma and shock. Am Rev Respir Dis 115:531–536
19. Von Wichert P, Kohl FV (1977) Decrease dipalmitoyllecithin content found in lung specimens from patients with so-called shock-lung. Eur J Intensiv Care 3:27–30
20. Hallman M, Spragg RG, Harrell JH, Moser KM, Gluck L (1982) Evidence of lung surfactant abnormality in respiratory failure. Study of bronchoalveolar lavage phospholipids, surface activity and plasma myoinosotol. J Clin Invest 72:656–666
21. Rauvala H, Hallman M (1984) Glycolipid accumulation in bronchoalveolar space in adult respiratory distress syndrome. J Lipid Res 25:1257–1262
22. Holm BA, Notter RH, Finkelstein JN (1985) Surface property changes from interactions of albumin with natural lung surfactant and extracted lung lipids. Chem Phys Lipids 38:287–298
23. Holm BA, Notter RH (1987) Effects of hemoglobin and cell membrane lipids on pulmonary surfactant activity. J Appl Physiol 63:1434–1442
24. Seeger WG, Stohr G, Wolf HRD, Neuhof H (1985) Alteration of surfactant function due to protein leakage: special interaction with fibrin monomer. J Appl Physiol 58:326–338
25. Fuchimukai T, Fujiwara T, Takahashi A, Enhorning G (1987) Artificial pulmonary surfactant inhibited by proteins. J Appl Physiol 62:429

26. Merritt TA, Revak S, Hallman M (1987) Elastolytic degradation of surfactant 35kD apoprotein. Pediatr Res 21:460A
27. Nieman GF, Bredenberg CE (1985) High surface tension pulmonary edema induced by detergent aerosol. J Appl Physiol 58:129–136
28. Bredenberg CE, Nieman GF, Paskanik AM, Hart KE (1986) Microvascular membrane permeability in high surface tension edema. J Appl Physiol 60:253–259
29. Richman P, Batcher S, Catanzaro (1986) Immunoregulation by pulmonary surfactant in vitro. Am Rev Respir Dis 133:A312
30. Ansfield MJ, Benson BJ, Kaltreider HB (1979) Immunosuppression by surface active material: lack of species specificity. Am Rev Respir Dis 120:949–952
31. LaForce FM, Kelly WJ, Huber GL (1977) Inactivation of staphlococci by alveolar macrophages with pulmonary observations on the importance of alveolar lining material. Am Rev Respir Dis 108:784–790
32. Jonsson S, Musher DM, Goree A, Lawrence EC (1986) Human alveolar lining material and antibacterial defenses. Am Rev Respir Dis 133:136–140
33. Bernard GR, Harris TR, Luce JE, Rinaldo JE, Sibbald WJ, Sprung CL, Tate RM, Higgins S, Kariman K, Bradley R, Brigham KL (1987) The effects of high dose methyl prednisolone in patients with the adult respiratory distress syndrome. Am Rev Respir Dis 125:A5
34. Luce JM, Marks JD, Montgomery AB, Turner J, Metz CA, Murray JE (1987) Methyl prednisolone does not prevent the adult respiratory distress syndrome or more mortality in patients with septic shock. Am Rev Respir Dis 135:A5
35. Russell JA (1987) Prostoglandin E_1 does not alter mortality of adult respiratory distress syndrome. Chest 92:59S
36. Wright JR, Clements JA (1987) Metabolism and turnover of lung surfactant. Am Rev Respir Dis 136:426–444
37. Lachmann B (1987) The role of pulmonary surfactant in the pathogenesis and therapy of ARDS. In: Vincent JL (ed) Update in intensive care and emergency medicine. Springer, Berlin Heidelberg New York, pp 123–134
38. Richman PS, Spragg RG, Merritt TA, Robertson B, Curstedt T (1987) Administration of porcine-lung surfactant to humans with ARDS: initial experience. Am Rev Respir Dis 135:A5

Surfactant Replacement in Acute Respiratory Failure: Animal Studies and First Clinical Trials

B. Lachmann

Introduction

The integrity of the surfactant system of the lung is essential for normal breathing. Decreasing the surface tension at the interface between alveoli and air ensures (a) mechanical stabilization of the lung; (b) equal ventilation of differently sized alveoli; (c) stabilization of fluid balance in the lung and protection against lung edema; and (d) adequate function of local defense mechanisms against air-borne, bacterial, and viral infections (for review see [1]). Furthermore, we have recently demonstrated that the pulmonary surfactant system may also be involved in protecting the lung against its own mediators (e.g., angiotensin II) and in protecting the cardiocurculatory system against mediators produced by the lung [2].

Besides its action primarily at the alveolar level, surfactant also plays an important role in the airways. It is essential for stabilization of the small airways [3] and for non ciliar and mucous transport [4]. Surfactant also acts as an anti-glue factor between mucous particles, as well as between mucous and the bronchial wall [5] through its masking of receptors of smooth muscle with respect to substances which induce contraction [3, 6].

In spite of this wide range of different functions of the surfactant system in the lung, to date (except for "idiopathic" respiratory distress syndrome, IRDS) there is very little information about the role of surfactant in other acute and chronic lung diseases. One reason for this may be that, although the chemical components of the surfactant system are more or less known, many questions remain unanswered concerning the interaction between components of the surfactant system and other chemical components which are present as a result of lung diseases (e.g., plasma in adult respiratory distress syndrome, or ARDS; special proteins in alveolar proteinosis; free radicals; proteinases; and bacterial membranes, etc. under special circumstances). Another reason may be that for in vivo diagnosis of a functionally disturbed surfactant system only a few parameters can be used: dynamic and static lung compliance and, with some limitation, blood gases and the functional residual capacity.

With these problems in mind we developed a simple but rational philosophy in order to investigate and solve these problems:

Hypothesis 1: If surfactant is necessary for keeping retractive forces as low as possible and also for optimal gas exchange, any disturbance in the surfactant system will result in abnormalities in lung distensibility and gas exchange.

Hypothesis 2: If hypothesis 1 is correct, surfactant replacement will restore lung distensibility and gas exchange to normal.

Under experimental and clinical conditions this means that in a lung disease in which decreased lung compliance and impaired gas exchange are present, if surfactant replacement improves these functions, the surfactant system should also be involved in the pathogenesis of this lung disease.

In order to confirm this hypothesis we developed different animal models, all characterised by decreased compliance and impaired gas exchange, in which most cases needed artificial ventilation to keep the animals alive. In the following animal models (available in our laboratory) it has been shown that the surfactant system improves the level of lung compliance and gas exchange and partly prevents bronchial obstruction in allergic asthma:

Respiratory failure due to surfactant depletion (rats, guinea pigs, adult rabbits) [7–12].

Respiratory failure due to viral pneumonia (mice) [13].

Respiratory failure due to bacterial pneumonia (unpublished observations).

Respiratory failure induced by free oxygen radicals following intratracheal instillation of xanthese oxidase (guinea pigs, rats) [14].

Respiratory failure after intravenous injection of anti-lung serum, capillary leakage syndrome (guinea pigs) [15, 16].

Respiratory failure caused by damaging the bronchial surfactant with a combination of artificial ventilation and tracheal instillation of a small amount of saline (rats, guinea pigs) [3].

Allergic bronchial asthma (guinea pigs) [6].

The following examples demonstrate that surfactant replacement improves lung function in acute respiratory failure, independently of its etiology.

Surfactant Replacement

Animal Studies

Regarding respiratory failure due to damaged bronchial surfactant, as early as 1970, Macklem et al. called attention to the significance of bronchial surfactant for stabilization of the peripheral airways; they suggested that its absence may cause airway obstruction or collapse of the small bronchi with air trapping [17]. Nevertheless, little is known about the importance of, for example, a transient disturbance of surfactant in the pathogenesis of asthma, although it has been known for some time that the most effective therapeutic agents in the asthmatic attack, glucocorticoids and β-adrenergic agents, stimulate synthesis and release of surfactant phospholipids by type II cells.

The results of investigations performed recently in our laboratory emphasize the importance of bronchial surfactant for the stabilization of peripheral airways as indicated by Macklem et al. We induced severe respiratory insufficiency in guinea pigs by a combination of ventilation (pressure-controlled ventilation;

inspiration time 50%; peak airway pressure 35 cm H_2O; positive end-expiratory pressure, or PEEP, 4–5 cm H_2O; rate 30/min) and tracheal saline instillation. In this in vivo model we have been able to trace the main cause for the massive decline in thorax-lung compliance to inactivation (damage?) of the bronchial surfactant.

Guinea pigs (body weight 324 ± 18 g) were anesthetized with urethane (1.3 g/kg), relaxed with pancuronium bromide (1.2 mg/kg), and subjected to pressure-controlled air ventilation with different peak airway pressures (20, 30, and 35 cm H_2O), a PEEP of 4–5 cm H_2O and a rate of 30/min. Immediately preceding saline instillation and 15, 60, and 120 min afterwards, a thorax-lung pressure-volume diagram of each animal was recorded, and blood for arterial P_aO_2 determination was drawn from a catheter in the carotid artery. With all peak airway pressures used, a significant reduction in thorax-lung compliance occurred within 15 min after saline instillation. Animals ventilated with a pressure of only +20 cm H_2O died from hypoxia (P_aO_2 31 ± 6 mmHg) after 1 h.

Analysis of the pressure-volume diagrams revealed that 15 min after saline instillation, the opening pressure was already elevated to 20 cm H_2O whereas the total compliance was only minimally reduced up to a pressure of 40 cm H_2O. After overcoming the opening pressure, the slopes of the pressure-volume diagrams before and after saline instillation were nearly identical; this seems to us to indicate that the total retractive forces, which always increase markedly with impairment of the alveolar surfactant, increased only slightly in our model

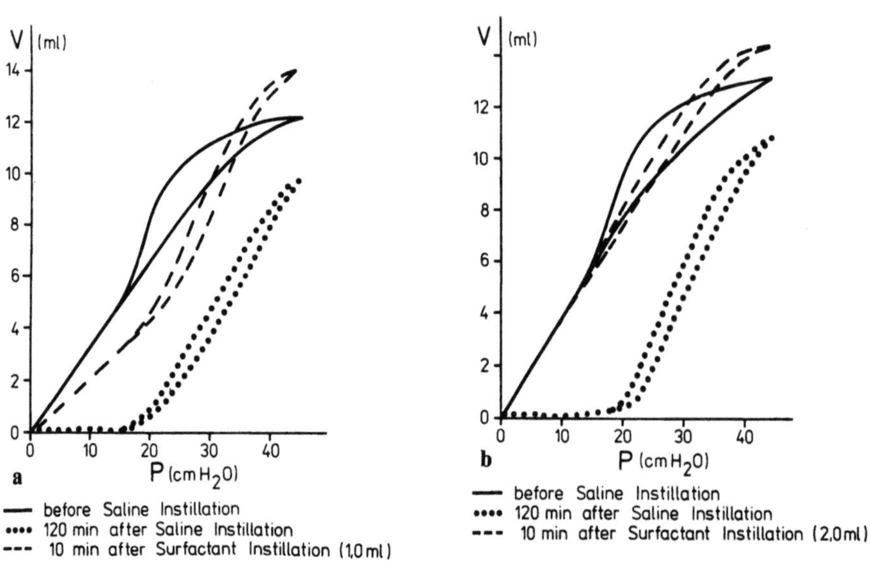

Fig. 1. Thorax-lung pressure-volume diagram of guinea pig before and 120 min after saline instillation and after subsequent instillation of 1 ml (**a**) and 2 ml (**b**) surfactant. Note that only the instillation of 2 ml surfactant results in complete restitution of the mechanical properties of the thorax-lung system. The minor volume increase after surfactant instillation might be explained by the opening of atelectatic regions. (From [3] with permission)

Table 1. Changes in thorax-lung compliance and arterial PO_2 induced by different amounts of surfactant applied intratracheally to animals in severe respiratory distress due to impairment of bronchial surfactant

	Before saline instillation	After saline instillation		
	120 min		Amount of surfactant (ml)	130 min (10 min after surfactant instillation)
Compliance (ml/cm H_2O kg)	1.21 + 0.34 (n = 5)	0.35 + 0.08	1	0.72 + 0.23
	1.27 + 0.28 (n = 4)	0.35 + 0.11	2	1.21 + 0.34
P_aO_2 (mmHg)	82 + 6 (n = 5)	32 + 6	1	75 + 16
	84 + 8 (n = 4)	34 + 5	2	79 + 14

Compliance was calculated at +20 cm H_2O on the deflation limb of the pressure-volume diagram. Blood was drawn at the ventilation pressure of +30 cm H_2O, and a PEEP of 4–5 cm H_2O was used throughout the experiment.

during the observed period. These findings therefore permit the conjecture that the functional changes obtained in this model are caused in the first place by collapse of peripheral airways due to impairment of the bronchial surfactant.

More evidence for the impairment of the bronchial (and alveolar?) surfactant in our model is provided by the following results. We showed that tracheal instillation of exogenous surfactant normalizes the impaired thorax-lung mechanics and gas exchange. Whereas the instillation of 1 ml surfactant (50 mg total phospholipid content) markedly improved thorax-lung mechanics, an almost complete correction of the functional changes was achieved in those animals which received 2 ml surfactant by tracheal instillation (Fig. 1; Table 1).

ARDS due to Surfactant Depletion

In experimental ARDS, induced by in vivo lung lavage, we were the first to demonstrate that tracheal instillation of surfactant lipids results in a striking improvement in gas exchange [7, 10] (Fig. 2). Other experiments (in animals receiving surfactant) have shown that the improved blood gas levels are stable for at least 5 h, whereas the P_aO_2 in control animals remains low, despite ventilation with PEEP and 100% oxygen. Histological examination of lung sections from surfactant-treated animals showed a uniform pattern of well-aerated alveoli, with only minimal intra-alveolar edema and hyaline membranes, whereas control animals had extensive atelectasis and prominent hyaline membranes.

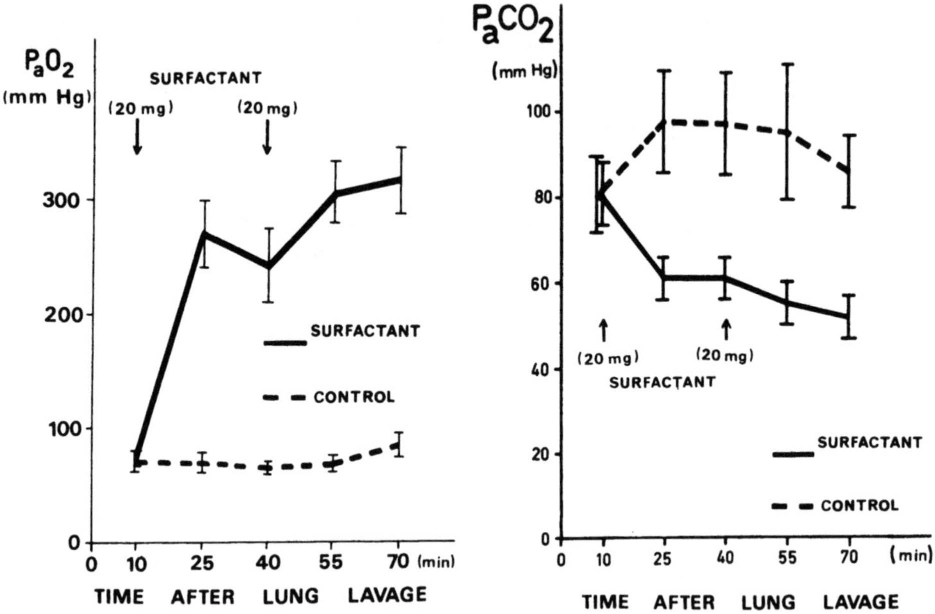

Fig. 2. Behavior of P_aO_2 and P_aCO_2 in adult guinea pigs with severe RDS after tracheal instillation of surfactant in comparison to controls. Animals were ventilated pressure-controlled with pure oxygen at an I:E ratio of 1:1, a frequency of 20/min, an insufflation pressure of 28 cm H_2O and a PEEP of 5 cm H_2O. (From [7, 10])

These results indicate that the ventilator per se is not harmful to the pulmonary parenchyma, provided that alveolar collapse is prevented by surfactant replacement, and shear forces are thereby avoided. One may argue that the use of the surfactant-depleted lung model for assessing the effects of surfactant replacement does not represent all the changes typical of ARDS lungs. This is why we also used other animal models with severe respiratory failure and treated them with tracheal instillation of surfactant (e.g., respiratory failure due to free oxygen radicals [14], virus pneumonia [13], and ARDS due to intravenous injection of anti-lung serum [16].

Acute Respiratory Insufficiency due to Free Oxygen Radicals

Free oxygen radicals play an important role in a variety of diseases. The hypoxanthine-xanthine oxidase system, which generates the superoxide radical, exerts a damaging effect on several organs, including the lung [18]. We have recently demonstrated that xanthine oxidase (XO) applied to the trachea of guinea pigs induces dramatic changes in lung-thorax compliance, in the course of a few minutes, by destroying the functional integrity of the bronchial and alveolar surfactant system. This effect probably results from the generation of free oxygen radicals by XO, but was only minimally prevented by superoxide dismutase (SOD) [19]. However, it was almost blocked by tracheal instillation

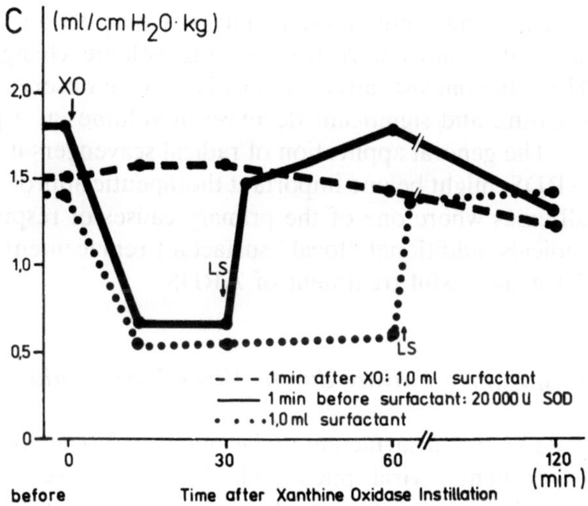

Fig. 3. Course of lung-thorax compliance *(C)* in three guinea pigs after tracheal instillation of 1 U xanthine-oxidase *(XO)*, natural surfactant *(LS)*, and superoxide dismutase *(SOD)*. C was measured at 30 cm H_2O. Note that C could be restored by surfactant replacement to the inital values, and that SOD in this animal with this protocol had no further therapeutic effect on improvement of C

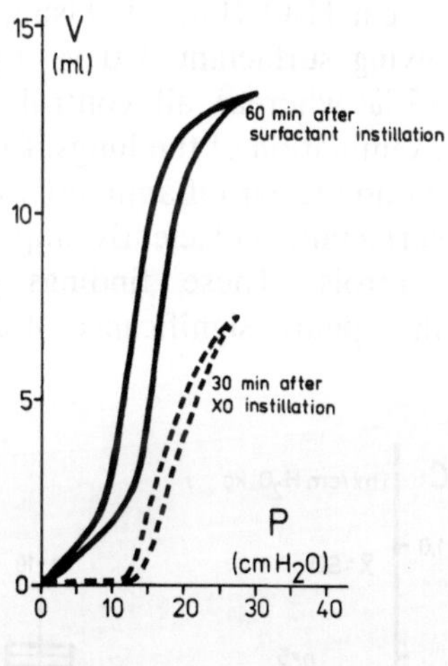

Fig. 4. Pressure-volume diagrams from a guinea pig 30 min after xanthine oxidase *(XO)* instillation *(broken line)* and 60 min after surfactant replacement *(solid line)*. Note that up to 60 min after surfactant replacement it was possible to record an almost normal pressure-volume diagram

of exogenous surfactant (Fig. 3). One possible explanation for the latter result could be that the lipid peroxidation by free oxygen radicals takes place with exogenous surfactant lipids, thus preventing the peroxidation of the lipids from surfactant system, which line the alveoli and airways.

Figure 4 shows that decreased lung-thorax compliance after XO instillation could almost completely be restored by tracheal instillation of surfactant. This effect must be related to the surfactant lipids and not to fluid administration, as the lung-thorax compliance deteriorated further in those animals receiving saline instead of surfactant.

In earlier investigations we demonstrated that the combination of fluid with artificial ventilation already leads to damage, mainly of the bronchial surfactant system [20]. Thus, also in this study, one factor of functional disturbance can be explained due to the fluid administration. However, the combination of fluid and artificial ventilation mainly led only to an increased opening pressure, without significant restrictive lung volume changes, at a pressure of 30 cm H_2O. In contrast, after XO instillation we observed both an increased opening pressure and significant decrease in volume, at a pressure of 30–35 cm H_2O.

The general application of radical scavengers in patients who might develop ARDS, might be one important therapeutic approach in the future. However, in all cases where one of the primary causes of respiratory failure is free oxygen radicals, additional "local" surfactant replacement might be an important step for a successful treatment of ARDS.

Respiratory Failure due to Virus Pneumonia

We found that surfactant replacement is also an effective therapeutic measure in experimental viral pneumonia [13]. In these studies, lung mechanics was recorded in mice 6 days after infection with influenza virus, before and after tracheal instillation of surfactant. Surfactant replacement resulted in a significant augmentation of thorax-lung compliance at insufflation pressure 26 cm H_2O (Fig. 5). Using the same model we could also demonstrate that giving surfactant 4 days after virus infection, the survival rate increased to 75 % whereas all control animals died between days 6 and 7. Histologic examination of the lungs, from all animals, showed the interstitial infiltration of mononuclear inflammatory cells, typical of viral pneumonia. Animals receiving surfactant had clearly improved lung aeration in comparison with untreated controls. These findings provide additional indication of the potential therapeutic significance of surfactant replacement in ARDS.

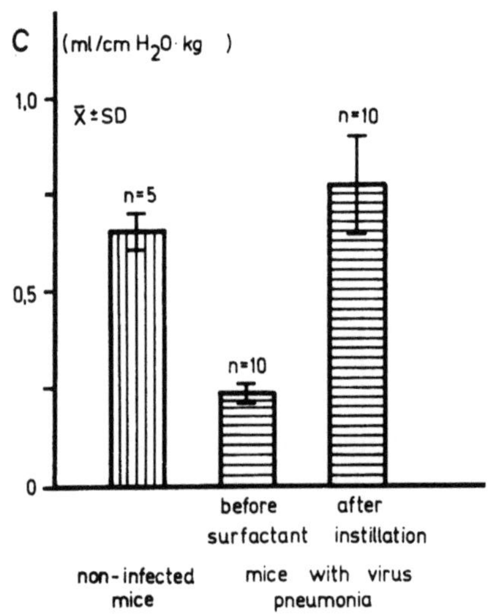

Fig. 5. Thorax-lung compliance per kilogram body weight in mice with viral pneumonia before and 10 min after tracheal instillation of 0.15 ml surfactant (60 mg phospholipids/ml) at 25 cm H_2O. *Left,* the thorax-lung compliance of noninfected mice

Experimental Capillary Leakage Syndrome

Intravenous injection of anti-lung serum in guinea pigs causes acute and fatal respiratory failure. Previous studies have shown marked decrease in thorax-lung compliance, increase in water content, decrease in phospholipid content,

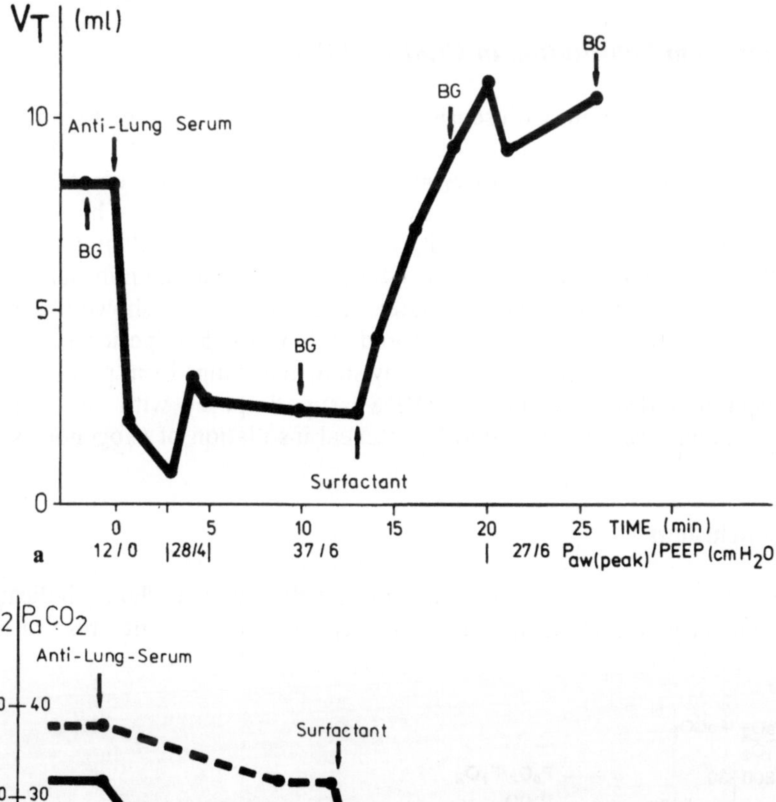

a

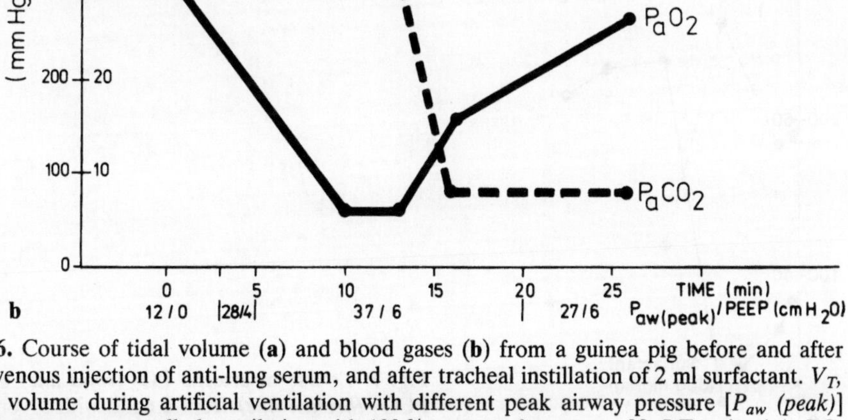

b

Fig. 6. Course of tidal volume (**a**) and blood gases (**b**) from a guinea pig before and after intravenous injection of anti-lung serum, and after tracheal instillation of 2 ml surfactant. V_T, Tidal volume during artificial ventilation with different peak airway pressure [P_{aw} *(peak)*] during pressure-controlled ventilation with 100 % oxygen; frequency, 30; I:E ratio, 1:1. *BG*, blood gases and diminished surface activity in these lungs, as well as severe morphological damage especially to the alveolar capillary membrane of these lungs [15]. Using this model – where the failure is clearly caused by damage to the capillary membranes – we could demonstrate that surfactant instillation also significantly improved blood gases and lung mechanics in these animals [16] (Fig. 6).

Surfactant Substitution in Clinical ARDS

In a terminal patient with sepsis and severe ARDS (P_aO_2 of 19 mmHg), despite pressure-controlled ventilation with an I:E ratio of 3:1 and peak airway pressure of 48 cm H_2O (FiO_2 = 1), tracheal instillation of natural surfactant (300 mg phospholipids/kg body weight) led within 4 h to a dramatic improvement in gas exchange (P_aO_2 from 19 mmHg to 240 mmHg; P_aCO_2 from 68 to 45 mmHg; see Fig. 7) [9]. Chest X rays made 20 min before surfactant instillation (Fig. 8a) and 4 h later (Fig. 8b) clearly showed that a nearly "normal" situation had been restored within this short period of time.

These first clinical results already show that lungs from patients with severe respiratory distress syndrome (RDS), superimposed with virus and bacterial pneumonia, can be reaerated by tracheal instillation of exogenous surfactant.

Conclusions

Results from these experimental and first clinical trials clearly indicate that the pulmonary surfactant system is at least involved in acute respiratory failure

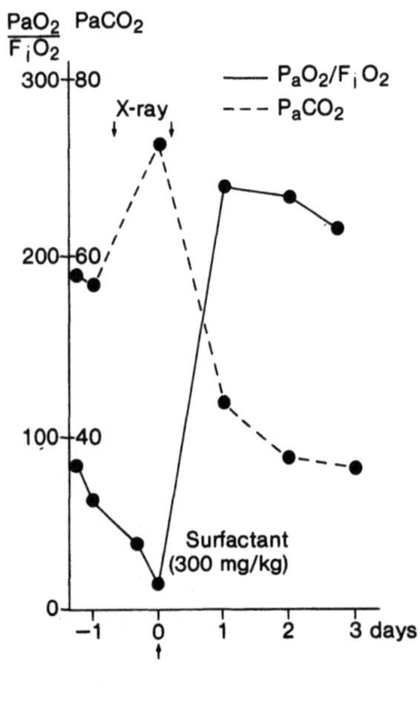

Fig. 7. Course of arterial PO_2 *(solid line)* and PCO_2 *(broken line)* from a child with severe ARDS, before and after surfactant replacement therapy

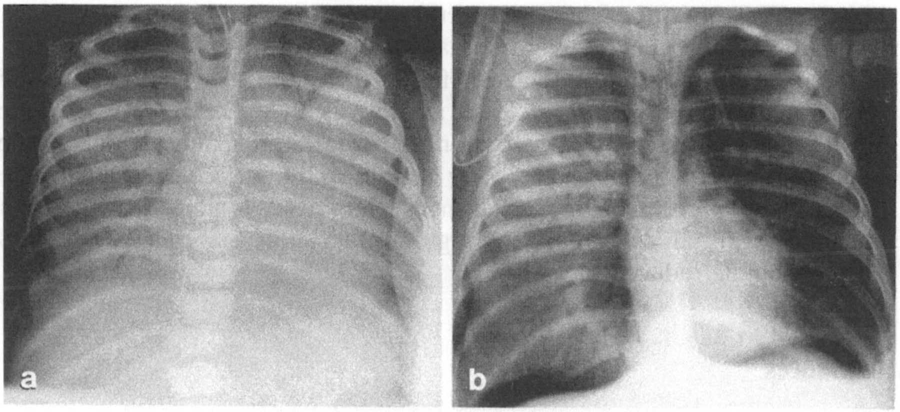

Fig. 8. X rays from a child with a severe ARDS immediately before (**a**) and (**b**) 4 h after surfactant replacement therapy. (From [9] with permission)

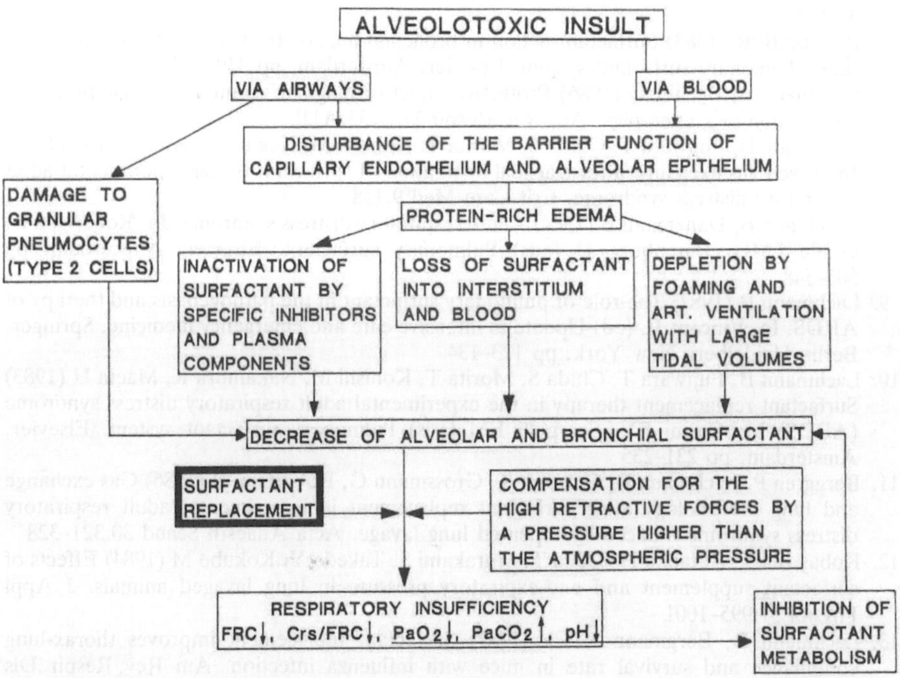

Fig. 9. Pathogenesis of ARDS with special reference to the surfactant system, including suggestions to compensate for a damaged surfactant system

such as ARDS, independently of its etiology. Thus, we conclude that after injury to the alveolar-capillary membrane, followed by capillary leakage, the surfactant system will be responsible for further pathophysiological changes

(Fig. 9). These well-documented functional disturbances in the lung result finally in the failure of the lung as a gas exchange organ.

If it becomes possible in the near future to control mechanisms of blood coagulation and complement release as well as to prevent or delay toxic influence of high oxygen concentration and free oxygen radicals by specific agents and if, on the other hand, early damage in the surfactant system of the lung can be prevented by surfactant replacement combined with "non traumatic" artificial ventilation, the high mortality rate associated with RDS can probably be reduced.

References

1. Lachmann B, Danzmann E (1984) Acute respiratory distress syndrome. In: Robertson B, Golde LMG, Batenburg JJ (eds) Pulmonary surfactant. Elsevier, Amsterdam
2. Hein T, Lachmann B, Armbruster S, Smit JM, Voelkel N, Erdmann W (1987) Pulmonary surfactant inhibits the cardiovascular effects of platelet activating factor (PAF), 5-hydroxytryptamine (5-HT) and angiotensin II. Am Rev Respir Dis 135:A506
3. Lachmann B (1985) Possible function of bronchial surfactant. Eur J Respir Dis 67:49–61
4. Green GM (1973) Alveolobronchiolar transport mechanisms. Arch Intern Med 31:109–114
5. Reifenrath R (1983) Surfactant action in bronchial mucus. In: Cosmi EV, Scarpelli EM (eds) Pulmonary surfactant system. Elsevier, Amsterdam, pp 339–347
6. Lachmann B, Becher G (1986) Protective effect of lung surfactant on allergic bronchial constriction in guinea pigs. Am Rev Respir Dis 133:A118
7. Lachmann B, Fujiwara T, Chida S, Morita T, Konishi M, Nakamura K, Maeta H (1981) Improved gas exchange after tracheal instillation of surfactant in the experimental adult respiratory distress syndrome. Crit Care Med 9:158
8. Lachmann B, Danzmann E (1984) Acute respiratory distress syndrome. In: Robertson B, Golde LMG, Batenburg JJ (eds) Pulmonary surfactant. Elsevier, Amsterdam, pp 505–548
9. Lachmann B (1987) The role of pulmonary surfactant in the pathogenesis and therapy of ARDS. In: Vincent JL (ed) Update in intensive care and emergency medicine. Springer, Berlin Heidelberg New York, pp 123–134
10. Lachmann B, Fujiwara T, Chida S, Morita T, Konishi M, Nakamura K, Maeta H (1983) Surfactant replacement therapy in the experimental adult respiratory distress syndrome (ARDS) In: Cosmi EV, Scarpelli EM (eds) Pulmonary surfactant system. Elsevier, Amsterdam, pp 231–235
11. Berggren P, Lachmann B, Curstedt T, Grossmann G, Robertson B (1986) Gas exchange and lung morphology after surfactant replacement in experimental adult respiratory distress syndrome induced by repeated lung lavage. Acta Anaesth Scand 30:321–328
12. Kobayashi T, Kataoka H, Ueda T, Murakami S, Takeda Y, Kokubo M (1984) Effects of surfactant supplement and end-expiratory pressure in lung lavaged animals. J Appl Physiol 57:995–1001
13. Lachmann B, Bergmann K Ch (1987) Surfactant replacement improves thorax-lung compliance and survival rate in mice with influenza infection. Am Rev Respir Dis 135:A6
14. Lachmann B, Saugstad OD, Erdmann W (1987) Effect of surfactant replacement on respiratory failure induced by free oxygen radicals. In: Schlag G, Redl H (eds) First Vienna shock forum. Part B: Monitoring and treatment of shock. Liss, New York, pp 305–313
15. Lachmann B, Bergmann KC, Winsel K, Müller E, Petro W, Schäffer C, Vogel J (1975) Experimental respiratory distress syndrome after injection of anti-lung serum. III. Chronic experimental trial. Pädiatrie und Grenzgebiete 14:211–233

16. Lachmann B, Hallman M, Bergmann K Ch (1987) Respiratory failure following anti-lung serum: study on mechanisms associated with surfactant system damage. Exp Lung Res 12:163–180
17. Macklem PT, Proctor DF, Hogg (1970) The stability of peripheral airways. Respir Physiol 8:191–203
18. Saugstad OD (1985) Oxygen radicals and pulmonary damage. Pediatr Pulmonol 1:167–175
19. Saugstad OD, Hallman M, Becher G, Oddoy A, Lachmann B (1984) Protective effect of superoxide dismutase on severe lung damage caused by xanthine oxidase. Pediatr Res 18:802
20. Lachmann B (1985) Possible function of bronchial surfactant. Eur J Respir Dis 67:49–61

Pulmonary Surfactant Effects and Replacement in Oxygen Toxicity and Other ARDS–Type Lung Injuries

B. A. Holm, S. Matalon, and R. H. Notter

Introduction

Surfactant in Respiratory Distress

In the thirty years since the discovery of surface active material in the alveolar region of the lung, and its fundamental importance for respiration, extensive studies have been directed at characterizing and understanding the pulmonary surfactant system (e.g. [20, 24]). A basic driving force for much of this lung surfactant (LS) research was the discovery in 1959 that the respiratory distress syndrome of the premature newborn (RDS) resulted directly from a lack of surfactant-producing type II pneumocytes in the immature lung, giving rise to a primary deficiency in alveolar surfactant at birth [1]. After the fundamental importance of LS for normal respiration was established, extensive research efforts were initiated to develop beneficial exogenous surfactant replacement therapy for neonatal RDS. The pathway to such therapy has been longer than many researchers expected, but since the beginning of this decade, it has become increasingly clear that exogenous surfactant replacement in RDS is highly efficacious in premature infants [e.g. 5, 20 (for review), 27, 29.

Another clinically relevant area of lung surfactant research that has recently attracted significant attention involves contributions from pulmonary surfactant dysfunction to the pathophysiology of the widespread adult respiratory distress syndrome (ARDS). In characterizing surfactant effects in ARDS, previous research on natural LS, and its biophysical and physiological activity in normal and abnormal respiration, can often be applied directly to enhance understanding of the lung injury that is present. It is this area of research which is the primary focus of the present paper. Specifically, surfactant system changes that can occur in the ARDS-type lung injury due to hyperoxia are discussed in detail, along with the potential efficacy of exogenous surfactant replacement therapy in mitigating some of the pathology found in this complex lung injury syndrome. In addition, the ramifications for more general ARDS lung injury and pathology are also noted and analyzed.

To aid the discussion here, ARDS is generally defined by a set of five clinical and pathophysiological features that are characteristic of acute injury of the alveolar-capillary unit, and this can occur from a variety of etiologies [13, 26]. However, regardless of specific cause, ARDS lung injuries exhibit the following common features: intrapulmonary shunting leading to arterial hypoxemia; decreased lung volumes such as functional residual capacity

(FRC); decreased lung compliance; diffuse but prominent alveolar infiltration on chest roentgenogram; and the absence of cardiogenic causes for this pathology [13, 26]. The clinical disorders which have been associated with ARDS lung injury are diverse in nature, and include shock, systemic sepsis, liquid aspiration, trauma, drug overdose, radiation pneumonitis, oxygen toxicity, and many others [13, 26].

Lung Surfactant Dysfunction in ARDS

Although a variety of etiologies are associated with ARDS, the underlying pathophysiology is felt to be relatively similar. As inferred above, the major early hallmark of ARDS is damage to the alveolo-capillary membrane, with a concomitant increased permeability to high molecular weight solutes. This permeability increase is thought to be the common pathophysiologic initiator in ARDS, and likely occurs through damage to the alveolar type I epithelial cell. This inevitably leads to effects on LS-producing type II pneumocytes, since they are the stem cell for the alveolar lining type I pneumocyte. Any effects on the alveolar type II cell resulting from epithelial injury could then alter the synthesis, storage, and/or secretion of natural LS. Furthermore, a second and potentially more direct means of LS dysfunction in lung injury could occur

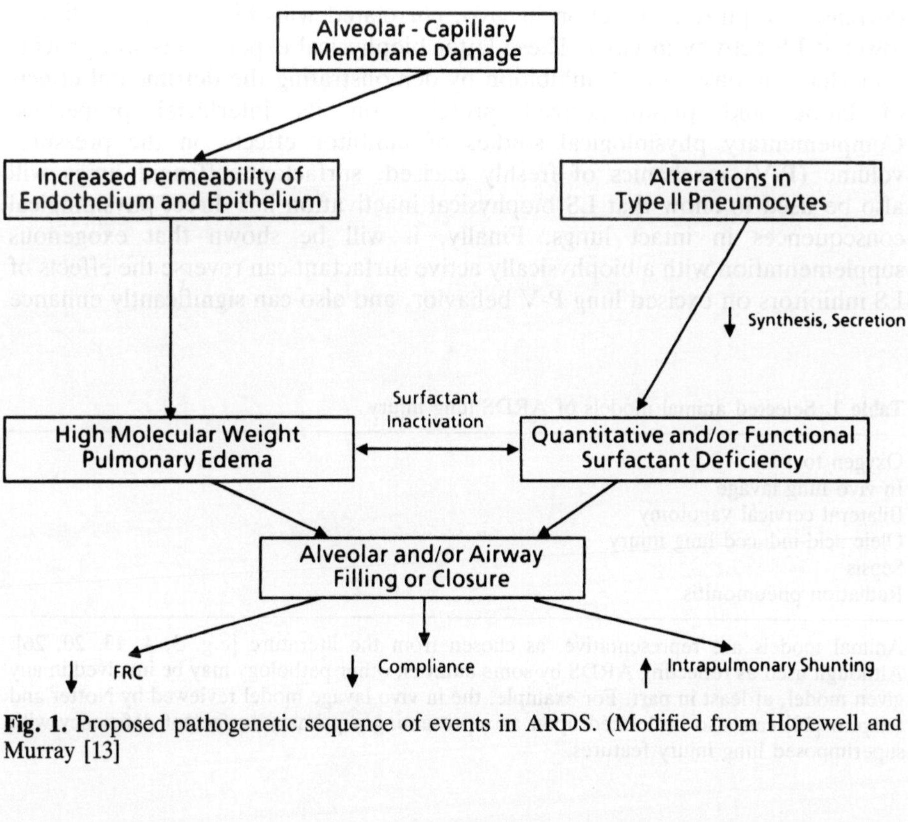

Fig. 1. Proposed pathogenetic sequence of events in ARDS. (Modified from Hopewell and Murray [13]

through biophysical inhibition of surfactant material already present in the alveolar hypophase, or at the air-hypophase interface (e.g. [8–12, 24]). A postulated sequence of events in ARDS lung injury, modified from Hopewell and Murray [13], is described in Fig. 1. As shown schematically in this figure, it is generally felt that pulmonary surfactant is affected secondarily in ARDS [26]. This does not mean, however, that LS alterations are unimportant in a quantitative sense. Indeed, any functional surfactant deficiency, whether primary or secondary, would contribute significantly to the pulmonary pathology present in ARDS.

Part of the difficulty in determining the role and importance of the pulmonary surfactant system in the pathophysiology of ARDS arises from a lack of precise and truly comparable animal models of such lung injuries. Even though similar pathologic features may be present in many ARDS models (e.g. increased alveolar permeability), differences in other primary and secondary lung injury variables, and in their time dependence, may also occur. In spite of these difficulties, animal models of ARDS can be useful and productive, particularly when designed and analyzed so as to maximize the comparability of experimental variables. Several different animals models which have been used previously to study ARDS are shown in Table 1, including the model of severe pulmonary oxygen toxicity which forms the major focus for discussion here.

In the remainder of this paper, lung injury due to hyperoxia will be described from a multidisciplinary perspective, including characterizations of decreased respiratory function in vivo, correlated with biophysical studies of lowered LS activity in vitro. These latter biophysical experiments will provide a mechanistic basis for LS inhibition by demonstrating the detrimental effects of blood and plasma-derived proteins on its interfacial properties. Complementary physiological studies of inhibitor effects on the pressure-volume (P-V) mechanics of freshly excised, surfactant-sufficient lungs will also be used to show that LS biophysical inactivation has direct physiological consequences in intact lungs. Finally, it will be shown that exogenous supplementation with a biophysically active surfactant can reverse the effects of LS inhibitors on excised lung P-V behavior, and also can significantly enhance

Table 1. Selected animal models of ARDS lung injury

Oxygen toxicity
In vivo lung lavage
Bilateral cervical vagotomy
Oleic acid-induced lung injury
Sepsis
Radiation pneumonitis

Animal models are representative, as chosen from the literature [e.g. 3, 4, 13, 20, 26]. Although used as reflecting ARDS by some authors, other pathology may be involved in any given model, at least in part. For example, the in vivo lavage model reviewed by Notter and Shapiro [20] can also be looked upon as representing a primary surfactant deficiency with superimposed lung injury features.

lung function and mechanics in vivo for animals with pulmonary oxygen toxicity. These results will also be discussed in terms of their relevance for future considerations of exogenous surfactant therapy for ARDS in humans.

Pulmonary Oxygen Toxicity (Hyperoxia)

Overview of Pathology

Prolonged exposure of animals to 100 % oxygen at atmospheric pressure causes progressive lung injury characterized by a gradual increase in permeability of the blood-gas barrier, leading to the formation of alveolar edema, pulmonary atelectasis and, eventually, death due to respiratory failure [4, 10, 11, 16, 17]. Previous work by Matalon and Egan [16] has indicated that for adult rabbits breathing 100 % O_2, alveolar permeability begins to increase after 48 hours of exposure, reaching a plateau after 64 hours. Moreover, these investigators also found that the permeability increase during the initial 64 hours of hyperoxic exposure occurred prior to the formation of alveolar edema, and at a time when the alveolar epithelium appeared normal histologically (although some interstitial edema might have been present near the end of this timecourse [17]).

If adult rabbits are allowed to continue breathing 100% O_2 for over 64 hours, their pulmonary function (assessed by arterial blood gas values) decreases rapidly. The onset of this decline is somewhat variable, although it does not generally occur prior to 72 hours of exposure [17]. Nonetheless, if such severe exposure continues, about 90% of the animals studied by Matalon and Egan [17] died from arterial hypoxemia over the next 10–15 hours in 100% oxygen. This lethal exposure protocol is too fulminant and rapid for accurate characterizations of several ventilatory variables and their timecourse. For this reason, Holm et al. [11] sought a severe but less lethal injury. They found that if rabbits were removed to room air following the initial 64 hour (100% O_2) exposure period, the ensuing respiratory distress occurred with a less fulminant onset, peaking at 24 hours (mean) post-exposure, and lasting for a day or more. Moreover, the mortality rate in these animals dropped to 35%–40% with the surviving animals showing complete recovery by 200 hours post-exposure. This in vivo model, characterized and described in detail by Holm and Notter [10] and Holm et al. [11], still has the characteristics relevant for clinical ARDS (e.g. increased alveolar permeability and edema), but at the same time allows for more extensive studies of hyperoxic lung injury through the stages of early development, acute respiratory distress, and eventual recovery. The discussion of pulmonary oxygen toxicity in this article will be centered, principally, around this model.

Indications of Surfactant Effects in Pulmonary Hyperoxia in Rabbits

Table 2 summarizes the arterial blood gas changes that occurred throughout the 64 hr hyperoxic exposure protocol (100% O_2), and after subsequent removal of

Table 2. Arterial blood gases in hyperoxic lung injury (from Holm et al. [11])

			PaO_2 after 64 h 100 % O_2 plus				
	Control	64 h 100 % O_2	0–12 h air	12–24 h air	24–48 h air	48–96 h air	96–200 h air
P_aO_2 (Torr)	79 ± 3	500 ± 18	81 ± 5	62 ± 4[a]	67 ± 3[a]	68 ± 2[a]	76 ± 2
P_aCO_2 (Torr)	31 ± 3	33 ± 2	27 ± 2	31 ± 2	31 ± 2	33 ± 1	38 ± 1
pH	7.4 ± 0.01	7.35 ± 0.03	7.45 ± 0.02	7.44 ± 0.02	7.47 ± 0.01	7.46 ± 0.01	7.44 ± 0.01
Lung wet/dry weight	4.1 ± 0.3	4.5 ± 0.2	–	6.2 ± 0.2[a]	6.2 ± 0.2[a]	–	4.4 ± 0.2

Data taken from Holm and Notter [10] and Holm et al. [11]; $PaCO_2$ are arterial blood gases (torr). Data are given as mean ± SEM.
[a] Significantly different from control rabbits ($p < 0.05$ or less).

animals to room air [10, 11]. As shown, pulmonary gas exchange (indicated by the arterial partial pressure of oxygen [PaO_2]) remained normal throughout the exposure period, and during the first 12 hours post-exposure. However, PaO_2 values decreased significantly by 24 hours post-exposure, and remained low for several days. This decrease in lung function did not give rise to significant changes in arterial partial pressures of carbon dioxide ($PaCO_2$) or arterial pH, nor was it associated with changes in other cardio-respiratory variables. A final variable assessed in Table 2 is the lung wet/dry weight ratios at various times during the hyperoxic exposure and post-exposure periods. Again, as found for arterial oxygenation, the maximal wet/dry weight ratios occurred at the 24 hour (mean) post-exposure point, indicating the presence of significant pulmonary edema at this peak period of lung injury.

In addition to the pulmonary functional changes and edema just noted, the 24 hour post-exposure animals also exhibited significant lung mechanical changes, as shown in Fig. 2. Such mechanical changes were not present earlier at the end of the 64 hr hyperoxic exposure period (Curve B, Fig. 2, top and bottom). However, at the 24 hr lung injury peak (Curve C, Fig. 2, top and bottom), the deflation limbs of pressure-volume (P-V) curves showed a 50 % decrease in total lung capacity (TLC) and a large decrease in compliance compared with controls [10, 11]. Importantly, P-V curves with saline-filling rather than air-filling in Fig. 3 showed no significant differences between lungs from control and 24 hour postexposure animals [11]. This indicates that the elastic properties of lung tissue remained unchanged during the hyperoxic exposure and post-exposure period, and implies that the differences in lung mechanics between control and 24 hour post-exposure animals (Fig. 2) reflect, at least in part, alterations in surface forces at the alveolar air-liquid interface.

The indication that surface tension abnormalities were present in lungs injured by hyperoxia was studied more directly by in vitro biophysical activity

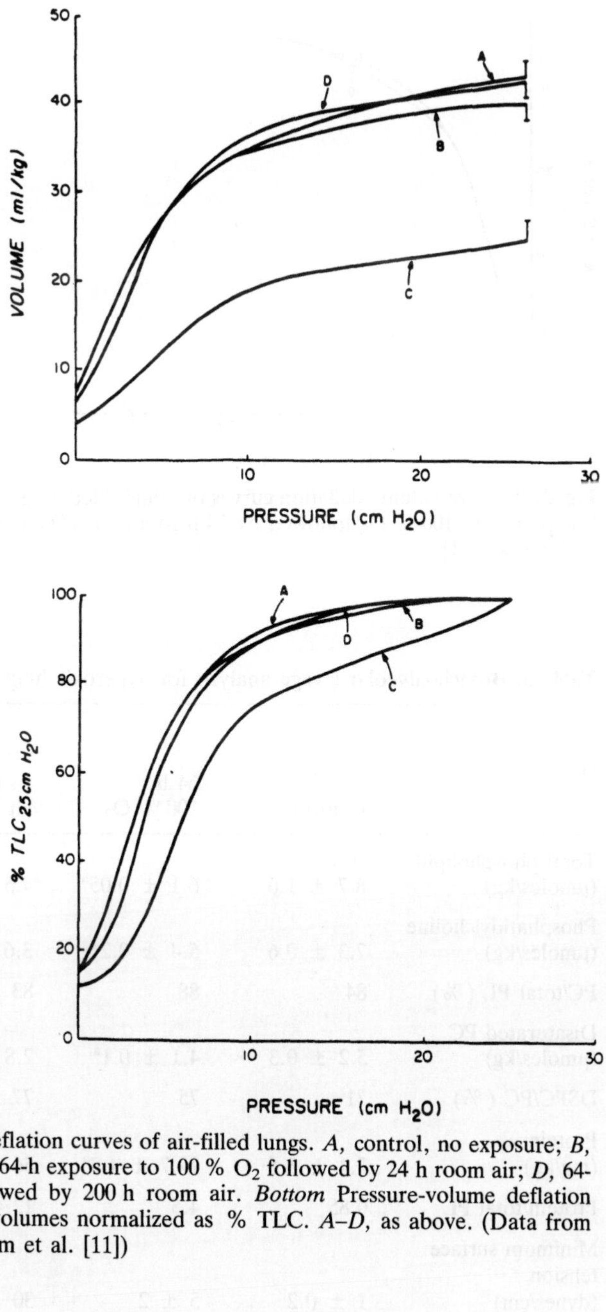

Fig. 2. *Top* Pressure-volume deflation curves of air-filled lungs. *A,* control, no exposure; *B,* 64-h exposure to 100 % O_2; *C,* 64-h exposure to 100 % O_2 followed by 24 h room air; *D,* 64-h exposure to 100 % O_2 followed by 200 h room air. *Bottom* Pressure-volume deflation curves of air-filled lungs with volumes normalized as % TLC. *A–D,* as above. (Data from Holm and Notter [10] and Holm et al. [11])

measurements on cell-free bronchoalveolar lavage (BAL) samples obtained from these animals. Table 3 shows that total surfactant phospholipid levels in cell-free BAL decreased 30 % by the end of the 64 hour hyperoxic exposure, and an additional 20 % by 24 hours post-exposure. Thin layer chromatographic analyses indicated that although the total phospholipid levels decreased, the

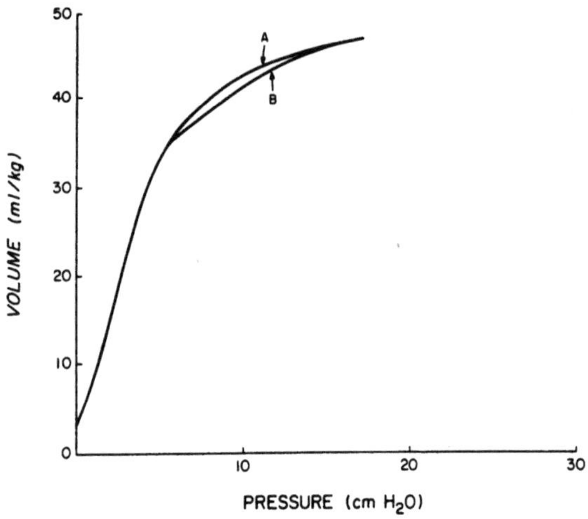

Fig. 3. Pressure-volume deflation curves of saline-filled lungs. *A*, control, no exposure; *B*, 64-h exposure to 100 % O_2 followed by 24 h room air. (Data from Holm and Notter [10] and Holm et al. [11])

Table 3. Bronchoalveolar lavage analysis for hyperoxic lung injury in rabbits

			64 h 100% O_2 plus	
	Control	64 h 100 % O_2	24 h air	200 h air
Total phospholipid (μmoles/kg)	8.7 ± 1.6	6.1 ± 0.05[a]	4.5 ± 0.6[a]	13.6 ± 2.9[a]
Phosphatidylcholine (μmoles/kg)	7.3 ± 0.6	5.4 ± 0.2[a]	3.6 ± 0.2[a]	11.8 ± 1.3[a]
PC/total PL (%)	84	88	83	87
Disaturated PC (μmoles/kg)	5.2 ± 0.3	4.1 ± 0.1[a]	2.8 ± 0.1[a]	8.8 ± 0.6[a]
DSPC/PC (%)	71	75	77	75
Protein (mg/kg)	7.4 ± 3.5	27.7 ± 4.7[a]	56.0 ± 3.5[a]	10.5 ± 2.6
Protein/total PL	0.85	4.5	12.4	0.80
Minimum surface tension (dynes/cm)	1 ± 0.2	5 ± 2	30 ± 2	1 ± 0.2

Data are taken from Holm and Notter [10] and Holm et al. [11], with units as given in the table, for material from bronchoalveolar lavage with 0.15 M NaCl. The minimum surface tension is that found at up to 30 minutes of pulsation in an oscillating bubble apparatus [6, 10, 11] at physical conditions relevant for the lung in vivo. See text for details.
[a] Data are expressed as mean ± SEM; significantly different from controls ($p < 0.05$ or less).

composition of surfactant in BAL remained normal [10, 11]. While the total amount of LS was decreasing, however, protein levels in BAL were increasing. The ratio of protein to LS-phospholipid in BAL was highest at the 24 hr post-exposure timepoint (Table 3), at the peak of lung injury. (As an aside for the present paper, further experiments by Holm et al. [9] indicated that the variations in alveolar phospholipid levels correlated well with hyperoxia-induced changes in the phospholipid biosynthetic abilities of type II pneumocytes, as shown in Fig. 4. However, detailed descriptions of the effects of hyperoxia on type II cell metabolism will not be presented here. For this discussion, the point is that there is minimum LS and maximum protein in BAL during the worst stage of lung injury.)

The amount of protein and LS in bronchoalveolar lavage is of critical importance because of the inhibitory effects that plasma proteins can have on the biophysical activity of pulmonary surfactant. The P-V mechanical analysis earlier (Figs. 2, 3) have suggested that LS-inhibition (biophysical) is contributing to the pathophysiology of severe pulmonary hyperoxia, at least at the peak of lung injury. The biochemical data in Table 3 are also consistent with conditions needed for such inhibition, and the biophysical activity studies of BAL in the same table suggest the same. Specifically, the maximal protein

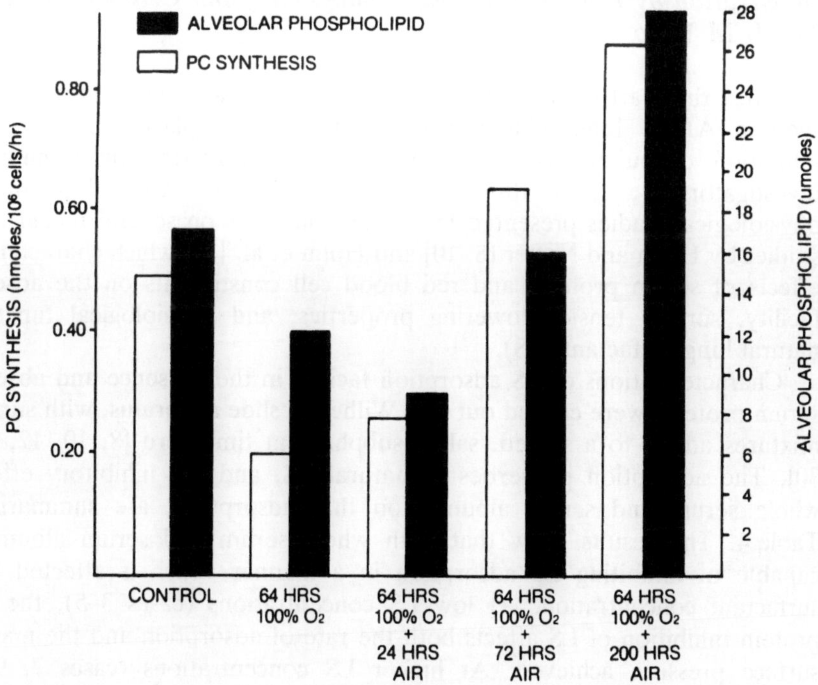

Fig. 4. Comparison of alveolar phospholipid values and phosphatidylcholine *(PC)* biosynthesis by type II cells at various times during and following exposure to 100 % O_2. (Data from Holm et al. [9])

levels and minimal LS-phospholipid levels in Table 3 are seen to correspond to increased BAL surface tensions, as measured on an oscillating bubble apparatus with phospholipid concentrations standardized to avoid dilution artifacts [6, 9–11]. The greatest effect occurs at 24 hours post-exposure, where the lavage minimum surface tension of 30 dynes/cm is clearly incompatible with a physiologically active surfactant.

In summary, the ARDS-like lung injury caused by severe hyperoxic exposure (64 hrs, 100% O_2) in adult rabbits was progressive in its pathology and reached its peak at 24 hours post-exposure. The pathophysiological mechanisms behind the respiratory changes in these animals covered a broad spectrum, including increased alveolar permeability, high molecular weight edema, and LS dysfunction. These mechanisms in turn led to pulmonary functional and mechanical deficits. The data strongly suggest that although alterations in LS metabolism (Type II cell) might contribute to this pathology of oxygen toxicity, at least part of the LS-related effects here were caused by biophysical inhibition of surfactant by alveolar edema constituents.

Specifics of Lung Surfactant Biophysical Inhibition

Inactivation by Plasma Proteins, Hemoglobin, and Cell Membrane Lipids In Vitro

The idea that part of the impaired lung functional and mechanical properties seen in ARDS lung injuries may be due to the plasma protein-induced inhibition of pulmonary surfactant has been suggested by a number of investigators [4, 7, 8, 10–12, 13, 26–28]. The in vitro biophysical and physiological studies presented in this section focus on several recent in vitro studies by Holm and Notter [8, 10] and Holm et al. [12] which characterize the effects of serum proteins and red blood cell constituents on the adsorption facility, surface tension lowering properties, and physiological function of natural lung surfactant (LS).

Characterizations of LS adsorption facility in the presence and absence of serum proteins were carried out in a Wilhelmy slide apparatus, with surfactant mixtures added to a stirred, saline subphase at time zero [8, 10, 12, 21–23, 30]. The adsorption properties of natural LS, and the inhibitory effects of whole serum and serum albumin on that adsorption, are summarized in Table 4. The results show that both whole serum and serum albumin are capable of inhibiting LS adsorption in a manner which is affected by the surfactant concentration. At low LS concentrations (cases 3–5), the serum protein inhibition of LS affects both the rate of adsorption and the maximum surface pressure achieved. At higher LS concentrations (cases 7, 9), the inhibition primarily affects adsorption rate, and the maximum surface pressure reached over time does not change. Indeed, if LS concentration is raised high enough, plasma protein inhibitory effects can be completely abolished [8, 12].

The protein inhibitory effects shown in Table 4 are not restrictive to plasma or serum proteins alone. For example, the representative adsorption isotherms

Table 4. Effects of serum albumin on LS adsorption in vitro

Case no.	Surfactants	Phospholipid concentration (mg/ml)	Protein concentration (mg/ml)	Adsorption surface pressure[a] (dynes/cm) 0 min	5 min	10 min	15 min	20 min
1	Albumin/serum	–	0.025–1.875	14	17	17	17	17
2	LS	0.063	–	17	46	47	47	47
3	LS + albumin/serum	0.063	0.375	17	22	34	43	45
4	LS + albumin/serum	0.063	1.875	18	22	25	38	45
5	LS + albumin/serum	0.063	1.875	18	21	23	24	27
6	LS	0.125	–	45	47	47	47	47
7	LS + albumin/serum	0.125	1.25–2.5	19	25	42	46	47
8	LS	0.300	–	46	47	47	47	47
9	LS + albumin/serum	0.300	2.5–5.0	38	45	46	47	47

Surfactant material was added (to the given final concentration shown) at time 0, with this initial time point representing the initial 15 s or less after addition to the stirred subphase [10, 12]. Adsorption results for albumin were essentially equivalent to those found for serum of equal protein concentration. All experiments were done in the presence of 1.4 mM $CaCl_2$ in the subphase, along with 0.15 M NaCl. For cases where a protein concentration range is given (cases 1, 7, 9), adsorption results listed were identical in all experiments performed within that range. Results shown are for 2–5 closely reproducible, independent samples (within 1–2 dynes/cm). See text plus ref. [8, 10, 12] for details.

[a] Data are from Holm and Notter [10] and Holm, Notter, Finkelstein [12], where methods are detailed. Surface pressure is the amount by which surface tension is lowered below that of the pure subphase (70 dynes/cm for water at 37° C).

in Fig. 5 show that hemoglobin, of interest as a hemorrhagic edema component, inhibits LS adsorption in the same manner described in Table 4 for serum and serum albumin. In principle, rapid adsorption to the air-water interface is just one of several biophysical properties necessary for physiologically beneficial effects of a surfactant in lungs [20, 24]. In natural LS, this property is dependent upon specific surfactant apoproteins [20, 21], and it appears from the data in Fig. 5 and Table 4 that plasma protein-apoprotein interactions may be responsible in part for protein-inhibitory effects on LS.

In addition to adsorption facility, the surface properties of LS films under dynamic compression are of critical importance for pulmonary surfactant to exert its physiologic functions. This is because, for example, that under equilibrium conditions, surfactant mixtures such as LS cannot reduce surface tension below the equilibrium spreading pressure of fluid phospholipids, i.e. approximately 20 dynes/cm (e.g. [24]). However, for full physiological activity, LS films must decrease surface tension further, certainly down into a range well-below 10 dynes/cm; this requires dynamic interfacial compression. Thus, the effects of plasma proteins and other potential inhibitory agents on LS *dynamic* surface tension lowering was measured using an oscillating bubble apparatus developed by Enhorning [6]. The oscillating bubble apparatus measures a combination of adsorption, dynamic surface tension lowering, and

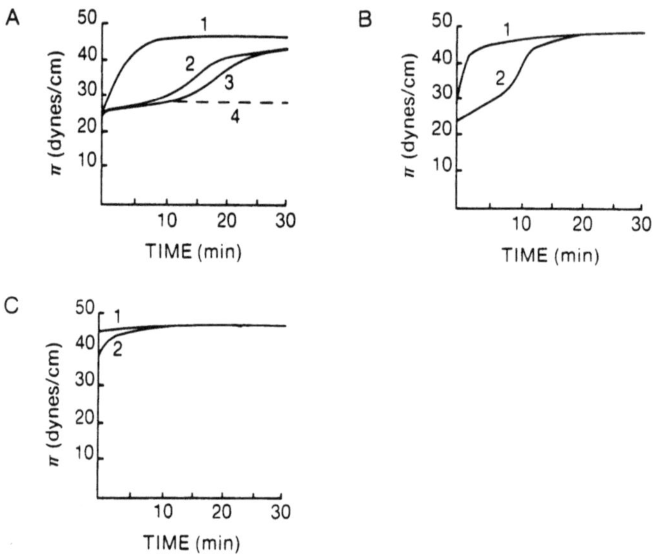

Fig. 5A–C. Effects of hemoglobin on LS adsorption properties (from Holm and Notter [8]). Representative isotherms are shown for the cases below. All experiments were done with an 80-ml subphase containing 0.15 M NaCl and 1.4 mM Ca^{++}. **A** LS, 5 mg phospholipid *(1)*; LS, 5 mg phospholipid plus 25 mg hemoglobin *(2)*; LS, 5 mg phospholipid plus 50 mg hemoglobin *(3)*; hemoglobin, 25–200 mg protein *(4)*. **B** LS, 10 mg phospholipid *(1)*; LS, 10 mg phospholipid plus 50–200 mg hemoglobin *(2)*. **C** LS, 25 mg phospholipid *(1)*; LS, 25 mg phospholipid plus 200–300 mg hemoglobin *(2)*

respreading at cycling rates, temperatures, and area changes relatively reflective of the lung in vivo.

Table 5 shows the minimum surface tension achieved after 20 minutes of bubble pulsation in aqueous dispersions of LS in the presence and absence of albumin, serum, or hemoglobin. As with adsorption inhibition, it is found that these proteins can inhibit LS dynamic surface activity dramatically, but that this inhibition can again be overcome by increasing LS concentration. The results in Table 5 display that the LS concentration to abolish inhibition varies depending upon the inhibitor protein used, but the pattern is conceptually identical. At low enough LS concentrations (e.g. 0.5 μmoles LS per ml), serum, hemoglobin, and albumin are all able to prevent the surfactant mixture from reaching surface tensions below 21 dynes/cm, a value which is incompatible with a physiologically active lung surfactant [20, 24]. However, at sufficiently high LS concentrations, the inhibitory effects of these blood and plasma constituents on LS dynamic surface activity were abolished, even when the molar ratio of protein to phospholipid was increased.

Pulmonary Mechanical Deficits in Intact Lungs

Because the effects of pulmonary surfactant on respiratory physiology are directly related to its biophysical properties, molecules which are able to inhibit

Table 5. Effects of exogenous proteins on LS dynamic surface tension lowering in vitro

Inhibitor	LS concentration (mg/ml)	Protein concentration (mg/ml)	Minimum surface tension (dynes/cm)
None	0.5	–	< 1
Albumin/serum	0.5	5–200	21
Hemoglobin	0.5	10–200	25
None	1.0	–	< 1
Albumin/serum	1.0	5–200	< 1
Hemoglobin	1.0	10–200	25
None	2.0	–	< 1
Albumin/serum	2.0	5–200	< 1
Hemoglobin	2.0	25–200	25
None	10.0	–	< 1
Albumin/serum	10.0	5–200	< 1
Hemoglobin	10.0	25–200	< 1

LS biophysical data are from Holm and Notter [8, 10] and Holm, Notter and Finkelstein [12]. The data express LS concentration and inhibitor (albumin, hemoglobin) concentration as dispersed in 0.15M NaCl prior to study for dynamic surface tension lowering under dynamic cycling (20 cycles/min, 50% area change on compression, 37° C and 90%–100% humidity) [8, 10, 12]. The minimum surface tension reached at end-compression is given; a value above 10 dynes/cm is indicative (conservatively) of a surfactant without high activity in lungs [e.g. 24]. See text and references for details.

LS biophysically (e.g. blood and plasma constituents) should also be directly detrimental to lung mechanical parameters. Moreover, given the concentration-dependent inhibition behavior above [8, 12], any such effects on lung mechanics should be reversible by providing sufficient exogenous surfactant supplementation within the lung. These postulated behaviors were thus studied directly by Holm and Notter [8] using the excised lung mechanical model developed by Bermel et al. [2].

Figure 6 shows the results of adding inhibitor substances (albumin, hemoglobin) to normal and partially surfactant-deficient excised lungs. Curve A of Fig. 6 (top) shows the quasi-static deflation limb of a pressure-volume (P-V) curve of a freshly excised, surfactant-sufficient lung, with volume normalized as percent TLC. Curve B then illustrates the detrimental effects of intratracheally-instilled protein on P-V mechanics. The decrease in mechanics (increased recoil forces) shown is not dramatic, but it was found to be reproducible and consistent [8]. Importantly, Curve C in Fig. 6 (top) shows that the instillation of exogenous surfactant [calf lung surfactant extract (CLSE)] into these lungs reverses almost completely the damaging effects of the proteins on lung mechanics. The CLSE preparation used in Fig. 6 has been shown in extensive previous studies to have high biophysical and physiological activity (e.g. 5, 8, 10, 12, 20–23, 30]. It has also been used in several successful clinical trials of surfactant replacement therapy in infants with RDS [e.g. 15, 29], and is currently in use in expanded RDS clinical studies.

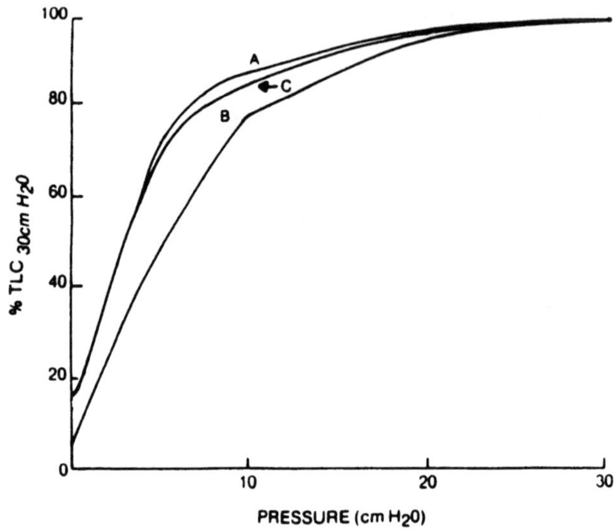

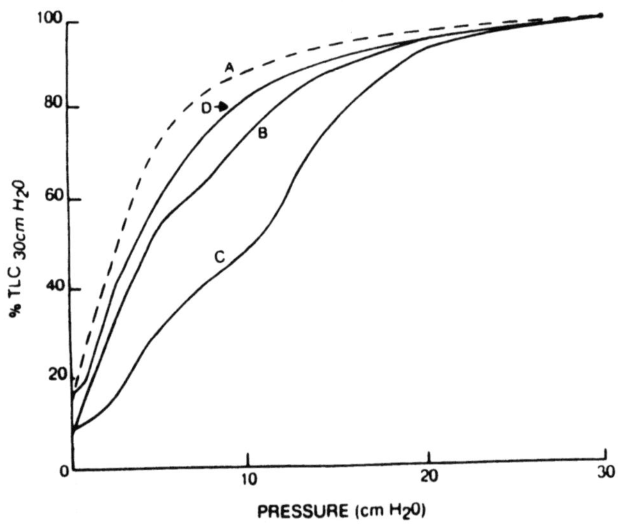

Fig. 6. *Top* Effects of albumin and hemoglobin on pressure-volume deflation curves of excised rat lungs, and reversal by exogenous surfactant treatment. (Data from Holm and Notter [8]). *A*, normal lung (surfactant sufficient, no inhibitors), *B*, lung instilled with 400 mg albumin or hemoglobin; *C*, lung instilled with 400 mg albumin or hemoglobin and then supplemented with 25 mg CLSE. *Bottom* Effects of albumin and hemoglobin on pressure-volume deflation curves of partially surfactant-deficient excised rat lungs. *A*, normal lung (surfactant sufficient, no inhibitors); *B*, partially surfactant-deficient lung (one lavage); *C*, partially surfactant-deficient lung instilled with 400 mg albumin or hemoglobin; *D*, partially surfactant-deficient lung instilled with 400 mg albumin or hemoglobin and then supplemented with 25 mg CLSE

To test the dependence of protein inhibitory effects on surfactant concentration, the (P-V) mechanical study described above was repeated using lungs that had been made partially surfactant deficient by uniformly lavaging one time with 15 ml saline. Curve A in Fig. 6 (bottom) shows the P-V deflation limb of a freshly excised surfactant-sufficient lung, while curve B is the reproducible partially-deficient P-V curve obtained after the single, standardized isovolumetric saline lavage [8]. Curve C then shows that the addition of albumin or hemoglobin to the partially surfactant deficient lung has an even greater detrimental effect on lung compliance than when they were instilled into the surfactant-sufficient lung. However, in spite of the larger mechanical change, curve D in Fig. 6 (bottom) shows that the deleterious effects of partial lung lavage and surfactant inhibition on P-V mechanics are almost completely reversed by 25 mg of exogenous surfactant (CLSE) supplementation.

The concept, presented above, that the protein-induced inhibition of lung surfactant activity is dependent upon the surfactant phospholipid (and apoprotein) concentration may help to explain the time course of the progressive pathology found in hyperoxic lung injury in rabbits as described in Tables 2, 3 and Figs. 2, 3 [10,11]. Specifically, a certain "threshold" needs to be reached where alveolar LS reaches a low enough level, and alveolar protein a high enough level, to induce a functional surfactant-deficient state in the presence of inhibitor molecules. Such a state then acts to exacerbate the existing pulmonary pathology, unless it is counteracted by some intervention such as exogenous surfactant replacement. The in vitro biophysical studies in this section [8, 12], as well as the complimentary intact lung mechanical studies [8], give a fundamental rationale for surfactant-replacement therapy in ARDS in vivo. Consequently, the following section examines directly the efficacy of exogenous surfactant supplementation in mitigating the pathological changes seen in the ARDS lung injury induced by severe pulmonary hyperoxia.

Exogenous Surfactant Supplementation in ARDS-Type Hyperoxic Lung Injury

One way in which the existence of LS dysfunction in ARDS-like hyperoxic lung injury is to see if exogenous surfactant supplementation can mitigate the pathological features present, and experiments of this type were carried out by Matalon, Holm and Notter [18]. For surfactant supplementation experiments, the same CLSE exogenous surfactant extract described in the previous section was utilized. For surfactant supplementation in ARDS, where significant alveolar edema may be present, it is necessary to design therapeutic protocols to optimize alveolar delivery. To accomplish this, supplemental CLSE was given after the formation of alveolar permeability changes, but prior to the development of gross alveolar edema [18]. This was done by two surfactant instillations in vivo, given at 0 and 12 hours post-exposure according to the protocol in Fig. 7; each dose consisted of 125 mg of CLSE dispersed in 6 ml sterile saline. Pulmonary physiological and bronchoalveolar lavage variables

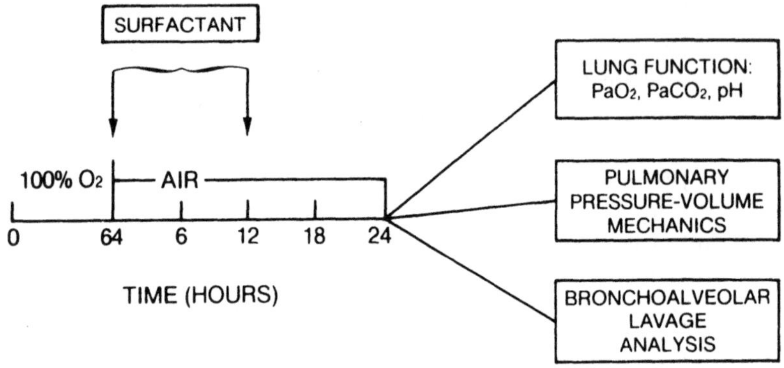

Fig. 7. Protocol for surfactant instillation in the in vivo hyperoxic lung injury model. One group of control animals received an equal volume of saline-placebo treatment at the same times, and this group was equivalent to a second control group with no treatment after hyperoxic exposure

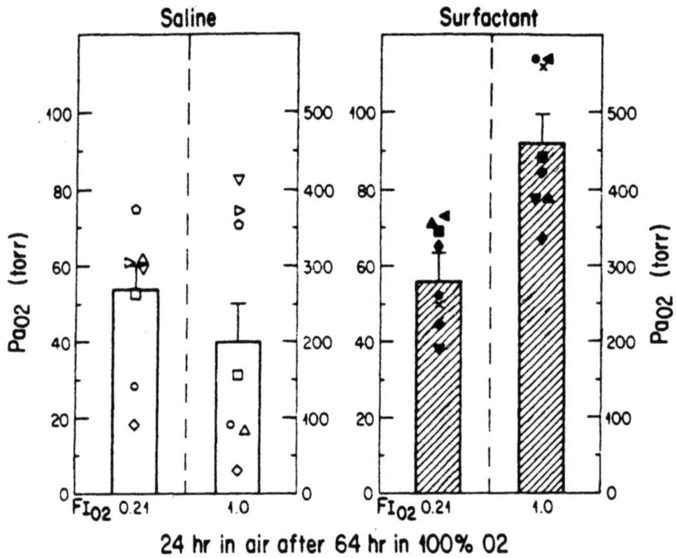

Fig. 8. Arterial O_2 partial pressure (PaO_2) in saline-*(left)* and surfactant-*(right)* treated rabbits. PaO_2 values were measured at 24 h after end of exposure while animals were breathing either air or 100 % O_2 for 30 min (latter only at sampling). Each *symbol* represents a different rabbit, with the group means and standard errors also shown. (Data from Matalon, Holm, and Notter [18])

were then assessed at 24 hours post-exposure, the time point at which the most severe detriments in pulmonary function and lung mechanics had been found in the absence of surfactant supplementation (Table 2, Fig. 2).

Figure 8 shows the PaO_2 values for surfactant and saline (placebo) treated rabbits, measured at 24 hours post-exposure, while animals were breathing either air or 100 % O_2 (the latter for 30 min during measurements only). Both surfactant- and saline-treated rabbits exhibited the same degree of arterial hypoxemia while breathing room air in Fig. 8. However, the surfactant treated animals had significantly higher PaO_2 values while breathing 100% O_2. One explanation for these results is that the hypoxemia in the surfactant-treated group resulted from ventilation-perfusion mismatching, whereas the hypoxemia in the saline-treated animals was more likely caused by a large degree of right-to-left shunt secondary to pulmonary atelectasis.

In addition to beneficial effects on atelectasis, the administration of exogenous CLSE surfactant was also found to decrease the amount of pulmonary edema and the amount of alveolar protein. As shown in Table 6, the surfactant-treated rabbits had an eight-fold increase in lavagable phospholipid, a 40% decrease in the amount of lavagable protein, and a 13% decrease in lung wet/dry weight ratios compared with saline (placebo) animals, or with untreated control animals exposed to 100% oxygen for 64 hours and returned to room air for 24 hours. Also given in Table 6 are minimum surface tension values, found with the pulsating bubble apparatus, for cell-free lavages from the various groups. As shown, the minimum dynamic surface tension for surfactant-treated animals was < 1 dyne/cm, compared to the high, physiologically incompatible, values for lavages from saline-placebo or untreated animals. It is probable that the greatly improved surface tension lowering ability in the CLSE-treated

Table 6. Effects of surfactant supplementation on bronchoalveolar lavage (BAL) parameters in hyperoxia

		64 h in 100 % O_2 + 24 h in air		
	Control (unexposed)	Surfactant-treated	Saline placebo	Untreated
Total phospholipid (µmoles/kg)	8.7 ± 1.6	34 ± 4[a, b]	6 ± 1	4.5 ± 0.6[b]
Total protein (mg/kg)	7.4 ± 3.5	36 ± 3[a, b]	60 ± 2[b]	56 ± 3.5[b]
Wet/dry weight	4.1 ± 0.3	5.6 ± 0.1[a, b]	6.3 ± 0.3[b]	6.2 ± 0.2[b]
Minimum BAL surface tension (dynes/cm)	1 ± 0.2	1 ± 0.2[a]	26 ± 2[b]	30 ± 2[b]

These data, from Matalon, Holm and Notter [18], show the beneficial effects of treating hyperoxia-injured rabbits with an active exogenous surfactant extract CLSE. Bronchoalveolar lavage (BAL) was obtained by methods detailed previously [11, 18]. Groups shown are normoxic controls (unexposed), hyperoxic exposed treated with CLSE surfactant, hyperoxic exposed saline (placebo) group, and hyperoxic exposed but untreated group. The latter two groups are statistically equivalent [18]. The minimum surface tension given is that reached on pulsating bubble analysis at a uniform phospholipid concentration of 1 µmole phospholipid/ml for all groups in the table. See text and references for details.

[a] Significantly different from hyperoxic exposed but saline-treated (placebo) group ($p < 0.05$ or less).

[b] Significantly different from unexposed (control) group ($p < 0.05$ or less).

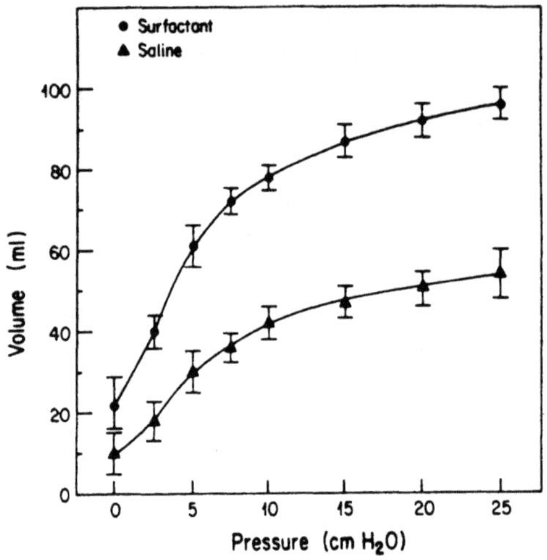

Fig. 9. Deflation limbs of pressure-volume curves for surfactant- and saline-treated rabbits after hyperoxic exposure. (Data from Matalon, Holm, and Notter [18]). The deflation curve for saline (placebo) rabbits after exposure was not significantly different from untreated, exposed rabbits. By contrast, the P-V curve for exposed rabbits, subsequently given exogenous CLSE surfactant, recovered to the point that it was not significantly different from unexposed, normoxic control animals. See text for details

animals reflects the reversal of alveolar surfactant inhibition, although the decreased alveolar protein content itself may also contribute.

The most dramatic result of exogenous surfactant supplementation in hyperoxic lung injury was the effect on pulmonary mechanics. As shown in the P-V deflation curves in Fig. 9, the lungs of surfactant-treated animals had significantly increased TLC and resting volumes compared to those from saline-treated animals. In fact, a comparison of Fig. 9 with Fig. 2 shows that the recovery of P-V mechanics with CLSE supplementation is almost equivalent to normal rabbit lungs. In contrast, the P-V curve from the saline-placebo group was not significantly different from that obtained from animals exposed to 64 hours in 100 % O_2 and returned to room air for 24 hours.

In summary, the instillation of exogenous CLSE surfactant in rabbits exposed to 100 % O_2 for 64 hours prevents the development of abnormal lung mechanics and alveolar collapse, and attenuates the degree of pulmonary edema formation normally seen at 24 hours post-exposure. The beneficial effects of surfactant treatment in ARDS appear to be related to its action in lowering alveolar surface tension. It is this effect which directly lowers the work of breathing and atelectasis, while raising lung compliance. Moreover, by reducing the pressure drop across the air-hypophase interface in alveoli, active surfactant lowers the driving force for pulmonary edema [19, 25]. Alveolar permeability may also be improved in surfactant-sufficient lungs due to more uniform alveolar expansion [14], and this too could help retard edema formation. Finally, the fact that exogenous surfactant supplementation has clearly beneficial effects in mitigating hyperoxic lung injury would seem to show that surfactant-related effects are important in the pathophysiology of this disease. However, further studies are needed to accurately assess the benefits of surfactant therapy in ARDS-like lung injuries in general, a subject which will be discussed further in the next section.

Conclusions and Considerations for Future Studies

The experiments, results, and analyses examined in this article have shown that significant changes in the pulmonary surfactant system occur in hyperoxic lung injury, and that these are important contributors to the progressive ARDS-type pathophysiology of oxygen toxicity. For hyperoxia specifically, the data presented and discussed [8–12, 16–19] clearly show that LS defects (cellular and direct biophysical inactivation) are present in the lung injury pathology. It was this functional deficiency of alveolar surfactant which led to the detrimental pulmonary mechanical changes, and part of the alveolar edema, present at the peak of hyperoxic lung injury. In turn, it was these factors (most specifically the P-V mechanical changes) that were reversed by exogenous surfactant supplementation. Results consistent with this were also found in complementary in vitro biophysical studies, which demonstrated that plasma-derived proteins could detrimentally affect both LS surface activity and excised lung mechanical properties, and that these could be reversed by increased surfactant supplementation. All of these varied aspects relevant for ARDS-type lung injury, involving in vitro biophysical studies, P-V mechanical studies in excised lungs, and physiological studies in adult animals in vivo, provide a self-consistent framework to help understand this complex lung injury, and the role of exogenous surfactant supplementation in its therapy.

Another important point involving surfactant replacement in ARDS is that the pathology of this lung injury is multifocal, and does not involve simply an LS deficiency (quantitative or qualitative) as its only component. Consequently, while surfactant replacement therapy early in the course of neonatal RDS may be providing specific therapy for a primary LS-deficiency, this is not the case in ARDS. As a result, therapeutic expectations need to be realistic: effective treatment of surfactant dysfunction in ARDS will have significant effects on some variables (e.g. P-V mechanics), but the effects on other variables may be less pronounced. Even with this caveat, however, the mitigation of hyperoxic lung injury found by Matalon, Holm and Notter [18] is impressive. The effects of CLSE supplementation on mechanics was essentially a return to normal levels. In addition, the reduction of edema (Table 6), while not complete to control levels, was still found to be significant. Similarly, the improvement in arterial oxygenation (Fig. 8) is promising. It would seem that as appropriate therapeutic protocols (e.g. doses, timing, etc.) are refined, exogenous surfactant replacement therapy in ARDS may prove surprisingly effective.

Although clinical surfactant replacement in ARDS may be promising, it is clearly for the future, at least in terms of developing rational therapy. The discussion to here has stressed the positive features of surfactant-replacement in a *specific* lung ARDS injury, in a *specific* animal species, at a *specific* age (e.g. adults, not neonates). It is particularly important to recognize that the success of surfactant replacement therapy in pulmonary oxygen toxicity followed from an understanding of the pathophysiology of this specific lung injury, and the biophysical behavior and inhibition pattern of natural LS and CLSE surfactants. Once defined [8, 10–12], these biophysical properties of lung surfactants, and their sensitivity and resistance to inhibition, should remain (relatively) predictable for future lung injury analyses. However, this is not necessarily true for the complexities of other ARDS-related animal models or clinical lung injuries, where significant pathophysiological differences may exist.

Although the combination of diverse etiologies to define a common clinical syndrome is attractive, it can be carried too far, and this consideration is relevant for ARDS. Almost certainly there are clinical conditions which have some ARDS characteristics, but also many differences. At least one example of this is in the selected "ARDS" models in Table 1. The in vivo lavage model of small animals (guinea-pigs) for instance, has characteristics of an induced surfactant deficiency, with superimposed lung injury pathology; it has been used as an ARDS model by some researchers, as a model of ventilated surfactant-deficient lungs by others, and so on [e.g. 20]. Indeed, even the hyperoxic lung injury here must be assessed with caution for relations to classic ARDS. For example, the ARDS pathology found in severe hyperoxia in adult rabbits is not identical in all its characteristics for other animal species, including even some variations (e.g. the timescale of effects) between neonatal vs. adult rabbits. Similarly, the basic pathology of hyperoxic lung injury will certainly vary with the oxygen exposure (duration and oxygen level) used. From these factors, it could be argued that before surfactant therapy is considered in any given ARDS-related animal model (or clinical illness if feasible), fundamental pathophysiological studies are first needed to obtain quantitative data to verify that surfactant-dysfunction is present. Alternatively, the existence of such pathology could be indirectly inferred, of course, by using exogenous surfactant replacement on a purely empirical basis in all ARDS-related lung injuries; if the therapy works, then surfactant dysfunction was probably present. Unfortunately, such an empirical approach generally provides no fundamental insights for extensions of knowledge.

In the studies analyzed here, exogenous surfactant replacement in rabbits was employed after the existence of surfactant-deficient behavior was inferred from P-V mechanics and BAL surface activity data. Its success thus served two purposes: first, the therapy mitigated the pathology that was present; and second, the efficacy of the therapy gave a confirmation that endogenous LS dysfunction was present in the severe hyperoxic lung injury. This latter interpretation, however, was itself stronger and more reliable because of its internal consistency with the other independent measures of decreased BAL biophysical activity, and pulmonary P-V mechanical deficits. Evidence that LS can become deficient in another model of ARDS has been supplied by Berry, et al. [3] who performed bilateral cervical vagotomies on adult rabbits to induce an ARDS-like lung injury. Their data also demonstrated an inhibition of surfactant function, and most importantly, a significant improvement in lung function and compliance in these vagatomized rabbits following treatment with natural surfactant.

Finally, as development of surfactant replacement therapy for ARDS proceeds, there will be many clinical questions to answer, after the fact that a component of surfactant-deficiency has been established for the specific lung injury at question. Aside from the choice of optimal exogenous surfactant (i.e.

one which is active plus resistant to inhibition, such as CLSE), such questions include determinations of the most effective surfactant dose amount and installation technique, of the necessity of repeat doses, the timing of initial and repeat treatments, and so on. Also, in terms of surfactant replacement in ARDS-linked lung injury syndromes, these involve more than just a pulmonary surfactant dysfunction. Therefore, as noted earlier, the effects of surfactant supplementation can be expected to mitigate, but not abolish completely, the multifaceted pulmonary pathology present. Moreover, it is likely that maximum benefits of surfactant therapy in ARDS will be seen if the surfactant is given early, before the multifocal pathology of lung injury becomes fulminant.

Acknowledgements. This work was supported in part by grant HL-37388 from the National Institutes of Health (NIH), and by a Pulmonary SCOR (HL-36543) at the University of Rochester (Neonatology).

References

1. Avery ME, Mead J (1959) Surface properties in relation to atelectasis and hyaline membrane disease. Am J Dis Child 97:517–523
2. Bermel M, McBride JT, Notter RH (1984) Lavaged excised rat lungs as a model of surfactant-deficiency. Lung 162:99–113
3. Berry D, Ikegami M, Jobe A (1986) Respiratory distress and surfactant inhibition following vagotomy in rabbits. J Appl Physiol 61:1741–1748
4. Clark JM, Lambertsen CJ (1971) Pulmonary oxygen toxicity: a review. Pharmacol Rev 23:37–133
5. Egan EA, Notter RH, Kwong MS, Shapiro DL (1983) Natural and artifical lung surfactant replacement therapy in premature lambs. J Appl Physiol 55:875–883
6. Enhorning G (1977) Pulsating bubble technique for evaluating pulmonary surfactant. J Appl Physiol 43:198–203
7. Gross NJ, Smith DM (1981) Impaired surfactant phospholipid metabolism in hyperoxic mouse lungs. J Appl Physiol 51:1198–1203
8. Holm BA, Notter RH (1987) Effects of hemoglobin and red blood cell membrane lipids on the biophysical and physiological activity of pulmonary surfactant. J Appl Physiol 63:1434–1442
9. Holm BA, Notter RH, Finkelstein JN, Matalon S (1987) Changes in type II pneumocytes during injury and recovery from hyperoxia. Am Rev Respir Dis 135:15A
10. Holm BA, Notter RH (1986) Pulmonary surfactant effects in sublethal hyperoxic lung injury. In: Taylor A, Matalon S, Ward P (eds) Physiology of oxygen radicals. APS, Bethesda, pp 71–86
11. Holm BA, Notter RH, Siegle JC, Matalon S (1985) Pulmonary physiological and surfactant changes during injury and recovery from hyperoxia. J Appl Physiol 59:1402–1409
12. Holm BA, Notter RH, Finkelstein JN (1985) Surface property changes from interactions of albumin with natural lung surfactant and extracted lung lipids. Chem Phys Lipids 38:287–298
13. Hopewell PC, Murray J (1977) The adult respiratory distress syndrome. In: Shibel EM, Moser KM (eds) Respiratory emergencies. Mosby, St Louis, p 101
14. Jobe A, Machiko I, Jacobs H, Jones S, Conaway D (1983) Permeability of premature lamb lungs to protein and the effect of surfactant on that permeability. J Appl Physiol 55:169–176
15. Kwong MC, Egan EA, Notter RH, Shapiro DL (1985) Double-blind clinical trial of calf lung surfactant extract for the prevention of hyaline membrane disease in extremely premature infants. Pediatrics 76:585–592

16. Matalon S, Egan EA (1981) Effects of 100 % O_2 breathing on permeability of alveolar epithelium to solute. J Appl Physiol 50:859–863
17. Matalon S, Egan EA (1984) Interstitial fluid volumes and albumin spaces in pulmonary oxygen toxicity. J Appl Physiol 57:1767–1772
18. Matalon S, Holm BA, Notter RH (1987) Mitigation of hyperoxic injury by exogenous surfactant supplementation. J Appl Physiol 62:756–761
19. Nieman GF, Bredenberg CE (1985) High surface tension pulmonary edema induced by detergent aerosol. J Appl Physiol 58:129–136
20. Notter RH, Shapiro DL (1987) Lung surfactants for replacement therapy: biochemical, biophysical and clinical aspects. Clin Perinatol 14:433–479
21. Notter RH, Shapiro DL, Ohning B, Whitsett J (1987) Biophysical activity of synthetic phospholipids combined with purified lung surfactant 6,000 dalton apoprotein. Chem Phys Lipids 44:1–17
22. Notter RH, Penney DP, Finkelstein JN, Shapiro DL (1986) Adsorption of natural lung surfactant and phospholipid extracts related to tubular myelin formation. Pediatr Res 20:97–101
23. Notter RH, Egan EA, Kwong MS, Holm BA, Shapiro DL (1985) Lung surfactant replacement in premature lambs with extracted lipids from bovine lung lavage: effects of dose, dispersion technique, and gestational age. Pediatr Res 19:570–577
24. Notter RH, Finkelstein JN (1984) Pulmonary surfactant: an interdisciplinary approach. J Appl Physiol 57:1613–1624
25. Raj JU (1985) Alveolar liquid pressure in excised air-inflated lungs of mature and immature fetal rabbits (abstr). Physiologist 28:349
26. Rinaldo JE, Rogers RM (1982) Adult respiratory distress syndrome, changing concepts of lung injury and repair. N Engl J Med 15:900–909
27. Robertson B, van Golde LMG, Batenburg JJ (eds) (1984) Pulmonary surfactant. Elsevier, Amsterdam
28. Seeger W, Stohr G, Wolf HRD, Neuhof H (1985) Alteration of surfactant function due to protein leakage: special interaction with fibrin monomer. J Appl Physiol 58:326–338
29. Shapiro DL, Notter RH, Morin FC, Deluga K, Golub LM, Sinkin R, Weiss K, Cox C (1985) A double-blind randomized clinical trial of a calf lung surfactant extract administered at birth to very premature infants for prevention of the respiratory distress syndrome. Pediatrics 75:593–599
30. Whitsett J, Ohning BL, Ross G, Meuth J, Weaver T, Holm BA, Shapiro DL, Notter RH (1986) Hydrophobic surfactant-associated protein in whole lung surfactant and its importance for biophysical activity in lung surfactant extracts used for replacement therapy. Pediatr Res 20:460–467

Surfactant Replacement in Respiratory Failure Induced by Aspiration of Hydrochloric Acid in Rabbits

T. Kobayashi, M. Ganzuka, T. Ueda, J. Taniguchi, and S. Murakami

Introduction

Aspiration of the acid gastric contents is a common clinical occurrence associated with high morbidity and mortality. Acids with a hydrogen ion concentration (pH) of less than 2.5 produce degeneration of the alveolar cells, infiltration of leukocytes, accumulation of proteinaceous intraalveolar edema, atelectasis, a decrease in lung compliance, and an impairment of gas exchange [8, 20, 21, 23, 28]. These latter pathophysiological events might reflect abnormalities of the lung surfactant system, at least in part, since surfactant is a substance which stabilizes the alveoli [3].

Data from several recent studies indicate the efficacy of surfactant replacement in surfactant-deficient prematurely delivered animals [14, 15, 19], in newborn babies with respiratory distress syndrome (RDS) [4, 7, 9, 17, 26], and in adult animals with surfactant depletion induced by repeated lung lavage [1, 18]. Under experimental and clinical conditions, exogenous surfactant replacement performed through the airway strikingly improved alveolar air expansion, lung compliance and gas exchange.

As far as we know, surfactant replacement has not been tested as a treatment for respiratory failure caused by the aspiration of acids. Moreover, participation of the surfactant system in this respiratory failure has yet to be clearly elucidated. In this study, we evaluated the effects of surfactant replacement in adult rabbits with respiratory failure induced by aspiration of hydrochloric acid into the lung. In addition, we examined the characteristics of the lung edema fluid, with particular reference to their surfactant inhibitory actions.

Materials and Methods

Preparation of Surfactant. In this study a type of porcine natural surfactant (Surfactant CK) was used. The isolation methods, chemical composition, surface properties and in vivo activity of this surfactant have been previously described [16]. In brief, the lungs of recently slaughtered pigs were lavaged with physiological saline. The combined lavage fluid was centrifuged for 10 min at 200 × g to remove cell debris, and the supernatant was further centrifuged for 1 hr at 2000 × g at 4° C. Pellets of crude lung surfactant were extracted by means of a chloroform-methanol mixture (2 : 1, v : v), and were washed in a 0.5% sodium chloride solution according to the method described by Folch et al. [5]. A phospholipids fraction, which also contains about a 1% low molecular weight (smaller than 16,000 Daltons) of hydrophobic proteins, was separated by acetone precipitation, and was used as Surfactant CK.

Preparation of Animals. Adult rabbits weighing between 2.8 and 3.5 kg were anesthetized with an intramuscular injection of ketamine hydrochloride (10 mg/kg). Pentobarbital sodium (single dose of 20 mg/kg) was intermittently given by intravenous injection when additional anesthesia was necessary. A tracheotomy was performed and an endotracheal tube was inserted, after which the animals were mechanically ventilated by use of a pump (respiration pump, model 618, Havard Apparatus, USA). The system, as shown in Fig. 1, delivered a time-cycled, pressure-preset ventilation. Throughout the experiment, pure oxygen was used for the inspiratory gas, and the conditions were set at a 20 cm H_2O peak inspiratory pressure, a 30 beats/min frequency, and a 0.66 inspiratory-to-expiratory ratio. Unless specified, these respiratory conditions were not changed until end of the experiment.

The femoral artery was cannulated to monitor blood pressure and to draw blood samples. A lactated Ringer's solution was continuously infused through an auricular vein at a rate of 2.0 ml/kg/hr. Rectal temperature was maintained at 35.6–37.8° C by the intermittent use of an electric blanket. The animals were

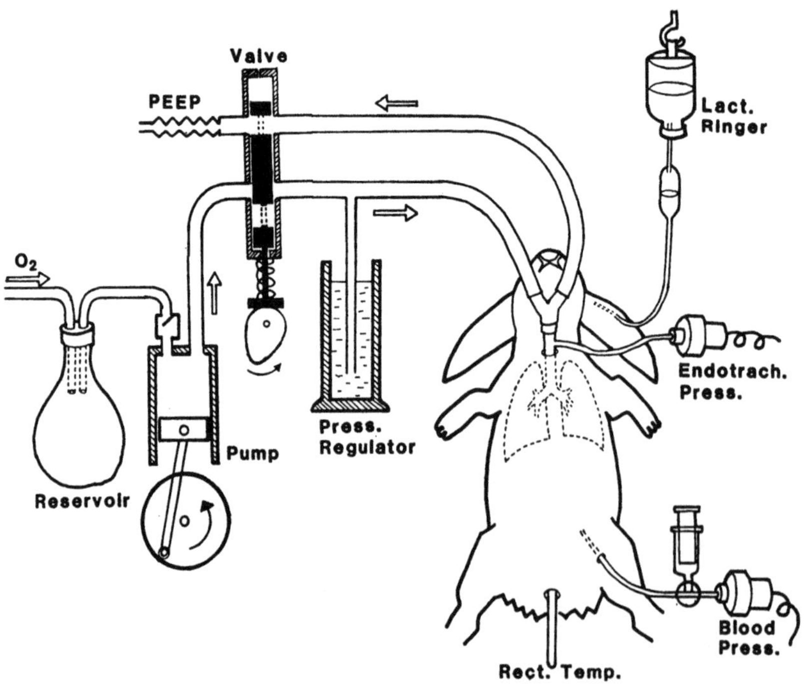

Fig. 1. Experimental design. Peak inspiratory pressure is regulated by submerging the side tube into a column of water. *PEEP,* positive end-expiratory pressure

kept paralyzed by an intermittent intravenous injection of a muscle relaxant (pancuronium bromide, single dose of 0.3 mg/kg). After these preparations, the arterial blood gases for oxygen tension (Po_2), the carbon dioxide tension (PCO_2), and the pH and base excess of the base line period were analysed by conventional methods (ABL 2, Radiometer, Denmark). The base line value of the lung-thorax compliance was also measured (see below).

Aspiration of Hydrochloric Acid. After determination of the base line values, respiratory failure was induced by aspiration of hydrochloric acid (0.1 N, pH = 1.0, aspirated total amount = 2.5 ml/kg). A polyethylene catheter (inside diameter, 1.0 mm) was inserted into the endotracheal tube, and the tip was placed about 1 cm above the carina. 1.5 ml/kg of hydrochloric acid, which was warmed to 37° C, was then injected into the lung through the catheter as the animals were held in a right lateral position. Two minutes later, the animals were held in a left lateral position and 1.0 ml/kg of this acid was injected in the same manner. After this aspiration procedure, thc animals were mechanically ventilated in a supine position. Positive end-expiratory pressure (PEPP) was not applied until 30 min after aspiration.

Treatment and Assignment of Animals. Twenty-five minutes after aspiration of the hydrochloric acid the edema fluid, which appeared in the trachea, was drawn by suction with a fine catheter. The volume and pH of this fluid was measured immediately, and the fluid stored in glass tubes at –20° C for further analysis. Thirty minutes after aspiration (5 min after the tracheal suction), arterial blood gases and lung-thorax compliance of the postaspiration period were measured. The animals were then randomly assigned to one of the three groups shown in Table 1.

In the Lavage+sact group, both alveolar lavage and surfactant replacement were performed. For the alveolar lavage, a physiological saline at 37° C was poured into the trachea at a volume of 25 ml/kg, which roughly corresponds to 80% of the total lung capacity, and this saline was withdrawn after 45–55 sec. Immediately after this lavage, 8 ml of Surfactant CK, suspended in a physiological saline at a concentration of 20 mg/ml, was instilled into the endotracheal tube through a polyethylene catheter (inside diameter, 1.0 mm) while holding the animals in many positions. The animals were then manually ventilated, using an anesthesia bag with pure oxygen for 1 min to facilitate entry of the surfactant. Peak inspiratory pressure and frequency of the manual ventilation were 30–35 cm H_2O and 45–50 beats/min, respectively.

Table 1. Treatment of rabbits

Group	*n*
Non-treated	7
Lavage only	7
Lavage plus surfactant replacement	7

In the Lavage-only group, alveolar lavage was performed, but no surfactant was replaced. The animals of this group were manually ventilated in the same manner as the former group.

In the Non-treated group, neither alveolar lavage nor surfactant replacement was performed, but the animals were manually ventilated in the same manner as the other two groups.

After these procedures, all animals were mechanically ventilated with a 5 cm H_2O PEEP, but other ventilatory conditions were not changed. The arterial blood gases in the post-treatment period were analyzed at 15, 30, 90, 120 and 150 min after treatment. The lung-thorax compliance was measured at 30 and 120 min.

Measurement of Lung-Thorax Compliance. The animals were disconnected from the ventilator for 6 sec, and the lungs were inflated by injecting 50 ml of air (full inflation) while monitoring the endotracheal pressure. After maintaining the full inflation volume for about 8 sec, half of the injected air (25 ml) was withdrawn (semi-deflation). The semi-deflation volume was also maintained for about 8 sec to obtain an equilibrated pressure. On the assumption that 25 ml of the injected air remained in the lung, the compliance was calculated by the following equation:

compliance = 25 ml/pressure (cm H_2O) at semi-deflation

Histological Examination. At the end of the observation period (150 min after the treatments), the lung edema fluid was drawn from the trachea, and the animals were killed by an overdose of pentobarbital sodium. The thorax was widely exposed by a resection of the anterior chest wall, and the lungs were expanded with air for 5 min at a transpulmonary pressure of 30 cm H_2O. The pressure was then lowered to 10 cm H_2O and this pressure was maintained while the lungs were perfused for 3 hr with 4% neutral formalin at a pressure of 65 cm H_2O. Paraffin sections of these fixed lungs were stained with hematoxylin and eosin.

Surfactant Inhibitory Action of Lung Edema Fluid. The pH of the lung edema fluid was analysed by the same method as that used to analyse the arterial blood. Protein concentration in the edema fluid, from which clots had been removed, was determined by the biuret test (Clinalyzer, JCA MS-24, Nihon Denshi, Japan).

To assess the surfactant inhibitory action, Surfactant CK was suspended in the lung edema, from which clots had been removed, at a concentration of 10 mg/ml. Subsequently, the surface properties of the mixture were determined at 37° C by the pulsating bubble technique (Surfactometer, model ASC, Surfactometer International, Canada). The radius of the bubble, pulsating in the mixture, oscillated between 0.40 and 0.55 mm at the rate of 30 cycles/min. The surface tensions at minimum and maximum bubble sizes (γ min and γ max) after 5 min of pulsation were calculated from the pressure gradient across the bubble (P) and the radius (r) according to the Laplace formula $P = 2\ \gamma/r$. The surface properties of the Surfactant CK, diluted with a physiological saline to a concentration of 10 mg/ml, and those of the edema fluid also were also assessed in the same way.

Statistical Analysis. The significance in the differences of the survial rate was determined by the Chi-square test. All other data are presented as means ± SD, and the significance of these differences were assessed by one-way or two-way analysis of variance. Scheffé's method for multiple comparison was used to determine where significant differences had occurred. P values less than 0.05 were considered significant.

Results

Volume and Character of Lung Edema Fluid. The volume of the lung edema fluid drawn at 30 min after aspiration of hydrochloric acid was 6.0 ± 1.6 ml with no significant differences between the three groups (Table 2). The volumes of the Lavage-Only and the Lavage+Sact (surfactant replacement) groups at the end of observation period (150 min after treatment) were 5.2 ± 1.6 ml and 6.0 ± 3.7 ml, respectively. There was no statistical difference in the volume between the two groups. The volume of the Non-Treated group at the end of the observation period was not measured as 6 out of the 7 animals died before that period.

The edema fluid showed a higher pH than that of the arterial blood. This was caused in part by loss of carbon dioxide during collection of the edema fluid by suction. Even if the Pco_2 is considered as being low, it is obvious that the hydrochloric acid had already been neutralized 25 min after the aspiration.

Protein concentration in the edema fluid was very similar to that of the serum (3.5 ± 0.4 g/dl). There were no statistical differences in the protein concentration between the groups.

Arterial Po_2 and Survival Rate. The mean arterial Po_2 at the base line period of all animals was 586 ± 28 Torr, and the value at 30 min after aspiration of hydrochloric acid (post-aspiration period) was 68 ± 23 Torr with no significant differences between the groups (Fig. 2).

In the Non-Treated group, the Po_2 did not increase with application of a 5 cm H_2O PEEP and 1 min of manual ventilation. The Po_2 remained at less than

Table 2. Lung edema 25 min after aspiration of hydrochloric acid

Drawn volume (n = 21)	pH[a] (n = 8)	Protein concentration (n = 9)
6.0 ± 1.6 ml	7.62 ± 0.15	3.5 ± 0.5 g/dl

Values are mean ± SD.
[a] PCO_2 of the samples was less than 10 Torr when the pH was measured.

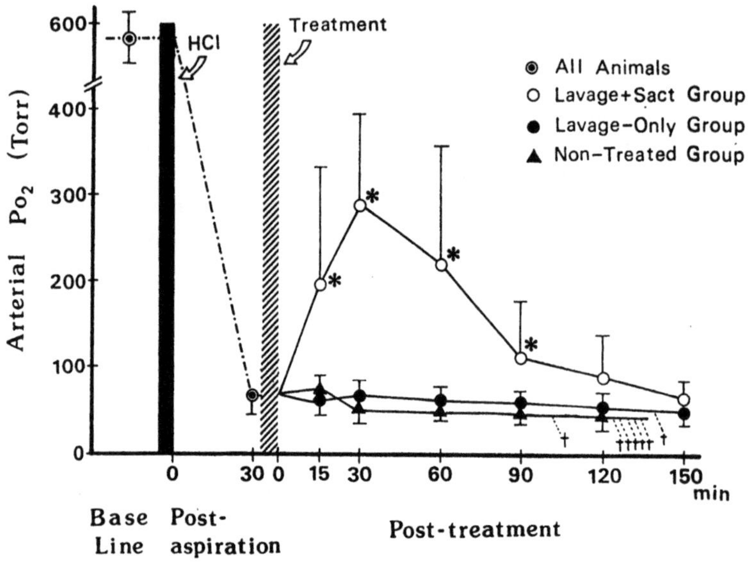

Fig. 2. Arterial Po_2 (means ± SD). *HCl,* aspiration of hydrochloric acid. Values of the base line and post-aspiration period are from all rabbits. A 5 cm H_2O positive end-expiratory pressure was applied after the treatment in all the groups. *Asterisk* indicates significant difference ($P < 0.05$) from the other groups. Value of the Non-Treated group at 150 min of post-treatment period was omitted because 6 of the 7 rabbits died. *Crosses* indicate rabbits that died. (Table 1 shows treatment groups)

70 Torr, and 6 of the 7 animals died during the observation period (150 min after treatment). In the Lavage-Only group, the Po_2 values of the post-treatment period were similar to those of the Non-Treated group, and 1 of the 7 animals died during the observation period. Conversely, the Po_2 of the Lavage+Sact group rose significantly to 196 ± 138 Torr at 15 min and 288 ± 108 Torr at 30 min after the surfactant replacement. Although the Po_2 tended to decrease 60 min after the replacement, values of the Lavage+sact group were significantly higher than those of the other groups until 90 min after replacement. None of the animals in the Lavage+Sact group died during the observation period. The survival rate of the Lavage-Only and the Lavage+Sact group was significantly better than that of the Non-Treated group.

Arterial Pco_2, pH and base excess. Mean arterial Pco_2 values in the Non-Treated and the Lavage-Only groups increased significantly with the time

Fig. 3. a Arterial PCO_2. **b** Arterial pH. **c** Base excess. (Values are means ± SD). *HCl,* aspiration of hydrochloric acid. Values of the base line and the post-aspiration period are from all rabbits. A 5 cm H_2O end-expiratory pressure was applied afer the treatment in all the groups. *Asterisk* indicates significant difference ($P < 0.05$) from the other groups. Value of the Non-Treated group at 150 min of post-treatment period was omitted because 6 of the 7 rabbits died. (Table 1 shows treatment groups) ▶

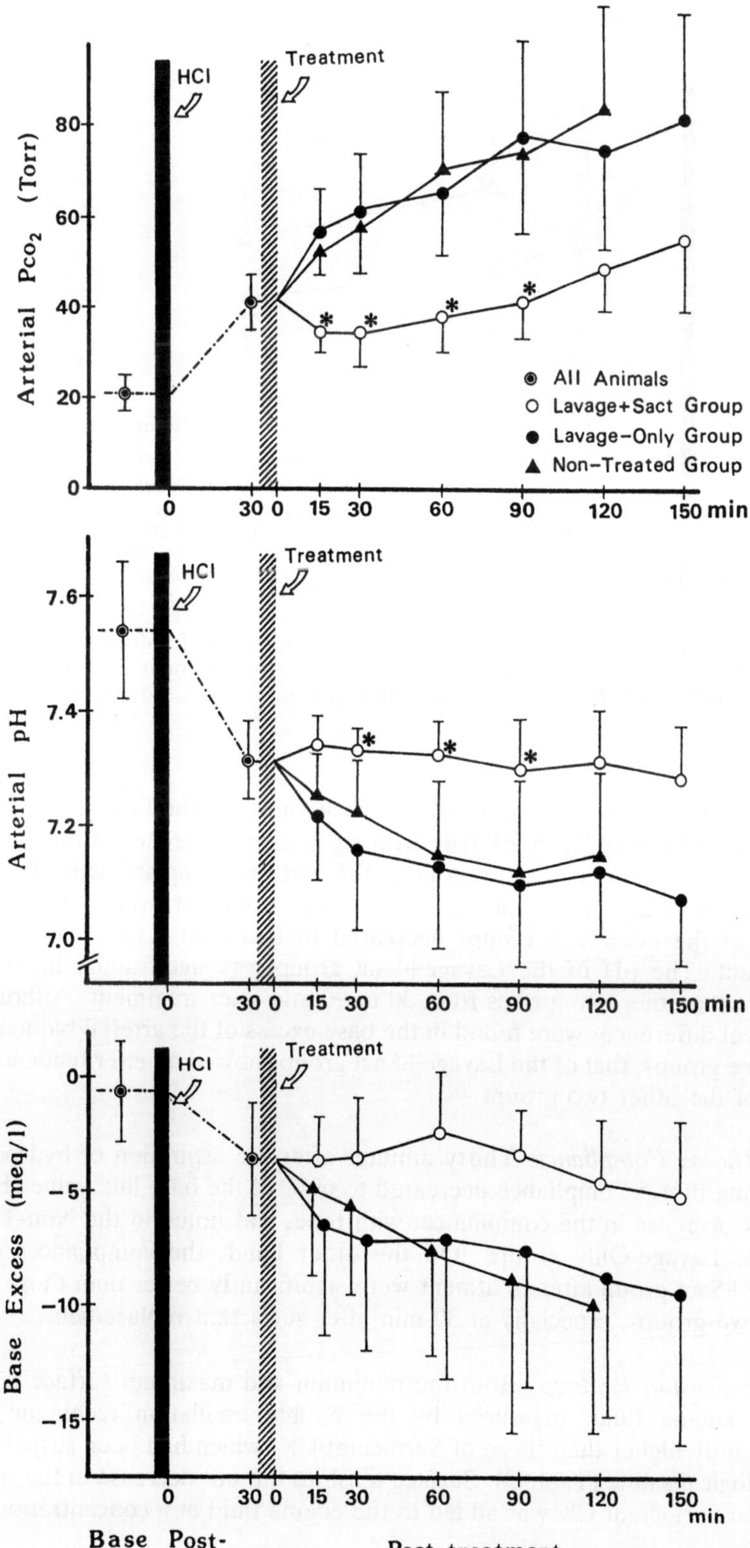
HCl
Treatment
Arterial P_{CO_2} (Torr)
80
60
40
20
0
All Animals
Lavage+Sact Group
Lavage-Only Group
Non-Treated Group
0
30 0 15 30 60 90 120 150 min
Arterial pH
7.6
7.4
7.2
7.0
Base Excess (meq/l)
0
−5
−10
−15
Base Line
Post-aspiration
Post-treatment

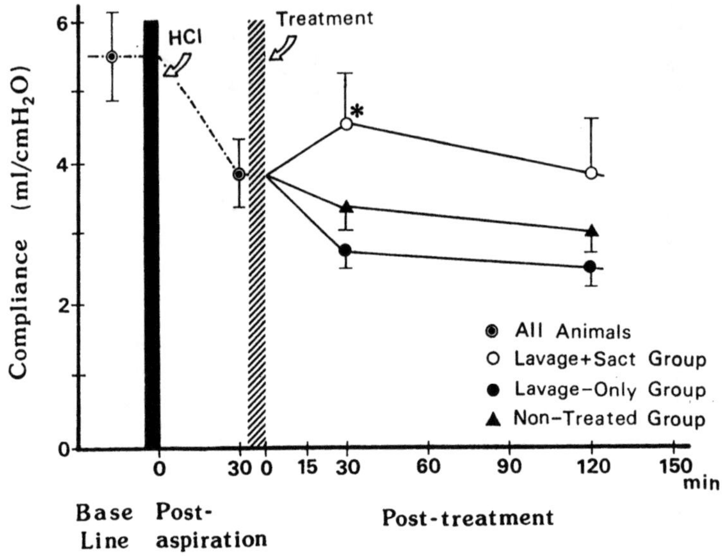

Fig. 4. Lung-thorax compliance (means ± SD). *HCl,* aspiration of hydrochloric acid. Values of the base line and the post-aspiration period are from all rabbits. *Asterisk* indicates significant difference ($P < 0.05$) from the other groups. (Table 1 shows treatment groups)

course (Fig. 3). On the other hand, the Pco_2 values of the Lavage+Sact group were lower by more than 20 Torr with significant differences from the other two groups, at least until 90 min after the surfactant replacement. The mean arterial pH values in the Lavage+Sact group remained over 7.28, while the values of the other two groups decreased to less than 7.15 at 90 min after treatment. The pH of the Lavage+Sact group was significantly higher than those of the other two groups from 30 to 90 min after treatment. Although no statistical differences were found in the base excess of the arterial blood among the three groups, that of the Lavage+Sact group showed a better tendency than those of the other two groups.

Lung-Thorax Compliance. Thirty minutes after the aspiration of hydrochloric acid, lung-thorax compliance decreased to 69% of the base line value (Fig. 4). Further decrease in the compliance, with time, was noted in the Non-Treated and the Lavage-Only groups. On the other hand, the compliances of the Lavage+Sact group after treatment were significantly better than those of the other two groups, especially at 30 min after surfactant replacement.

Surface Tension Findings. Both the minimum and maximum surface tensions of the edema fluid, measured by the bubble oscillation technique, were significantly higher than those of Surfactant CK, which had been suspended in physiological saline (Table 3). Surface tensions did not decrease in the samples in which Surfactant CK was added to the edema fluid at a concentration of 10 mg/ml.

Table 3. Minimum and maximum surface tension (γ min and γ max), mN/m

	Surfactanct CK in saline (n = 12)	Surfactant CK in lung edema (n = 8)	Unmixed lung edema (n = 8)
γ min	1.2 ± 0.4	21.3 ± 0.7*	22.8 ± 1.7*
γ max	31.5 ± 2.0	49.5 ± 4.2*	50.3 ± 5.1*

Values are mean ± SD. Lung edema fluid used in these measurements was obtained at 25 min after aspiration of hydrochloric acid. Concentration of Surfactant CK was adjusted to 10 mg/ml.
* $P < 0.01$ vs Surfactant CK in physiological saline. The surface tensions of Surfactant CK in the lung edema and the unmixed edema were statistically insignificant.

Histological Findings. In the lungs obtained from all animals, an accumulation of edema fluid in the alveoli, a migration of granulocytes into the alveoli and interstitium, intraalveolar hemorrhage, and necrosis of the airway epithelium were observed (Fig. 5). Although no statistical analysis was performed, we obtained a distinct impression that the alveolar air expansion of the Lavage+Sact group was better than that of the other two groups.

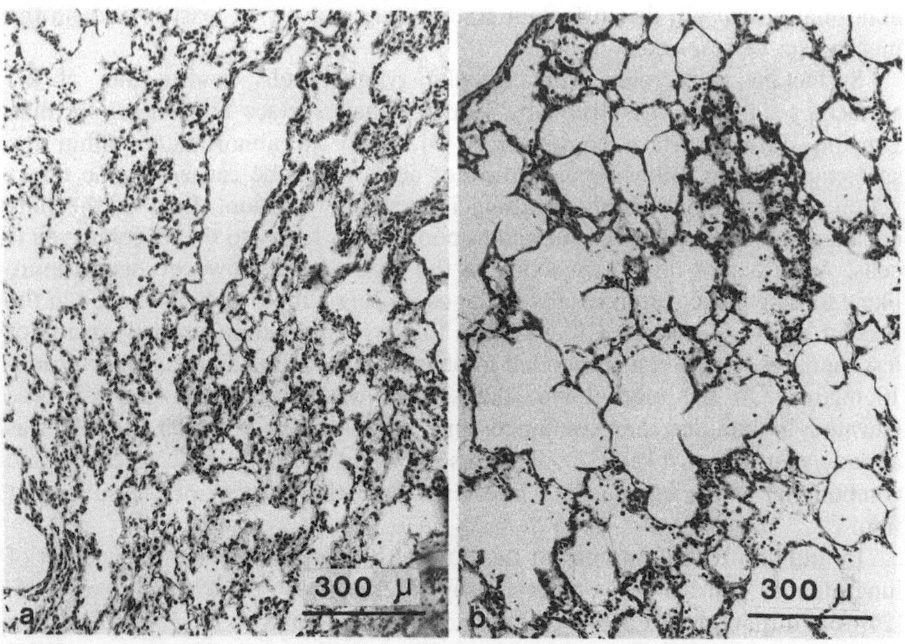

Fig. 5a, b. Histological appearance of lungs after aspiration of hydrochloric acid. The lungs were fixed at 10 cm H_2O of transpulmonary pressure. **a** Right lower lobe of the Lavage-Only group. **b** Right lower lobe of the Lavage+Sact group. Migration of granulocytes, intraalveolar edema are demonstrated in the lungs of both groups, but the alveolar air expansion is better in the Lavage+Sact group

Discussion

Results of the present study indicate that the combination of alveolar lavage and surfactant replacement improves the gas exchange, compliance, alveolar air expansion, and the survival rate in rabbits subjected to aspiration of hydrochloric acid.

The primary change in the lung after aspiration of acids is supposed to be due to the direct action of the acid, and this original insult may secondarily lead to a local accumulation of leukocytes and platelets which can further damage the lung [8]. Confirming earlier studies, the accumulation of these inflammatory cells were noted in the present study, as shown in Fig. 5. Activated neutrophils release noxious products, such as granular substances, toxic species of reduced oxygen, and products of arachidonic acid metabolism [29]. These products may enhance the permeability of the pulmonary capillaries and the alveolar epithelium [20, 27]. It was demonstrated in the present study that a considerable amount of lung edema fluid (6.0 ± 1.6 ml) was drawn from the trachea at 25 min after aspiration of the hydrochloric acid (Table 2).

The mechanisms involved in lung injury caused by aspiration of acids have been extensively studied, however, little attention has been paid to the participation of the lung surfactant system. This study has shown that, following surfactant replacement some favourable effects on blood gases, on compliance, and on alveolar air expansion took place. Thus, some abnormalities within the surfactant system are involved, at least in part, in the mechanism of respiratory failure.

Surfactant is secreted from alveolar type II cells, and a film of the surfactant stabilizes the alveoli by reducing their surface tension to less than 2 mN/m during surface compression [3]. Such an abnormality within the surfactant system following aspiration of acids might be caused by the direct action of the acid, possibly resulting in a simple dilution effect of the lung edema and a decrease in the surfactant pool size by injury to the alveolar type II cells. A surfactant inhibitory action of the lung edema, however, seems more likely to take place. From results of the oscillating bubble technique used in the present study, Surfactant CK was seen to have a minimum surface tension of less then 2 mN/m, when suspended in physiological saline at a concentration of 10 mg/ml. On the other hand, the surface tension of Surfactant CK was elevated in samples that contained lung edema fluid in which the pH was already neutralized (Table 2). This result would seem to indicate that some components of the edema fluid interfered with the stability of the surfactant film.

In addition to the findings in this study, the surfactant inhibitory action of lung edema fluid has been suggested by several other researchers. Said et al. [24] found that surface activity of extracts from edematous lungs was less than that of normal lungs. Recently, Seeger et al. [25] and Fuchimukai et al. [6] have shown that proteins in serum, especially fibrinogen, are potent inactivators of lung surfactant. Another serum protein that inhibits surfactant function has been isolated in premature babies [12] and in premature lambs [11]. This protein, which has a molecular weight of 110,000 daltons, seems to leak into the airspaces of the immature lung after the onset of ventilation, and might be involved in the pathogenesis of neonatal RDS. This same inhibitory protein was also found in tracheal fluid of patients with adult RDS [10].

In our previous observations, the effective duration of surfactant replacement was more than 11 hr in surfactant-depleted rabbits, which were rendered by repeated lung lavage [18]. On the other hand, differences in arterial Po_2 between the Lavage+Sact group and the other groups of this study disappeared only 120 min after the surfactant replacement. This rather short effective duration could be explained in part by the reappearance of the proteinaceous edema fluid, since a considerable amount of the fluid (6.0 $\pm$ 3.7 ml) was again obtained from the animals of the Lavage+Sact group at the end of the experiment. In our preliminary study on the aspiration of hydrochloric acid, we tested the surfactant replacement without alveolar lavage. However, we could not obtain any improvement in arterial blood gases (data not shown). This phenomenon i.e., surfactant replacement not being effective in cases in which the alveoli are already flooded with proteinaceous edema fluid, could also be explained by the surfactant inhibitory action of the edema fluid.

A vicious circle will arise between the lung edema and inactivation of the alveolar surfactant. Bredenberg et al. [2] have demonstrated in experiments on dogs that displacing the alveolar surfactant with a detergent, dioctyl sodium sulfosuccinate, increased the flux of fluid into the airspaces. Jobe et al. [13] and Robertson et al. [22] have shown that the bidirectional flux albumen across the lung epithelium is large in premature lambs and rabbits with surfactant deficiency. A high alveolar surface tension contributes to a negative hydrostatic pressure in the pulmonary interstitium and tends to increase the transport of fluid from the vascular lumen into the interstitium and, further, into the air spaces. The fluid that appears in the alveoli further deteriorates the surfactant, as already discussed here.

Given the results of this study and the above considerations, the prevention of edema formation or the removal of the fluid from the alveoli might be important therapy in respiratory failure. The administration of a cyclo-oxygenase antagonist, which inhibits the production of prostaglandins and thromboxanes in the arachidonic acid cascades, was reported to reduce the accumulation of edema fluid in dogs that had aspirated hydrochloric acid [28]. Positive end-expiratory pressure (PEEP) has also been known to prevent the formation of lung edema. These procedures, however, cannot prevent inactivation of the alveolar surfactant after the edema fluid has appeared in the alveoli. In the Non-Treated group of the present study, PEEP, which was applied after the appearance of the lung edema, did not improve the findings of blood gases or compliance.

In cases where gastric contents were aspirated, even the foodstuff itself is involved in the pathogenesis of a respiratory failure, as is the acidity [30]. Alveolar lavage can reduce the protein concentration of edema fluid and other noxious substances. The lavage procedure, however, also washes out the alveolar surfactant. In such cases, surfactant replacement may be useful in treatment and may help to break the vicious circle between the production of lung edema and inactivation of the alveolar surfactant.

The concentration of exogenous surfactant used in this study (20 mg/ml) is rather small. For experiments with premature animals, a surfactant concentration of over 50 mg/ml has been frequently used [15, 22]. Perhaps the replacement might have been more effective if the animals of the present study had been tested with a surfactant of a higher concentration. Further studies are required to find a more effective mode and dose of surfactant administration and a better method to remove the lung edema fluid. Identification of the inhibitory components of edema fluid is also necessary. We conclude, however, that surfactant inactivation by the lung edema is one of the mechanisms that hastens respiratory failure induced by the aspiration of acids, and that the combination of alveolar lavage and surfactant replacement has a therapeutic affect on the respiratory failure in the rabbit.

References

1. Berggren P, Lachmann B, Curstedt T, Grossmann G, Robertson B (1986) Gas exchange and lung morphology after surfactant replacement in experimental adult respiratory distress syndrome induced by repeated lung lavage. Acta Anaesthesiol Scand 30:321–328
2. Bredenberg CE, Nieman GF, Paskanik AM, Hart AK (1986) Microvascular membrane permeability in high surface tension pulmonary edema. J Appl Physiol 60:253–259
3. Clements JA (1977) Function of the alveolar lining. Am Rev Respir Dis 115:67–71
4. Enhorning G, Shennan A, Possmayer F, Dunn M, Chen CP, Milligan J (1985) Prevention of neonatal respiratory distress syndrome by tracheal instillation of surfactant: A randomized clinical trial. Pediatrics 76:145–153
5. Folch J, Ascoli I, Less M, Meath JA, LeBaron FN (1951) Preparation of lipid extraction from brain tissue. J Biol Chem 191:833–841
6. Fuchimukai T, Fujiwara T, Takahashi A, Enhorning G (1987) Artificial pulmonary surfactant inhibited by proteins. J Appl Physiol 62:429–437
7. Fujiwara T, Maeta H, Chida S, Morita T, Watabe Y, Abe T (1980) Artificial surfactant therapy in hyaline-membrane disease. Lancet 1:55–59
8. Greenfield LJ, Singleton RP, McCaffree Dr, Coalson JJ (1969) Pulmonary effects of experimental graded aspiration of hydrochloric acid. Ann Surg 170:74–86
9. Hallman M, Merritt TA, Jarvenpaa A-L, Boynton B, Mannino F, Gluck L, Moore T, Edwards D (1985) Exogenous human surfactant for treatment of severe respiratory distress syndrome: a randomized prospective clinical trial. J Pediatr 106:963–969
10. Ikegami M, Berry D, Kanata M, Nozaki M (1985) Surfactant inhibition in adult respiratory distress syndrome. J Jpn Soc Biol Interface 16:135–141
11. Ikegami M, Jobe A, Jacobs H, Lam R (1984) A protein from airways of premature lambs that inhibits surfactant function. J Appl Physiol 57:1134–1142
12. Ikegami M, Jobe A, Berry D (1986) A protein that inhibits surfactant in respiratory distress syndrome. Biol Neonate 50:121–129
13. Jobe A, Ikegami M, Jacobs H, Jones S, Conaway D (1983) Permeability of premature lamb lungs to protein and the effect of surfactant on that permeability. J Appl Physiol 55:169–176
14. Jobe A, Jacobs H, Ikegami M, Jones S (1983) Cardiovascular effects of surfactant suspensions given by tracheal instillation to premature lambs. Pediatr. Res 17:444–448
15. Kobayashi T, Grossmann G, Robertson B, Ueda T (1984) Effects of artificial and natural surfactant supplementation in immature newborn rabbits. J Jpn Med Soc Biol Interface 15:125–131
16. Kobayashi T, Izumi K, Ganzuka M, Murakami S, Kokubo M (1985) Correlation between in vitro property and in vivo function of artificial lung surfactant. J Jpn Med Soc Biol Interface 16:51–57

17. Kobayashi T, Kataoka H, Murakami S, Haruki S (1981) A case of idiopathic respiratory distress syndrome treated by newly developed surfactant (Surfactant CK). J Jpn Med Soc Biol Interface 12:1–6
18. Kobayashi T, Kataoka H, Ueda T, Murakami S, Takada Y, Kokubo M (1984) Effect of surfactant supplement and end-expiratory pressure in lung-lavaged rabbits. J Appl Physiol 57:995–1001
19. Lachmann B, Berggren P, Curstedt T, Grossmann G, Robertson B (1982) Combined effects of surfactant substitution and prolongation of inspiration phase in artificially ventilated premature newborn rabbits. Pediatr Res 16:921–927
20. Manny J, Manny N, Lelcuk S, Alexander F, Feingold H, Kobzik L, Valeri CR, Shepro D, Hechtman HB (1986) Pulmonary and systemic consequences of localized acid aspiration. Surg Gynecol Obstet 162:259–267
21. Morgan JG (1978) Pathophysiology of gastric aspiration. Prim Care 5:1–11
22. Robertson B, Berry D, Curstedt T, Grossmann G, Ikegami M, Jacobs H, Jobe A, Jones S (1985) Leakage of protein in the immature lung; effect of surfactant replacement. Respir Physiol 61:265–276
23. Royston D, Fleming JS, Krausz T (1986) The non-steroidal anti-inflammatory agent, indoprofen, does not protect against hydrochloric acid induced lung injury in the rat. Acta Anaesthesiol Scand 30:533–537
24. Said SI, Avery ME, Davis RK, Banerjee CM, El-Gohary M (1965) Pulmonary surface activity in induced pulmonary edema. J Clin Invest 44:458–464
25. Seeger W, Stöhr G, Wolf HRD (1985) Alteration of surfactant function due to protein leakage: special interaction with fibrin monomer. J Appl Physiol 58:326–338
26. Shapiro DL, Notter RH, Morin FC, Deluga KS, Golub LM, Sinkin RA, Weiss KI, Cox C (1985) Double-blind, randomized trial of a calf lung surfactant extract administered at birth to very premature infants for prevention of respiratory distress syndrome. Pediatrics 76:593–599
27. Tahamont MV, Malik AB (1983) Granulocytes mediated the increase in pulmonary vascular permeability after thrombin embolism. J Appl Physiol 54:1489–1495
28. Utsunomiya T, Krausz MM, Valeri CR, Shepro D, Hechtman HB (1981) Treatment of aspiration pneumonia with ibuprofen and prostacyclin. Surgery 90:170–176
29. Weissmann G, Smolen JE, Korchak HM (1980) Release of inflammatory mediators from stimulated neutrophils. N Engl J Med 303:27–34
30. Wynne JW, Reynolds JC, Hood CI, Auerbach D, Ondrasick J (1979) Steroid therapy for pneumonitis in rabbits by aspiration of foodstuff. Anesthesiology 51:11–19

Efficacy of Exogenous Surfactant in Experimental Aspiration Trauma

W. Strohmaier, H. Redl, and G. Schlag

Introduction

The use of exogenous surfactant derived from varying sources to treat premature babies who develop respiratory distress syndrome (RDS) has been proven to be effective by several groups (for review see [1–3]). Encouraged by these results, investigations were extended to the treatment of adults suffering from RDS. In contrast to prematurity, respiratory failure in the adult is a consequence of the activation of a wide spectrum of biochemical and pathobiochemical cascades (review [4]). Numerous animal models have been designed to induce lung damage [5] but only a small number actually respond to treatment [6–8]. The objective of this study was to investigate the effects of a semi-synthetic surfactant preparation in an adapted aspiration trauma model, thus providing a more appropriate reflection of the clinical situation.

Material and Methods

Animal Model. Lung damage was induced in 22 healthy New Zealand rabbits (weighing 2850–3400 g) by intratracheal administration of 2 ml/kg body weight (BW) Betain.HCl-pepsin solution (pH 1.78). During this procedure the animals were anesthetized (50 mg/kg Ketalar with 0.3 ml/kg Rompun i.m.) and orally intubated.

Exogenous Surfactant. After 24 hours the rabbits were anesthetized and again orally intubated. 5 animals (group T) received a sterile surfactant preparation [240 mg lyophilisate : 120 mg phospholipid (PHL), dipalmitoylphosphatidylcholine (DPPC) : Egg phosphatidylglycerol (PG) 4 : 1; Nattermann, Köln, FRG] suspended in 7.5 ml Ringer's solution containing 4.6 mg rabbit surfactant pool-PHL and 0.5 mg protein per ml. The surfactant pool was isolated from lavage fluid by discontinuous sucrose-gradient centrifugation (0.68 M / 0.25 M; 65.10^3 g until $w^2t = 1.2 \times 10^{10}$) (for details see [9]). 7 animals (group ST) were administered an equal volume of Ringer's solution. The animals received no additional treatment. 10 healthy animals served as controls (group C).

Lung Mechanics. Twenty-four hours after surfactant, or Ringer's solution instillation, the anesthetized animals were tracheotomized and intubated. After killing the animals with an i.v. overdose of Thiopental (Sanabo, Vienna, Austria) the chest was opened by a midline incision and the lungs were inflated once with twice the tidal volume (calculated as in [10]). The tracheal tube was then connected to a gas-tight syringe (Hamilton, Bonaduz, Switzerland), driven by a pump (Harvard Apparatus, 907A, Millis., Mass., USA). The pressure transducer (Gould Electronics, TH1017, Bilthoven, Netherlands) was adjusted to zero before closing the system. The lungs were inflated in 10 ml increments, allowing 1 minute equilibrium pauses until a stable plateau pressure of 20 mm Hg (= 26.3 cm H_2O) was reached. This volume was defined as total lung capacity above functional residual capacity (TLC°). Deflation was performed in the same rhythm. Static compliance was calculated as the steepest slope from at least 3 consecutive data points of the deflation limb. All data were taken from the first curve.

Statistical Analysis. Analysis of variance (ANOVA) and group comparison were performed by linear contrast according to Scheffe [11] for static compliance (C_{stat}) and according to Linder [12] for TLC°. Significance was accepted at $p < 0.05$. Results are presented as means ± SEM.

Results

The administration of exogenous semi-synthetic surfactant 24 hours after hydrochloric acid-pepsin instillation increases specific static compliance of damaged rabbit lungs from 2.12 ± 0.69 ml/cm H_2O/kg (54.1 %, group ST) to

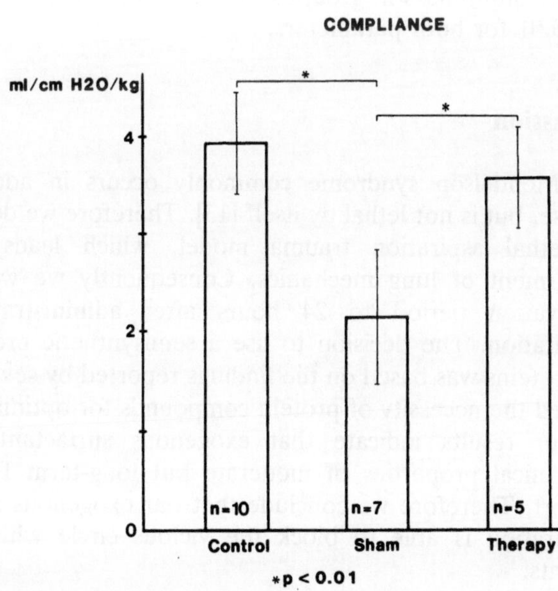

Fig. 1. Static compliance. *Sham,* group ST; *Therapy,* group T

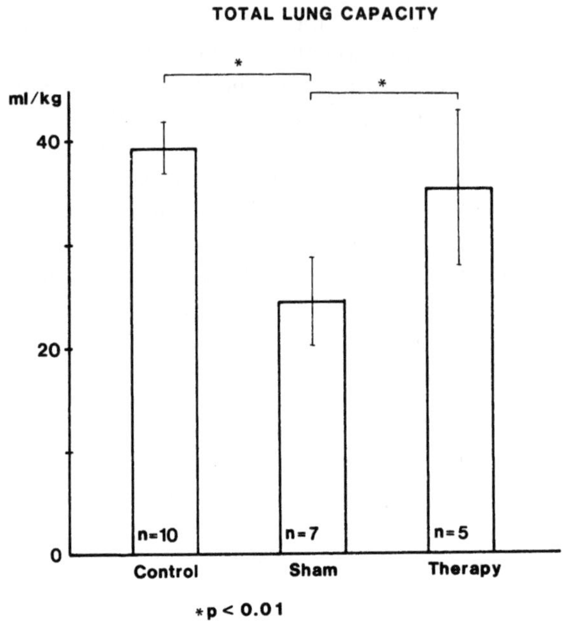

Fig. 2. Total lung capacity. *Sham,* group ST; *Therapy,* group T

3.58 ± 0.52 ml/cm H_2O/kg (91.3 %, group T) (Fig. 1), compared to healthy controls (group C: 3.92 ± 0.53 ml/cm H_2O/kg = 100 %).

Figure 2 shows the values for TLC°. A 37.7 % reduction (39.3 ± 2.59, group C) to 24.49 ± 4.33 ml/kg in group ST is compensated in group T (35.28 ± 7.58 ml/kg = 89.8 %).

The differences in groups C vs ST and ST vs T are significant at a level of $p < 0.01$ for both parameters.

Discussion

The Mendelson syndrome commonly occurs in addition to an underlying disease, but is not lethal by itself [13]. Therefore we decided to use an adapted non-lethal aspiration trauma model, which leads to RDS with severe impairment of lung mechanics. Consequently we were able to expand the observation period to 24 hours after administration of our surfactant preparation. The decision to use a semisynthetic preparation containing the apoproteins was based on the findings reported by several groups [14–17], who showed the necessity of protein compounds for optimal surfactant function.

Our results indicate that exogenous surfactant can restore the lung mechanical properties of moderate but long-term RDS without ventilatory support. Therefore we conclude that our exogenous semi-synthetic surfactant preparation is able to block the vicious circle which exists in severely ill patients.

References

1. Taeusch HW, Clements J, Benson B (1983) Exogenous surfactant for human lung disease. Am Rev Respir Dis 128:791–794
2. Robertson B (1984) Surfactant replacement in neonatal and adult respiratory distress syndrome. Eur J Anaesthesiol I:335–343
3. Ikegami M, Agata Y, Ewady T, Hallmann M, Berry D, Jobe A (1987) Comparison of four surfactants: in vitro surface properties and responses of preterm lambs to treatment at birth. Pediatrics 79:36–46
4. Hammerschmidt DE (1983) Activation of the complement system and of granulocytes in lung injury: the adult respiratory distress syndrome. Adv Inflamm Res 5:147–172
5. Lachmann B, Danzmann E (1984) Adult respiratory distress syndrome: In: Robertson B, Van Golde LMG, Batenburg H (eds) Pulmonary surfactant. Elsevier Science, Amsterdam, p 505–548
6. Berry D, Ikegami M, Jobe A (1986) Respiratory distress and surfactant inhibition following vagotomy in rabbits. J Appl Physiol 61:1741–1748
7. Lachmann B, Hallmann M, Bergmann K (1987) Respiratory failure following anti-lung serum: study on mechanisms associated with surfactant system damage. Exp Lung Res 12:163–180
8. Winn R, Stothert J, Nadir B, Hildebrandt J (1983) Lung mechanics following aspiration of 0.1 N hydrochloric acid. J Appl Physiol 55:1051–1056
9. Redl H, Strohmaier W, Stachelberger H, Schlag G (1984) Morphological appearance (REM and TEM) of surfactant derived from lavage, from lamellar body fraction and from lung tissue in dogs. Prog Respir Res 18:64–67
10. Kleinman IL, Redford EP (1964) Ventilation standards for small mammals. J Appl Physiol 29:360–362
11. Scheffe H (1953) A method for judging all contrasts in the analysis of variance. Biometrika 40:87–104
12. Linder A (1963) Trennverfahren bei qualitativen Merkmalen. Metrika 6:76–83
13. Wynne JW (1982) Aspiration pneumonitis. Correlation of experimental models with clinical disease. Clin Chest Med 3:25–34
14. Robertson B, Van Golde LMG (1984) Postscript: surfactant research: current concepts and perspectives for the future. In: Robertson B, Van Golde LMG, Batenburg H (eds) Pulmonary surfactant. Elsevier Science, Amsterdam, p 549
15. Taeusch HW, Keough KMW, Williams M, Slavin R, Steele E, Lee AS, Phelps D, Kariel N, Floros J, Avery ME (1986) Characterization of bovine surfactant for infants with respiratory distress syndrome. Pediatrics 77:572–581
16. Fujiwara T (1984) Surfactant replacement in neonatal RDS. In: Van Golde LMG, Batenburg J (eds) Pulmonary surfactant. Elsevier Science, Amsterdam, p 479
17. Tanaka Y, Takei T, Masuda K (1983) Lung surfactant: III. Correlations among activities in vitro, in situ, and in vivo, and chemical composition. Chem Pharm Bull (Tokyo) 31:4110–4115

Acute Effects of an Artificial Surfactant (Exosurf) Administered as an Aerosol in Sheep with Oleic Acid-Induced Lung Injury*

M. Zelter, B. J. Escudier, J. M. Hoeffel, and J. F. Murray

Introduction

Several investigators have found evidence of surfactant abnormalities in patients with ARDS (Asbaugh et al. 1967; Hallman et al. 1982; Petty et al. 1977). This raises the possibility that lowering intra-alveolar surface tension by surfactant administration might improve the gas exchange disturbances, abnormal mechanics and increased permeability that characterize acute lung injury.

To test this hypothesis we administered an exogenous surfactant prior to and during oleic acid injury in the sheep, a well known animal model of acute lung injury (Griffith et al. 1979; Julien et al. 1986).

Methods

We measured total resistance (R), respiratory system compliance (C) and blood gases (PaO_2, $PaCO_2$, pH) in 3 groups of anesthetized (pentobarbital sodium, 25 mg/kg), paralyzed (pancuronium bromide, 1 mg/h) and ventilated (15 ml/kg tidal volume, 12 breaths/min) sheep: Group 1 = oleic acid and aerosolized Exosurf (n = 5); Group 2 = oleic acid and aerosolized NaCl (n = 5); Group 3 = no lung injury (n = 6, 2 received Exosurf alone, 2 received NaCl alone, 2 received neither). The site of Exosurf deposition was studied in an additional 3 sheep.

R and C were measured hourly in duplicate, at randomized tidal volumes, 500–700 ml or 700–500 ml, by interrupting a pressure ramp, constant flow ventilation (Bates et al. 1985).

We administered either Exosurf, an artificial protein free surfactant [dipalmitoilphosphatidylcholine (DPPC), hexadecanol, tyloxapol, NaCl] or 0.9 % NaCl, with a Retec nebulizer (MMD 3.6 μm) using the method of Rizk et al. (1984). Aerosol delivery was started 1 h before oleic acid or sham injury and continued until the end of the experiment (4 h). The amount of aerosolized surfactant was 1200 ± 200 mg DPPC. Oleic acid was infused intravenously at the rate of 0.03 ml/kg over 45 min.

* Supported by Pulmonary Vascular SCOR HL-19155. M. Zelter and B. Escudier were supported in part by Philippe Foundation Inc.

Table 1. Results (X ± SD) of PaO_2, total lung resistance and respiratory system compliance during baseline (B) and 3 h after injury for the three groups of sheep

		Exosurf oleic-acid	NaCl oleic-acid	Control no oleic-acid
PaO_2	B	101 ± 6.2	102 ± 10.1	102 ± 3.0
(mm Hg)	3 h	76 ± 12.2	68 ± 8.1	102 ± 3.3
Compliance	B	71 ± 15.0	67 ± 8.0	57 ± 5.0
(ml/cm H_2O	3 h	53 ± 14.5	45 ± 13.5	56 ± 6.0
Resistance	B	8 ± 2.5	8 ± 2.3	9 ± 2.1
(cm H_2O/ml/sec)	3 h	16 ± 6.6	14 ± 7.2	10 ± 3.4

A multi-group, repeated measurements design of analysis of variance was used to test for overall differences over time between groups (Wallenstein et al. 1980).

Results

The results for baseline (B) and 3 h after intervention are shown in Table 1.

Tidal volume groups (700–500 ml and 500–700 ml) were identical and therefore pooled. R and C measurements were independent of volume within our tidal volume range in the aerosolized groups.

In the acute experiments the two groups that received oleic acid did not differ from each other but differed from control for R, C and PaO_2 after injury.

Discussion

Exosurf administered as an aerosol clearly does not mitigate the physiological abnormalities produced during the acute phase of oleic acid-induced injury. Surfactant abnormalities may indeed be unimportant in this type of injury as previously found (Oyarzun et al. 1984; Grossmann et al. 1980). Insufficient delivery of active material by aerosol inhalation may also explain this negative finding. However deposition experiments of dye-labeled surfactant demonstrated detectable deposition in the peripheral lung with little evidence of staining in the larger airways as early as one hour after beginning the aerosolization. Surfactant abnormalities may only become significant after the first five hours during which our animals were monitored. Aerosol inhalation at onset of injury would thus have limited effectiveness because of the short biological half-life of surface active material (Jacobs et al. 1982). However, by five hours lung mechanics, permeability and gas exchange abnormalities have already developed (Grossman et al. 1980; Julien et al. 1986) and appear largely attributable to alveolar flooding. Treatment with surfactant in this phase of injury is thus unlikely to be beneficial.

More likely endogenous and exogenous surfactant may have been partially or totally inactivated by protein leakage (Ikegami et al. 1983) or oleic acid (Oyarzun et al. 1984).

Finally, Exosurf, a protein-free compound may not include whatever components may be critically needed to ensure normal biophysical function of surfactant in the diseased adult lung (Magoon et al. 1983). Thus an active compound in pre-term lambs (Durand et al. 1985) and infants may not be useful in the adult.

The delayed effects of surfactant administration after acute lung injury are being studied. It may partially protect alveolar type II cells or provide an important increase in the alveolar phospholipid pool in poorly perfused parts of the injured lung; thus Exosurf may help de novo surfactant synthesis by providing substrate, despite reduced blood supply (Baritussio and Clements 1981). We suggest that surfactant replacement therapy be monitored long after the initial lung injury in adult animal models of acute lung injury in order to fully evaluate its effect.

References

Asbaugh DG, Bigelow DB, Petty TL, Levine BE (1967) Acute respiratory distress in adults. Lancet II:319–323

Baritussio A, Clements JA (1981) Acute effects of pulmonary arterial occlusion on the pool of alveolar surfactant. Respir Physiol 45:323–331

Bates JHT, Rossi A, Milic Emili J (1985) Analysis of the behavior of the respiratory system with constant inspiratory flow. J Appl Physiol 58:1840–1848

Durand DJ, Clyman RI, Heymann MA, Clements JA, Mauray F, Kitterman J, Ballard P (1985) Effects of a protein free synthetic surfactant on survival and pulmonary function in preterm lambs. J Pediatr 107:776–780

Griffith BP, Caroll RG, Hardesty RL, Peel RL, Borovetz HS (1979) Selected lobar injury after infusion of oleic acid. J Appl Physiol 47:706–711

Grossman RF, Jones JG, Murray JF (1980) Effects of oleic acid induced pulmonary edema on lung mechanics. J Appl Physiol 48:1045–1051

Hallman M, Spragg R, Harell JH, Moser KM, Gluck L (1982) Evidence of lung surfactant abnormality in respiratory failure. J Clin Invest 70:673–683

Ikegami M, Jacobs H, Jobe A (1983) Surfactant function in the respiratory distress syndrome. J Pediatr 102:443–447

Jacobs H, Jobe A, Ikegami M, Jones S (1982) Surfactant phosphatidylcholine source, fluxes and turnover time in 3 day old, 10 day old and adult rabbits. J Biol Chem 257:1805–1810

Julien M, Hoeffel JM, Flick MR (1986) Oleic acid lung injury in sheep. J Appl Physiol 60:433–440

Magoon MW, Wright JR, Baritussio A, Williams MC, Goerke K, Benson BJ, Hamilton RL, Clements JA (1983) Subfraction of lung surfactant. Implications for metabolism and surface activity. Biochim Biophys Acta 750:18–31

Oyarzun MJ, Cabezas E, Donoso P, Quijaoa D (1984) Effects of free fatty acid infusion on rabbit pulmonary surfactant. Influence of corticosteroids. Bull Eur Physiopathol Respir 20:105–111

Petty TL, Reiss OK, Paul GW, Silvers GW, Elkins ND (1977) Characteristics of pulmonary surfactant in adult respiratory distress syndrome associated with trauma and shock. Am Rev Respir Dis 115:531–536

Rizk NW, Luce JM, Hoeffel JM, Price DC, Murray JF (1984) Site of deposition and factors affecting clearance of aerosolized solute from canine lungs. J Appl Physiol 56:723–729

Wallenstein S, Zucker CL, Fleiss J (1980) Some statistical methods useful in circulation research. Circ Res 47:1–9

Discussion: What is the Surfactant of Choice for Replacement Therapy?

Chairman: A. Jobe

A. Jobe (Los Angeles, USA): It is a very difficult task to decide which surfactant to use since each group of investigators has their own opinion and uses their own surfactant. One of the major difficulties in this field is that to conduct research in surfactant it is necessary to make one's own surfactant. Thus almost everyone at this meeting has been using different surfactants in clinical trials and there is, consequently, an inherent difficulty in comparing preparations. I would like to ask Dr. Bloom which surfactant he would advise as he is the only one who has directly compared two different surfactants.

B. Bloom (Kansas, USA): I do not yet have a definite answer as to which surfactant is the best as there are so many confusing factors involved varying from single-dose and multi-dose trials. We need a much larger population on which to base our judgement.

A. Jobe: Would anybody like to comment on whether it is possible to define differences in terms of in vitro testing between surfactants? Most investigators have tested their surfactant in relation to controls but not against other surfactants. Is there a possibility to identify the best surfactant in vitro, or via animal models, or do we have to compare surfactants in clinical trials to ultimately determine which is the best surfactant?

M. Hallmann (Helsinki, Finland): I think that animal studies are extremely important, as they can provide a large amount of very important information that we could never get from clinical studies alone. But there are limitations to animal studies, for example, it is difficult to conduct long-term studies with animals and follow-up is also difficult. There are, of course, also inherent differences between clinical and experimental studies; patients may have underlying disorders which can, for example, precipitate pre-term labour etc. It is very important to conduct clinical studies. I do not yet have an answer as to which is the best surfactant to use.

A. Merritt (San Diego, USA): Clearly one of the reasons why we do not have an answer is that the surfactants are "first generation" and perhaps they are all at this stage "non-optimal". When new and second generation surfactants are developed, which are more reproducible, they might provide a better answer. As there is such great clinical variability this means that each study must in fact be exactly comparable. Until we can efficiently and rapidly define which babies indeed have the disease we are supposed to be treating, and realise that studies that are prophylactic without documentation of lung maturity status cannot be compared to studies in which there is careful attempt to document pulmonary maturity status, we cannot evaluate the success or failure of the treatment. The problem is that given the knowledge base we have today it is not possible to make a direct meta-analysis of the existing trials and draw meaningful conclusions. I would like to know why we have not been able to conduct studies which enable us to take surfactant A, B and C and compare them simultaneously to each other and, perhaps simultaneously, to a control group.

C. Morley (Cambridge, UK): First of all one has to define what outcome one wants; you can have increased oxygen at 1 hour or pneumothoraces, patent ductus or mortality and "pick your winner" depending on which outcome you want. Currently there is no overall winner with a meta-analysis even taking into account all the problems involved. Secondly, comparison trials are virtually impossible. The results discussed at this meeting indicate that surfactant treatment reduces mortality by about 30% (average), this is just taking mortality as the outcome, which is after all absolute. Assuming this is so, how can you compare two surfactants to see if one is better than the other? We estimate that about 3000 babies will be needed to make such a comparison. If you try and run a clinical trial of 3000 babies of less than 30 weeks gestation it will collapse, people will lose interest, change the treatment etc. I am sorry to be pessimistic, but I think that effective randomised trials with well-defined outcomes to compare surfactants will be almost impossible.

A. Jobe: That seems to be a fair summary of the situation! We are in the position of having multiple surfactants and if we examine the studies presented at this meeting, and others, the results are fairly consistent – bearing in mind it does depend on which standpoint you pick and the patient population studied. Obviously it is very difficult to ultimately clearly define which surfactant is most efficacious. Maybe we should consider another interesting issue that many people have mentioned – that of non-responders. In all studies there are babies that respond (the majority) and those that fail to respond. I think that if they do not respond to surfactant they probably do not have surfactant deficiency disease. It is an important issue to consider because possibly this subgroup could be used to help resolve the question as to which surfactant is better. Or, more importantly, we could use surfactant as a diagnostic tool to differentiate babies with RDS from those with other problems. This could be a major challenge for the coming years – to investigate what these babies have and to find solutions to their problems. I have listed some of the items participants have mentioned in relation to non-response, including: pulmonary edema, surfactant inhibition, ductus arteriosus, pulmonary hypoplasia, congenital heart disease, the question of dosage of the agent, cardiovascular shock related to surfactant therapy or tiny infant size – the problem discussed by Drs. Hallman and Clarke concerning tiny babies who give initial clinical response to surfactant but with no overall improvement in mortality in this subgroup of babies. Dr. Fujiwara also mentioned the issue of the learning curve basically involving how to use and administer surfactant to babies. It seems that non-responders do exist – maybe someone could give there experiences, as to how many babies under 1000 g are non-responders?

J. Kendig (Rochester, USA): We have not yet completed all the analysis from our current randomised trial, but there is clearly a subgroup that does not respond. I cannot give an exact percentage at this point nor have we identified any specific factors which are predictors for non-response, this remains to be investigated. We particularly want to compare non-responders in our current study to see if there are any significant differences between the preventilatory and post-ventilatory arms of our study.

A. Jobe: Dr. Fujiwara, do you have any thoughts about nonresponders? Do you think that if surfactant is used appropriately most babies will respond?

T. Fujiwara (Iwata, Japan): In our nursery center we analysed a total of 96 patients with birthweight ranging from 500–2500 g and had 5 % non-responders. Two of these were obviously due to persistent fetal circulation (PFC) associated with surfactant deficiency. Surfactant was effective in clearing the lung but PAO_2/PaO_2 did not decrease – thus resulting in PFC. Other factors for non-response can include cardiogenic shock or transient myocardial dysfunction. In total, including randomised and non-randomised multicenter studies in Japan, we have 350 patients of which about 5 % are non-responders.

A. Jobe: Dr. Robertson do you have comments about non-responders from your European trial?

B. Robertson (Stockholm, Sweden): Maybe my Belfast colleagues would like to say a few words, as they basically did this analysis.

H. Halliday (Belfast, N. Ireland): Of 35 patients treated so far, five have been non-responders (about 14%). The non-responders had a mortality rate of 80% – only one out of five survived, whereas those that responded well had oxygen concentrations of less than 30% with four hours and all survived. Those admitted with 30–60% oxygen concentration had a 95% survival, so even a fair response indicated which babies had a better chance of survival. The differences between the good responders and non-responders seems to be severity of illness, judged by oxygen concentration on admission.

A. Jobe: It seems to me that over the last few years the incidence of non-responders has decreased, which may be related to increased knowledge concerning how to use surfactant. I am surprised that the Belfast group had 14% non-response in their population – if we look at anatomic studies (for example from Dr. Hislop in the UK) and look at the difference in alveolar number and calculated surface area of lungs one can anticipate that there are many babies with very small lungs, not because they are hypoplastic but because they have not yet developed anatomically. Certainly the histologists have claimed that when an infant dies from severe RDS then the lung is too immature to survive, that there is not enough surface area to support gas exchange. The proof seems to be that most of these lungs are potentially treatable with surfactant. This is surprising to many people who assumed that the anatomy of the lung would not permit survival of these babies, and non-responders would be a group of babies with anatomically immature lungs. I would be interested to hear Dr. Enhorning's comments on this issue.

G. Enhorning (Buffalo, USA): We undertook some experiments several years ago with human fetal lungs and found that at 16 weeks the lungs expanded surprisingly well – one lung was used as control lung and the other was expanded by surfactant. Thus, in spite of the fact that the anatomy seems immature, there is a good chance with surfactant to open the lung sufficiently for gas exchange to take place.

B. Robertson: Firstly one has to differentiate between pulmonary hypoplasia and immaturity. An immature lung can be one with "primitive" air cells but can still be suitable for gas exchange if that lung produces adequate surfactant. I do not think you need mature alveolisation of the lung to be able to breathe properly – you cannot detect from the structure of the lung whether it is an RDS lung or not, unless you see the lesions, epithelium disruption, hyaline membranes etc. There is also the problem of permeability that may be related to surfactant deficiency or to something else – as indicated by the fact that corticosteroids may tighten the epithelium, which is another effect that surfactant also produces. Thus, it is a complicated issue, one cannot just look at the lung and say that, anatomically, this is an RDS candidate. It is a very complex situation.

A. Jobe: Would anyone like to tackle the issue of what is the minimum gestational age, or size of baby, we should be treating with surfactant? Particularly bearing in mind the reports from Tom Clarke and Mikko Hallman concerning very tiny infants with very severe disease.

T. Clarke (Dublin, Ireland): I think that in every country the approach to neonatal intensive care has to be taken gradually and certainly in Ireland we generally would not be aggressive with babies under 700 g (except in special circumstances). Judging by the overall results in the world literature, excluding certain centers of excellence, it seems that certainly one should not recommend widespread use of surfactant for babies under about 700 g. In my opinion, bearing in mind the survival and handicap rates, I would personally question use of surfactant in such tiny babies.

C. Morley: Our 10 center trial stratified the babies of 25–26 weeks in one strata and 27–29 weeks in another strata. There was no doubt that the babies in the latter group responded best, and there was a measureable difference between the groups. In spite of the fact that the younger babies had more immature lungs, surfactant was nevertheless having some effect. If you do not treat these tiny babies they will die or become handicapped and it becomes a self-fulfilling prophecy. I think neonatolgy has felt this and people involved will be treating every baby that is admitted because that infant needs the best treatment available.

A. Jobe: Are there any questions from the audience?

Participant: What is the latest experience and information concerning nebulisation?

A. Jobe: Dr. Adams spent a number of years in the early days of the development of surfactant therapy trying to apply nebulisation. One of the problems is the dose and the time it takes to deliver a dose, another problem is that phospholipids are difficult to nebulise – they foam and probably change their structure. Thus there are significant technical problems, at least with traditional methods of nebulisation. Perhaps there are strategies using dry powder nebulisation, but one of the considerations for people not in the surfactant field is that they are wary about the concept of giving babies 3–5 ml saline tracheally. This produced some resistance in the early days, but it works and seems to be harmless. Converted to adults that means that 5–10 g surfactant in about 200 ml fluid is instilled – this is in fact what we are doing with the babies. I am not sure whether the instillation technique is not the most appropriate technique, but it is worth thinking about other approaches. Dr. Kendig what was your experience when you worked with some of the aerosolisers in Rochester?

J. Kendig: Our initial interest with aerosols diminished after we saw the very effective response we had to our postventilatory doses, which were simply instilled. We were very impressed with how well the babies seemed to tolerate the instillation and thus our enthusiasm with aerosols diminished. But nebulisation may have some role in repeat or maintenance doses in the future. As far as administering the initial dose is concerned, the instillation and handbagging that we use works extremely well and may even open up some atelectic areas and encourages a better distribution than would be achieved with an aerosol.

C. Morley: I would like to make two practical comments on that. Firstly the surfactant on the whole has been too precious to let it disappear into the air and, secondly, as far as clinical trials are concerned we wanted to know whether it was getting into the lungs. If you just put it into a headbox or face mask you have no idea what proportion the baby is actually getting. These are two practical reasons why we have steered clear of nebulisation.

M. Obladen (Berlin, FRG): During these trials, has anyone looked at the structural morphology of the preparation simultaneous to the administration? My feeling is that it is not very homogeneous from batch to batch and some of the differences in response may be due to heterogenity in the preparation. Does anyone have experience with the structural morphology of exogenous surfactant?

A. Marini (Milan, Italy): When people are working with different types of natural surfactant the surfactant is usually prepared at different times, thus it is never totally consistent. Differences exist between individual batches and how can the differences be tested? Is there a standard animal model?

B. Disse (Ulm, FRG): I think that in the absence of experience there is no other choice but to test every batch in premature animal models.

A. Jobe: That is one possibility. If one is going to test in animals, or on a surface balance, or in babies one has to at least consider dose-response differences. If one is going to choose a dose of 200 mg per kg perhaps all the preparations will behave similarly even though they are quite different – for example, you may be using a very high dose. But if one is going to select preterm animals, (you have selected the ventilated preterm rabbit, others such as Dr. Fujiwara selected the pressure volume curve of the preterm rabbit and others are using lavaged rat lungs for the standard model), this means that even selection of an animal model does not solve the problem as there are multiple animal models. I suppose from a pharmacological point of view for a drug company that is making a massive dose one could consider animal testing but when we are using human surfactant the batches are so small it would all be used in testing before we had an opportunity to give it to the babies.

Are there other comments about animal models and how to approach this from a practical viewpoint?

C. Morley: For my purposes animal models, as they stand, are not suitable as they are too short-term. The effect I demonstrated with my surfactant was becoming apparent after several hours – about 6 hours. Very few animal models would have picked that up. We did show a difference in lung compliance in preterm rabbits but if looking for a longer-term effect it would have been missed in the animal experiments which are currently being conducted. As a neonatologist I am not so interested in what happens in 1 or 2 hours, but I want to know what happens in the course of several days and weeks.

M. Obladen: Colin, I am not sure if this is right. The major purpose why we are involved with surfactant substitution is that surfactant lowers surface tension at the air/water interface and by this action improves compliance. If someone tells me he substitutes surfactant I would expect compliance to increase immediately. If compliance does not go up it may be beneficial to the baby but he is not substituting that which is primarily a surfactant deficiency – he is not abolishing surfactant deficiency.

C. Morley: I agree that good surfactant should form a monolayer, lower surface tension and improve compliance straight away. But if the surfactant we are using is so surface active why do we need 100 mg when probably 1, 2 or 3 mg should do? If you need 100 mg of surfactant for certain dose responses it is obviously not very surface active.

A. Jobe: I would like to make a comment concerning the dose. The initial dose is very interesting – empirically we seem to use relatively large doses and across all the studies the average is about 100 mg per kg and in multiple doses even higher. For reasons we may not understand the normal newborn has a pool size of about 100 mg per kg, or higher. So why do babies need so much surfactant? I do not think we really understand this aspect. But it is clearly more than required to form a monolayer, and babies need a lot more to form a monolayer. This needs more investigation. I would like to ask Dr. Enhorning, particularly after seeing the studies from our Rochester colleagues, why do we not have to treat these babies in the delivery room?

G. Enhorning: Because it is important to give the surfactant before the first breath – I think this is most important. Dr. Merritt has pointed that out that it is also important to make sure that the tube is correctly placed by seeing that the lungs ventilate properly. But then it is too late – the lungs are already aerated and the surfactant is not likely to be correctly spread within the lungs.

A. Jobe: Yet a number of the clinical studies seem to demonstrate that there really is no difference in efficacy whether the babies are treated at 2 or 6 hours – I believe these were the figures from the recent Rochester study versus treatment in the delivery room. I think my bias would be that you really need delivery room treatment for optimal therapy – but some of the data that is coming in suggests that this may not be true.

G. Enhorning: But I think your own data demonstrates that it is true.

A. Jobe: Well the animal model data does, but I am not sure that this is so with the infant data. I think this is an important issue in terms of how we design our future studies.

A. Merritt: What evidence is there that 1 or 2 breaths will cause epithelium necrosis – or in fact do any harm? We certainly do not ventilate the babies, with a hand ventilation bag we determine that we can hear a breath sound in the left thorax. What evidence of harm is there?

G. Enhorning: I think there will be some damage. But the point I want to make is that if you have air in the airway it is less likely that you can get the surfactant in there. Another thing I would like to comment on is that we have here a panel of experts representing various types of surfactant and I do not agree that it is impossible to compare the surfactants. I agree with Dr. Morley that it is probably very difficult to determine a difference in mortality between the various preparations but certainly one could test them physically, or with animal experiments. I think that Dr. Lachmann would be interested in heading such testing so perhaps we could all submit our surfactant to him for comparative trials in animal models?

H. Halliday: Regarding the timing of replacement: recently we had a beautiful response at 31 hours with a baby on 100 % oxygen and the baby was off the ventilator within 24 hours. So babies do seem to respond somewhat later than we might think.

R. Rüfer (Mannheim, FRG): Continuing the discussions on different surfactant preparations I feel that we focus too much on the main function of lowering of surface tension and too little on the second function of stabilising the system of alveoli having different diameters. I feel we should acquire more information on inflation and deflation characteristics. It is clear that lowering surface tension increases compliance and decreases the opening pressure but we lack information on the effect of surfactant on, for example, deflation curves. I feel we should remind our colleagues that pressure volume curves are an excellent way to assist analysis.

T. Fujiwara: Our preparation is very homogenous and quality control is conducted before we give it to humans. If we test surfactant in animal experiments e.g. using immature rabbits it's harder to control other parameters than lung mechanics, for example, circulation etc. So in assisting therapeutic response using such an animal you have to kill the animal to avoid the effects of circulation. The deflation curve is very important especially at lower inflation pressure. Our surfactant shows residual volume of 42–53 ml per kg (average 45 ml per kg) at pressure of +5 cm H_2O. I think this is a very reasonable residual volume for newborns. This is one criteria and also in vitro on the Wilhelmy balance a small compression of the surface area of only 20% should reduce the surface tension to below 10 or 15 mN/m and I believe that this is another important criteria for an effective surfactant.

A. Jobe: To interpret your comment a little bit differently: I think your point is very important in terms of lung volumes. One of the problems with infants is knowing what the functional residual capacity is and a major problem with all the mechanics measurements in babies is that we may be able to measure compliance but we do not know where we are in the pressure volume curve, and if we are up too high up in the curve we do not improve compliance. My clinical impression has been that when using surfactant therapy we are sometimes actually way up on our pressure volume curve and we have air trapping and/or very large FRCs and therefore low compliances – this can be a potential problem. It surprises me that, under these circumstances, we do not often see pneumothoraces. Clinically this does not seem to be much of a problem. But we should not take all the mechanics measurements too literally when we do not know the lung volumes.

B. Lachmann (Rotterdam, The Netherlands): This is why it is important to undertake mechanics studies during spontaneous breathing, as well as during mechanical ventilation, only in connection with FRC measurements, to get specific lung compliance.

H. Halliday: When giving surfactant to babies there is generally an immediate, visible and dramatic improvement in chest movement noted by our nursing staff. I am surprised that this is not measured by the pulmonary function test of compliance. I do not understand this because one would expect a change in compliance.

Panel: The nurses are sometimes better than some of our machines.

A. Jobe: My thanks to everybody for this most stimulating discussion.

Part V
What Happens After Surfactant Replacement Therapy?

What Happens to Treatment Doses of Surfactant?

A. Jobe, A. Pettenazzo, S. Seidner, and M. Ikegami

Variables Affecting Surfactant Clearance

To start this discussion of what happens to treatment doses of surfactant, perhaps the conclusion should be stated first: we know very little about the fate of exogenously administered surfactants. Almost all the data available are from animal models and there is no quantitative information from the human. To further complicate the issue, multiple variables relating to the surfactant tested and the animal or human receiving the surfactant probably change overall clearance measurements. The pathways responsible for clearance also may change under different conditions. Some of the variables that have some experimental foundation are listed in Table 1. Each investigator has used different surfactants at different doses that have been resuspended in different water based vehicles for administration [10, 43–45, 53]. The aggregate sizes and lipid structures will differ based on the calcium, surfactant protein, and suspension techniques selected [39]. Total phosphatidylcholine or saturated phosphatidylcholine have been studied almost exclusively, and other components of surfactant may have different fates [53]. It is clear that catabolic rates change with lung development, as do macrophage numbers and the distributions of other cell types in the lungs. Surfactant treatments will be used clinically only for lung immaturity or in lung injury. The severity of lung immaturity probably influences surfactant clearance rates [26]. In the adult lung, the type of injury will no doubt affect the fate of exogenous surfactants [3, 4, 37]. A primarily epithelial injury with type II cell damage will change surfactant catabolism differently than will an endothelial injury with complicating alveolar edema. Mechanical ventilation in the developing or adult

Table 1. Variables affecting surfactant clearance

Surfactant-dependent variables	Subject-dependent variables
Type of surfactant [10, 43–45]	Developmental stage of lung [21, 22, 26]
Dose of surfactant [43–45]	Type of lung injury [3, 4, 37]
Component of surfactant to be evaluated [53]	Severity of lung injury [18, 47]
Aggregate form of surfactant [39, 43, 53]	Mechanical ventilation [11]
	Nutritional effects [5]

lung may also change surfactant clearance kinetics [11]. Also the health of the lung in terms of basic metabolic and healing processes will be influenced by nutritional status [5]. The results to be summarized in this review relate primarily to healthy lungs in animals as very little information is available that is directly applicable to the injured lung.

Surfactant Phosphatidylcholine Metabolic Pathways

Surfactant Synthesis, Secretion and Recycling. Normal surfactant metabolic pathways are complicated and not completely understood. Figure 1 summarizes the major directions of movement of surfactant phosphatidylcholine from synthesis in the endoplasmic reticulum to secretion. Surfactant phosphatidylcholine is synthesized in the same intracellular location in the type II cells as are other phospholipids that are not part of the surfactant complex. By presently unknown mechanisms, possibly involving transfer proteins, the predominantly saturated and monoenoic phosphatidylcholines are transferred via the golgi to vesicular forms thought to represent "new" or "growing" lamellar bodies [7, 25, 50, 51]. Probably very little synthetic activity occurs in these vesicular forms which then mature into lamellar bodies capable of basal or stimulated secretion to the alveolus [25, 32]. The kinetics of movement of surfactant phosphatidylcholine from synthesis to secretion is very slow, an observation leading to the conclusion that *de novo* synthetic pathways will not be responsive in a timely fashion to an acute surfactant deficiency [26]. For example, following the intravascular injection of radiolabeled palmitic acid into

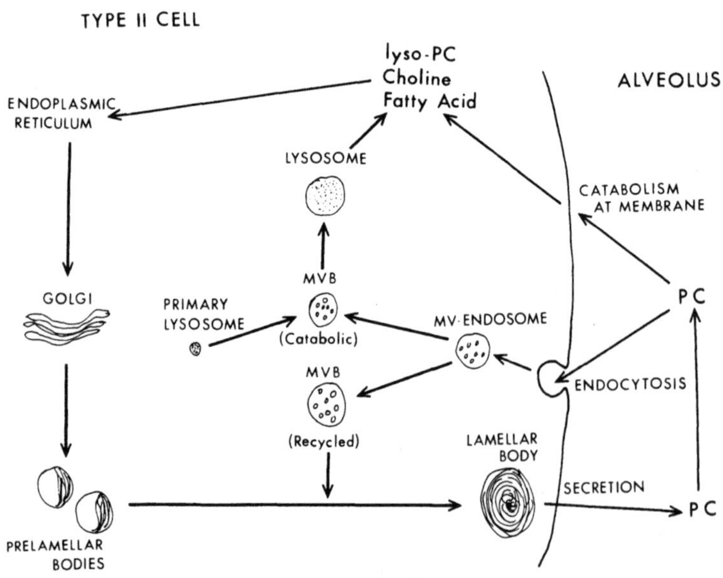

Fig. 1. Pathways of surfactant phosphatidylcholine from synthesis to catabolism. *PC,* phosphatidylcholine; *MV,* multivesicular; *MVB,* multivesicular bodies

rabbits or sheep, maximal incorporation of the labeled precursor into lung phosphatidylcholine occurs within about 10 min [19, 23, 31]. However, recovery of labeled surfactant phosphatidylcholine from the alveolar pool does not reach its peak until many hours later [23, 31, 34] (Fig. 2). The times to peak activity were on the order of 12 hrs for adult rabbits and 30 hrs for newborn rabbits [23]. The times were even longer in the sheep. As a generalization, the time delays from synthesis to maximal activity in the alveolar pools are longer in developing than adult animals.

Once secretion has occurred, surfactant seems to have a complex "alveolar life cycle [1, 33, 36, 52]. The surfactant is secreted as a large aggregate – the lamellar body, which is in the order of one micron in diameter. Tenuous evidence points to the unraveling of the lipid layers to form tubular myelin, a loose lattice of phospholipids and surfactant specific protein that easily adsorbs and spreads to form the monolayer [42]. With normal respiration, the monolayer is thought to squeeze out some components that were part of the secreted form of surfactant and to be "used" to then re-enter the hypophase as small phospholipid aggregates that have a phospholipid composition similar to whole surfactant but with less of the SP-35 protein [52]. Thus, there are a series of possible precursor-product relationships for the different aggregate forms of alveolar surfactant. The recovery of surfactant phosphatidylcholine by sequential differential centrifugations illustrates the complex size characteristics of alveolar surfactant, assuming the alveolar wash procedure recovers material representative of the material present in the alveolus [26] (Fig. 3). The surfactant aggregates isolated between about 100 xg and 48,000 xg are the

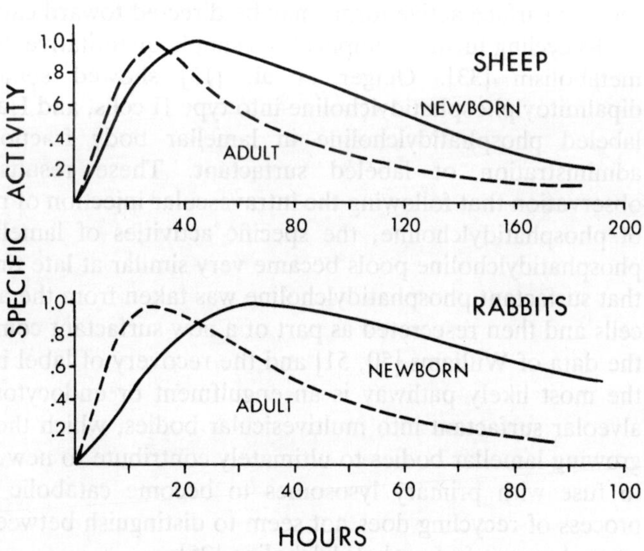

Fig. 2. Specific activity-time relationships for alveolar surfactant phosphatidylcholine. The adult and newborn sheep and rabbits received intravascular injections of radiolabeled palmitic acid and the radioactivity per mol phosphatidylcholine was measured in alveolar samples at multiple subsequent times. The curves were normalized such that the maximal specific activities were given a value of 1.0, as summarized in [26]. (Reproduced with permission)

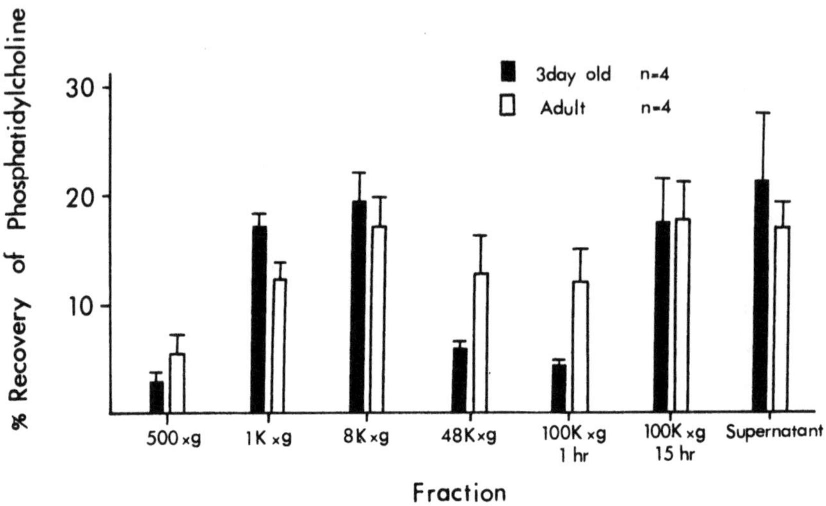

Fig. 3. A separation of alveolar surfactant aggregates by sequential differential centrifugation. Alveolar washes from healthy 3-day-old and adult rabbits were centrifuged at the indicated g forces and the percent recovery of phosphatidylcholine in each fraction was measured. The multiple aggregate sizes of the surfactant phospholipids in solution, are illustrated. (Reproduced with permission [26])

surface active forms, while the smaller aggregates do not adsorb or lower surface tension nearly as effectively [36, 53]. In the separation in Fig. 3, only about 45% of surfactant phosphatidylcholine is in a highly surface active form; the less surface active forms may be directed toward catabolism and recycling.

Recycling further complicates a simple quantitative description of surfactant metabolism [33]. Geiger et al. [13] showed some uptake of labeled dipalmitoylphosphatidylcholine into type II cells, and Hallman et al. [16] found labeled phosphatidylcholine in lamellar body fractions following tracheal administration of labeled surfactant. These results, together with the observation that following the intravascular injection of radiolabeled precursors of phosphatidylcholine, the specific activities of lamellar body and alveolar phosphatidylcholine pools became very similar at late times, strongly indicated that surfactant phosphatidylcholine was taken from the alveolar pool by type II cells and then resecreted as part of a new surfactant complex [2, 23]. Based on the data of Williams [50, 51] and the recovery of label in lamellar bodies [16], the most likely pathway is an engulfment or endocytosis of a subfraction of alveolar surfactant into multivesicular bodies, which then either fuse with the growing lamellar bodies to ultimately contribute to new lamellar bodies and/or to fuse with primary lysosomes to become catabolic vesicles (Fig. 1). The process of recycling does not seem to distinguish between phosphatidylcholine or analogues of phosphatidylcholine [25].

Estimates of the efficiency of the recycling process have varied [1, 16, 21, 22]. Measurements of efficiency are very indirect if intravascular precursors of phosphatidylcholine are used to label the surfactant pools via *de novo* synthesis [1]. Intratracheally injected labeled surfactant can be used to measure

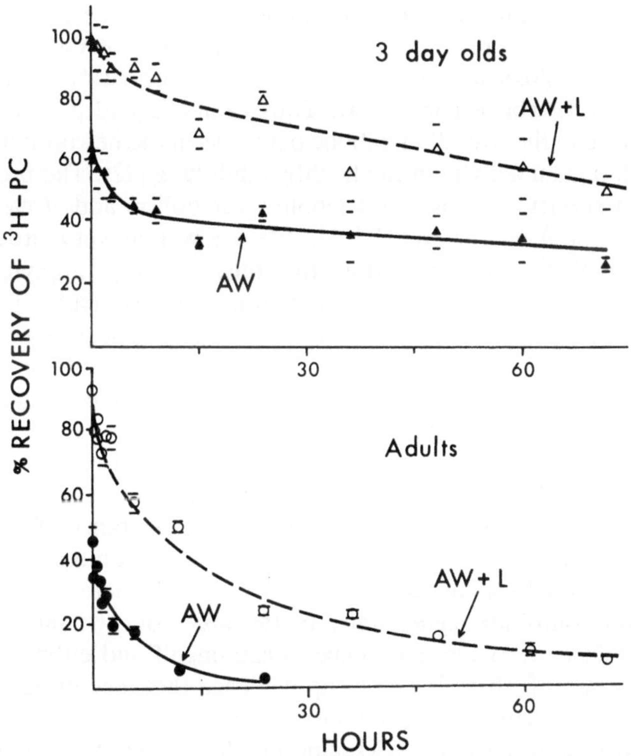

Fig. 4. Clearance of radiolabeled natural surfactant following airway injection into rabbits. Natural surfactant that was radiolabeled with [^{3}H]choline was recovered from the lungs of 3-day-old rabbits by alveolar wash. This surfactant then was injected into the airways of both adult and 3-day-old rabbits and the % recovery of labeled phosphatidylcholine with time was measured in the alveolar wash and lung tissue after alveolar wash. The *curves* show the recoveries for the alveolar wash *(AW)* and the lung compartment *(AW+L)*. (Reproduced with permission [26])

recycling efficiency only if the administered surfactant mixes with and is metabolized equivalently to the stable endogenous pool. Given that assumption, we have measured recycling efficiencies in 3-day-old and adult rabbits [21, 22]. The calculations depend on exponential curve fitting, but the essence of the results can be appreciated directly from the primary data (Fig. 4). Following the tracheal injection of labeled surfactant, about 50% of the labeled phosphatidylcholine becomes lung associated in that it cannot be recovered by alveolar wash in both adult and young rabbits. Subsequently the amount of label recovered by alveolar wash persists for many hours in young animals, while virtually all the label has disappeared from the alveolar wash of adult rabbits. These curves yield recycling efficiencies of about 90% for the young rabbits and about 20% for the adult rabbits [21, 22].

Surfactant Catabolism. The curves in Fig. 4 also point out another striking difference between the adult and developing animal – the amount of catabolic activity. Labeled surfactant phosphatidylcholine is rapidly lost from the adult lung while only 30% is lost from the lungs of the young animal over 72 hrs. The biological half-life of surfactant phosphatidylcholine was about 6 days in term newborn lambs [14]. Thus catabolic and clearance activities contribute little to the overall metabolic pathways in the newborn lung but may represent a dominant activity in the healthy adult lung [22]. The precise sites and pathways of surfactant phosphatidylcholine catabolism and/or clearance in the lungs are not well characterized [33]. We know that very little surfactant is lost via lymphatic drainage from the lung [8, 48], implying little movement of surfactant into the lung interstitium. A corollary of this lack of lymphatic clearance is that since lymphatics drain approximately the same space that exchanges across the vascular endothelium, little intact surfactant should move from the alveolus to the vascular space. We found that some surfactant phosphatidylcholine derived label was recovered from the liver, but the labeled phosphatidylcholine was from *de novo* synthesis using labeled precursors presumably released from the lungs [43].

In recent experiments we found that clearance of surfactant up the airways accounted for no more than 2 to 5% of the amount of labeled surfactant lost from the lungs of adult rabbits over a 24 hr period. Airway clearance also did not contribute significantly to the loss from the lungs of treatment doses of surfactant. While macrophages can engulf and either catabolize or synthesize surfactant, there is no strong evidence that macrophages contribute to a large extent to surfactant catabolism [9, 38]. As can be seen from Fig. 4, the labeled surfactant phosphatidylcholine rapidly became lung associated and presumably catabolic activity was a property of the lung tissue itself. In unpublished studies we have noticed that the diether analogue of phosphatidylcholine was cleared from the alveolus equivalenty to surfactant phosphatidylcholine, but it was not catabolized effectively by the lung. Since the ether label did not accumulate in macrophages, we do not think they are a major catabolic pathway for surfactant.

Presumably catabolism occurs at the alveolar and/or small airway epithelial surface or after epithelial uptake. The proposed uptake pathways for type II cells were outlined above [51]. Since both adult and developing lungs contain lysosomal enzymes, although at a lower level in the developing lung [17], the different recycling rates in developing and adult lungs may result from differential trafficking of endocytolic multivesicular bodies (Fig. 1). On the other hand, different cell types may be more or less important to catabolism at different stages of development. For example, the contribution of type I cells to surfactant metabolic pathways is unknown. Chander et al. [6] and Fisher et al. [12] recently characterized the catabolism of liposome associated phosphatidylcholine by type II cells from rat lungs and by isolated perfused rat lungs. In both systems the phosphatidylcholine was degraded and some of the degradation products were reused for surfactant phosphatidylcholine synthesis. They thought that the degradative activity might be at the plasma membrane since agents known to interfere with lysosomal pathways did not alter the catabolic activity [6]. In summary, catabolic pathways are active in the adult animal, but their characteristics and cellular locations remain to be defined.

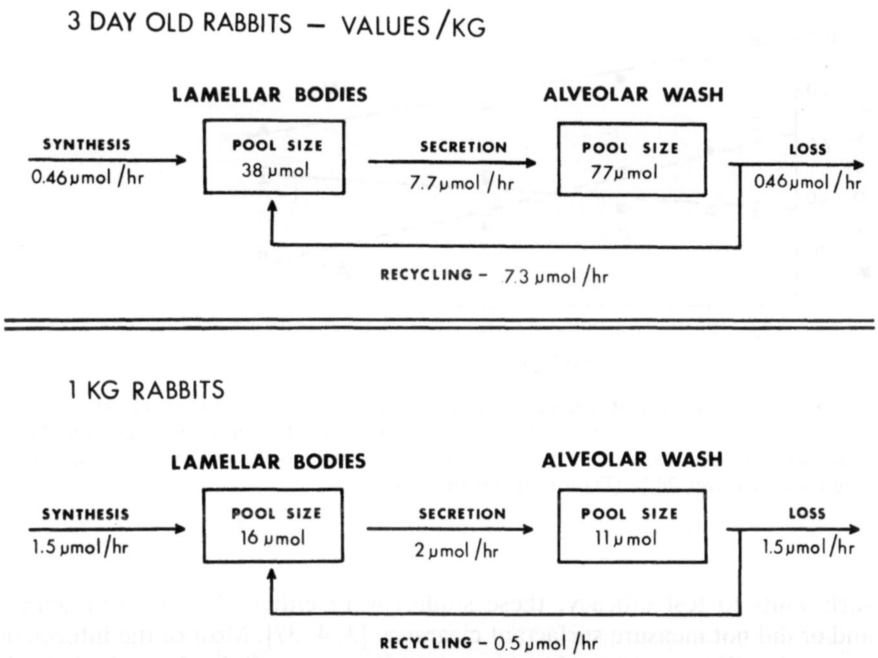

Fig. 5. Estimates of pool sizes and flux rates for surfactant phosphatidylcholine. Values for the 3-day-old and adult rabbits were converted to values/kg body weight. The numbers were derived from our studies of surfactant metabolism [21–23]

Surfactant Pool Sizes. Another striking difference between the term newborn and the adult lung is the surfactant pool size. Adult rabbits, sheep, and monkeys have alveolar surfactant pool sizes of about 10–15 μmol phosphatidylcholine/kg (about 10–15 mg total surfactant lipid/kg body weight), while the newborns have pool sizes of about 100 mg/kg [20, 28–30). In contrast, preterm animals with an RDS-like syndrome have alveolar pool sizes that approximate those of the adult [20, 27, 30]. The surfactant pool of the preterm is deficient relative to the healthy term animal but not the adult. The reason for the large pool size in the term animal is unknown. The quantitative relationships between lamellar body and alveolar phosphatidylcholine pools in adult and 3-day-old rabbits emphasize the differences that occur during development [21–23] (Fig. 5). This figure is an oversimplification in that it does not consider the precursor-product relationships within the maturation of lamellar bodies or within the alveolar pool. However, these metabolic pathways and pool size relationships are integral to trying to understand how surfactant treatments might interact with endogenous surfactant metabolic pathways.

Clearance and Catabolism of Treatment Doses of Surfactant

Adult Rabbits. While adult animals with different lung injuries such as vagotomy, oxygen toxicity, and saline lavage have been treated with

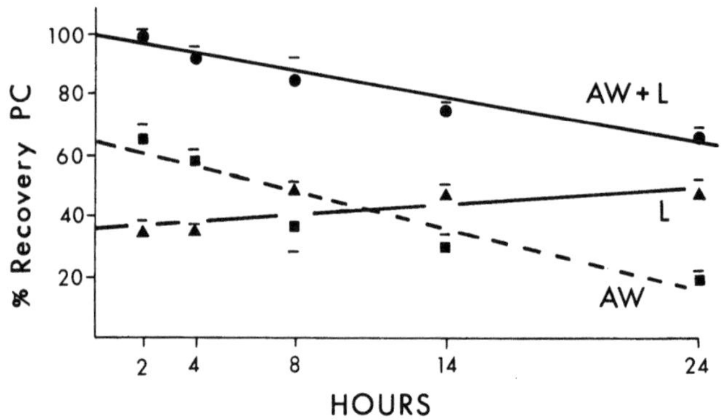

Fig. 6. Percent recovery of labeled phosphatidylcholine from calf surfactant. The treatment dose of surfactant was lost linearly from the alveolar wash *(AW)* and accumulated in the lung tissue after alveolar wash *(L)*. Overall about 30 % of the labeled PC was lost from the total lung *(AW+L)* over 24 h. (Data redrawn from [43])

surfactants to test efficacy, these studies were either of very short duration and/or did not measure surfactant clearance [3, 4, 37]. Most of the information concerning the clearance of exogenous surfactants or lipids from the lungs has been collected using doses that were small relative to the endogenous pool. These studies were designed to probe normal catabolic and clearance pathways. For example, Oyarzun et al. [41] showed that the clearance of liposomes of dipalmitoylphosphatidylcholine increased with hyperventilation and the addition of phosphatidylglycerol to the liposomes also increased clearance rates. Despite the predictions that the small aggregate forms of surfactant were the "catabolic forms" [1, 36], Wright et al. [53] found more rapid lung association of the large than the small aggregates in adult rabbits. As pointed out above, small amounts of exogenous surfactants will mix with the alveolar pool and apparently enter recycling pathways [22].

We recently measured clearance rates and some aspects of catabolism following the intratracheal injection of large doses of surfactants in rabbits [43]. The injections were made via a small catheter passed through the procedure channel of a bronchoscope that had been exteriorized about 8 cm from the tip of the bronchoscope. The catheter was advanced through the cords to just above the carinas of lightly ether anesthetized 1 kg rabbits, and the surfactant suspension in 4 ml of 0.45% NaCl was injected. The technique allowed us to inject large volumes with no visible loss up the airways. We injected natural calf surfactant and natural rabbit surfactant that had been isolated by centrifugation of alveolar washes and a lipid solvent extracted surfactant, Surfactant-TA, (Ross Laboratories). The surfactants were radiolabeled with dipalmitoylphosphatidylcholine and recoveries of the injected doses were quantified over 24 h in alveolar washes, the lung tissue after alveolar wash, and the total lung. An example of a clearance curve for calf surfactant injected at an average dose of 128 mg/kg is shown in Fig. 6. These animals have an endogenous alveolar surfactant pool size of about 12 mg/kg; therefore, the dose exceeded the

Table 2. Comparison of recoveries 24 h after treatment doses of three surfactants in 1-kg rabbits

Surfactant no.	Surfactant	*n*	Alveolar wash	Lung tissue	Total lung
1	Rabbit	5	6.9 ± 1.0	24.5 ± 1.51	31.4 ± 1.5
2	Calf	6	18.5 ± 2.9	45.6 ± 4.3	64.1 ± 3.6
3	Surfactant-TA	5	9.7 ± 0.9	46.2 ± 3.4	55.9 ± 3.8
	$p < 0.01$		2 > 1,3	2,3 > 1	2,3 > 1

Data abstracted from Pettenazzo et al. [43].

endogenous pool by about 10-fold and is approximately the dose used to treat preterm infants with respiratory distress syndrome. This curve was characteristic for the different surfactants in that a large percent of the labeled phosphatidylcholine was not recoverable by alveolar wash even at early times and most of the labeled phosphatidylcholine was lost from the alveolar wash over 24 h. The overall clearance from the total lungs was quite linear at a rate of 1.5%/h. The clearance rates from total lung for rabbit surfactant and Surfactant-TA were 3.1% and 1.8%/h, respectively. The recoveries in the alveolar wash, lung tissue, and total lung differed for the three surfactants at 24 h (Table 2). More of the calf surfactant and Surfactant-TA remained in the lung tissue than did rabbit surfactant, and more calf surfactant was recovered by alveolar wash than was either rabbit surfactant or Surfactant-TA. The lipid solvent extraction used to prepare Surfactant-TA removed SP-35 from the cow lung surfactant and the resuspension steps changed the aggregate sizes and forms of Surfactant-TA relative to the natural surfactant; however, overall clearance rates were not dramatically altered. The reasons for the differences between calf and rabbit surfactants are not known and could result from species differences in the surfactant specific proteins. However, it is perhaps better to emphasize similarities rather than differences: all three surfactants were rapidly lost from the alveoli over 24 h and the majority of the labeled phosphatidylcholine had become lung tissue associated.

A particularly remarkable characteristic of surfactant clearance in the adult rabbit is the nonsaturability of clearance of phosphatidylcholine [43] (Fig. 7). Rabbit surfactant was given to groups of five rabbits at six different doses ranging from about 1.5 mg/kg to 70 mg/kg (about 12% to 600% of the endogenous pool), and the amount of the labeled surfactant phosphatidylcholine recovered after 24 h was measured. The amount recovered was a relatively constant percent of the amount injected in alveolar wash (5.4 ± 0.6%), lung tissue (19.6 ± 1.8%), and total lung (24.9 ± 2.2%), indicating that overall about 75% of the labeled phosphatidylcholine administered was lost, independent of dose. This result indicates that the lung has clearance and catabolic pathways that can accommodate surfactant doses in the range tested.

We measured the recovery of labeled phosphatidylcholine in organs distant from the lungs following the tracheal injection of labeled rabbit surfactant [43]. While about 75% of the labeled surfactant phosphatidylcholine was lost from

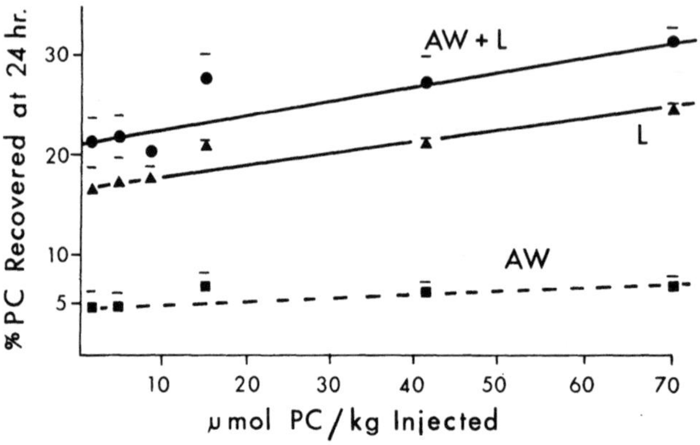

Fig. 7. Percent recovery of different doses of rabbit surfactant phosphatidylcholine from adult rabbit lungs 24 h after airway injection. The % recoveries are shown for the alveolar wash *(AW)*, lung tissue after alveolar wash *(L)*, and the total lung *(AW+L)*. There were comparable percent recoveries in the lung compartments, independent of dose. (Data redrawn from [43])

the lungs over 24 h, very little was detected anywhere except the liver, where about 8% of the label was recovered in phosphatidylcholine. Treatment of the labeled phosphatidylcholine from the liver with OsO_4 destroyed almost all the phosphatidylcholine indicating that it was unsaturated. Therefore, we assume the label that was found in the liver came from catabolic products of phosphatidylcholine derived from degradation in the lungs. This interpretation is consistent with the catabolic activity described by Chander et al. [6] and Fisher et al. [12].

In a further experiment we tried to detect effects of 100 mg/kg doses of rabbit surfactant, calf surfactant, or Surfactant-TA on the endogenous metabolic pathways of surfactant [43]. The surfactants seemed to have no effect on the total synthesis of phosphatidylcholine by the lung. Also, the percent secretion of the labeled phosphatidylcholine did not change for rabbits that received the surfactants or vehicle control. We interpret these experiments to indicate no large effect of the surfactants on normal metabolic pathways; however, the experiments were of low resolution because of variability in labeling using intravascular precursors in the different rabbits. In summary, we found treatment doses to about 100 mg/kg were not harmful to adult rabbits, and there were no indications of potential toxicity that resulted from the metabolic or clearance studies.

Three-Day-Old Rabbits. We have given 3-day-old rabbits very large amounts of natural surfactant and liposomes to measure clearance rates and metabolic effects of the intratracheal lipids on the lungs of young animals [40, 44, 45]. The 3-day-old rabbits have endogenous alveolar surfactant pool sizes of about 75 mg lipid/kg and the pool size changes very little over the 2–6 day period of life. The animals also can easily be given tracheal injections or intravascular injections. Since clearance rates are quite slow at this stage of development, the

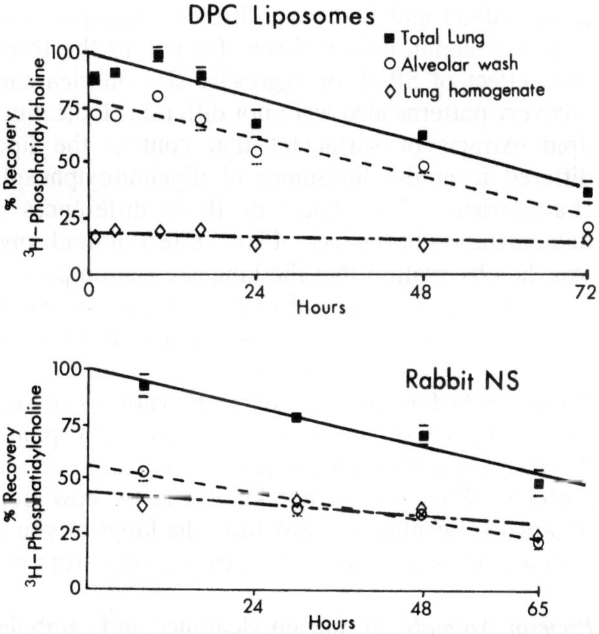

Fig. 8. Percent recovery of phosphatidylcholine from treatment doses of liposomes of dipalmitoylphosphatidylcholine *(DPC)* and rabbit natural surfactant *(NS)*. The 3-day-old rabbits were given the labeled DPC liposomes or rabbit NS by tracheal injection and recoveries measured. The distribution of the labeled PC between alveolar washes and lung tissue differed. (Data redrawn from [40])

measurements are made over 72 h. Labeled rabbit surfactant phosphatidylcholine was cleared from the lungs of 3-day-old rabbits at a constant percent of the injected dose, demonstrating the same nonsaturable clearance kinetics seen with adult rabbits [45]. However, the rate of clearance was 16%/24 h, a rate 4.7 times slower than for rabbit surfactant in adult rabbits. In the case of the 3-day-old rabbits, the highest dose tested, about 160 mg/kg, was about two times the endogenous pool because of the high endogenous pool size in these young rabbits. In other experiments, clearance rates did not change with species source of the surfactant except perhaps for a somewhat slower rate of clearance for sheep surfactant [45]. We found that while the rate of clearance from the lungs of small liposomes of dipalmitoylphosphatidylcholine was the same as that of rabbit surfactant when both were given at doses of over 300 mg lipid/kg, the percent recoveries by alveolar wash were different [40]. At all times most of the labeled phosphatidylcholine was recovered with the liposomes, while only about 50% of the labeled phosphatidylcholine could be recovered by alveolar wash from the rabbit surfactant treated animals (Fig. 8).

To more critically evaluate the effects of aggregate form on rates of clearance, we treated 3-day-old rabbits with about 100 mg/kg sheep surfactant, sheep surfactant that had been lipid solvent extracted and resuspended as large aggregates, or lipid solvent extracted sheep surfactant in the form of 175 ± 50 nanometer diameter liposomes [44]. The two lipid solvent extracted surfactants lacked SP-35 and had very different aggregate sizes. The rates of clearance were not significantly different for any of the three surfactants, documenting little effect of SP-35 or aggregate size on clearance rate. The alveolar wash recovery patterns also were not different, indicating that liposomes made from lipid extracts of surfactant that contain the lipophilic surfactant proteins differed from the liposomes of dipalmitoylphosphatidylcholine in clearance characteristics. The cause of these differences is not known. A further demonstration of the lack of large effect of lipid solvent extraction of surfactant was the observation that the lung association patterns and lung clearance rates for Surfactant-TA and calf surfactant were similar in 3-day-old animals [44].

As with the adult animals, we found no indication of major effects of surfactant treatments on endogenous synthesis or secretory pathways [40]. Rabbit surfactant had no effect on synthesis or secretion of phosphatidylcholine. Sheep surfactant did not effect lung phosphatidylcholine synthesis, but it did increase somewhat the percent secretion of the labeled surfactant to the alveolus. While we do not know exactly how treatment doses of surfactant phosphatidylcholine are lost from the lungs of young healthy animals, the rate is slow and characteristic of clearance rates from endogenous pools.

Preterm Animals. Surfactant clearance and catabolism needs to be thought of as a two stage process in the preterm with RDS just as in the newborn or adult: an initial lung association process followed by loss from the lungs. A significant factor in evaluating the effect of a treatment dose of surfactant on alveolar pool size is the rapid and large lung association of the surfactant with the lung tissue [18, 24]. About 50% of labeled surfactant could not be recovered by alveolar wash from preterm ventilated rabbits soon after treatment. While not formally proved, our suspicion is that the amount of lung association increases as immaturity and severity of respiratory failure increases (Table 3). In comparing results from two different studies, we found much less

Table 3. Surfactant recoveries in preterm lambs treated with 50 mg/kg, ^{3}H-labeled natural sheep surfactant

	Study I	Study II
n	8	6
Gestational age (days)	125 ± 1.8	135 ± 0.2
Age at treatment (h)	2.3 ± 0.6	5
Time from treatment to death (h)	3.0 ± 0.4	5
Recovery of labeled phosphatidylcholine (%)		
Alveolar wash	19 ± 3	44.2 ± 2.0
Lung tissue	55 ± 4	41.3 ± 1.5
Alveolar wash plus lung tissue	74 ± 4	85.5 ± 2.0

Data abstracted from Ikegami et al. [18] and Jacobs et al. [24].

recovery of treatment doses of natural sheep surfactant by alveolar wash in more immature lambs despite a shorter interval from treatment to sacrifice and alveolar wash [18, 24]. The data also suggest less recovery in the total lung with increased severity of lung disease.

We found in a series of lambs delivered at 135 days gestational age and ventilated by either conventional ventilation or high frequency oscillation that alveolar recovery of a trace dose of labeled surfactant decreased strikingly as the endogenous alveolar surfactant pool size decreased and severity of disease increased [47] (Fig. 9). Some of these large changes in percent recovery could result from difficulties with alveolar wash in the more immature lambs; however, total recoveries in the lungs also decreased significantly with increased disease severity. This effect was detected over only a 3 h period of ventilation and probably would be cumulative with time. In unpublished experiments we have found that preterm lambs delivered at about 138 days

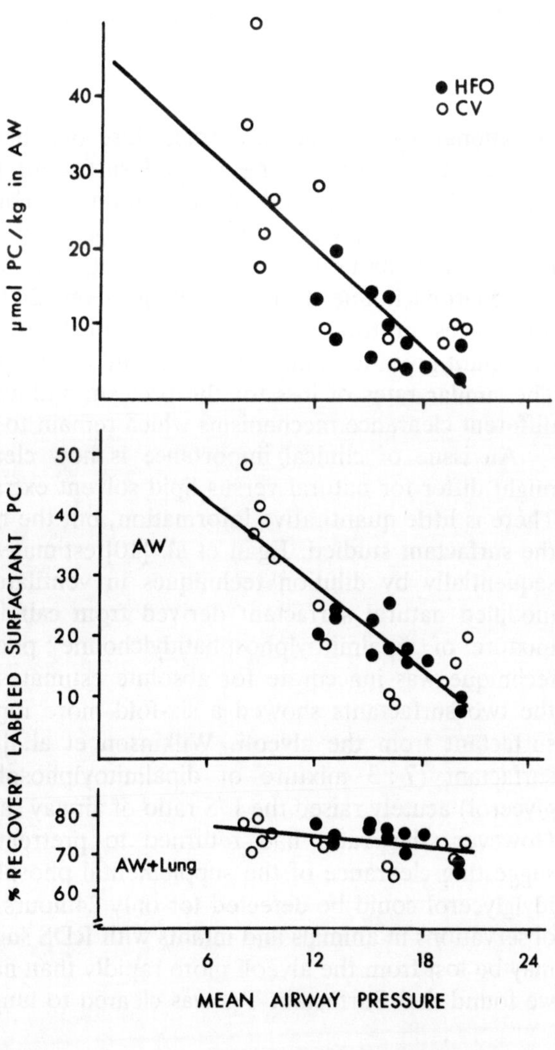

Fig. 9. Surfactant pool sizes and percent recovery of labeled phosphatidylcholine *(PC)* in preterm lambs ventilated for 3 h by either conventional ventilation *(CV)* or high-frequency oscillation *(HFO)*. The mean airway pressure needed to normalize pCO_2 values is shown versus alveolar wash *(AW)* pool size and surfactant recovery percentages. (Data redrawn from [47])

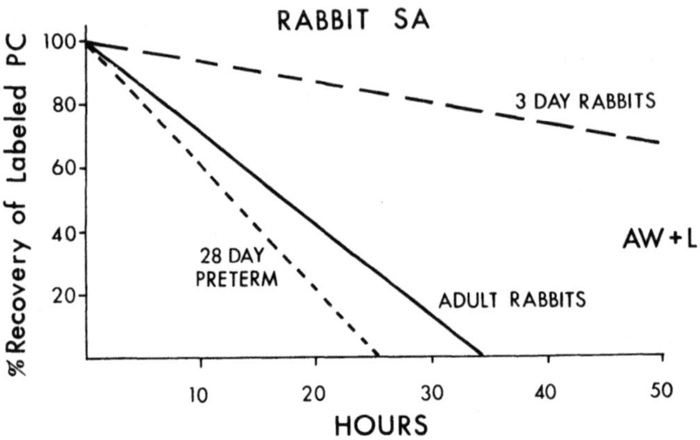

Fig. 10. Clearance of labeled rabbit surfactant following airway injections into adult, 3-day-old, or 28-day preterm ventilated rabbits. Clearance rates in the total lungs differed at different developmental stages. Curves derived from data presented in [44–46]. The curve for the 28-day rabbits was approximated by extrapolation from a 6-h experiment

gestational age and given a trace dose of labeled natural surfactant lost no measurable amount of the labeled surfactant phosphatidylcholine from the lungs over a 24 h period of ventilation. In contrast, more immature lambs delivered at 132 days gestational age and treated with 100 mg/kg of radiolabeled natural sheep surfactant lost about 30% of the labeled phosphatidylcholine from the lungs over 24 h. Figure 10 summarizes the differences in total lung recoveries of treatment doses of natural rabbit surfactant given to adult, 3-day-old, and 28 day preterm ventilated rabbits [35]. The similar rates of loss for the preterm and adult animals likely result from different clearance mechanisms which remain to be defined.

An issue of clinical importance is how clearance and tissue association might differ for natural versus lipid solvent extracted or synthetic surfactants. There is little quantitative information, but the result probably will depend on the surfactant studied. Egan et al. [10] estimated alveolar phospholipid pools sequentially by dilution techniques in ventilated lambs treated with either modified natural surfactant derived from calf lung alveolar wash or a 7 : 3 mixture of dipalmitoylphosphatidylcholine : phosphatidylglycerol. While the technique was inaccurate for absolute estimates, the relative comparisons of the two surfactants showed a six-fold more rapid clearance of the synthetic surfactant from the alveoli. Wilkinson et al. [49] found that 25 mg of dry surfactant (7 : 3 mixture of dipalmitoylphosphatidylcholine : phosphatidylglycerol) acutely raised the L/S ratio of airway samples from infants with RDS. However, this ratio had returned to pretreatment levels within 20 hours suggesting clearance of the supplemental phosphatidylcholine. The phosphatidylglycerol could be detected for only 24 hours in the airway samples. These observations in animals and infants with RDS suggest that synthetic surfactants may be lost from the alveoli more rapidly than natural surfactants. In contrast, we found that Surfactant-TA was cleared to lung tissue equivalently to rabbit surfactant from the alveoli of preterm ventilated rabbits but loss from the lung tissue was less rapid [46].

Following intratracheal administration of either tracer or treatment doses of natural surfactant to preterm animals with RDS, at least 50% of the phosphatidylcholine associated with the natural surfactant rapidly becomes "lung associated" in that the phosphatidylcholine cannot be recovered by alveolar wash. If the alveolar wash relatively accurately recovers the alveolar surfactant pool, only half of the treatment dose of natural surfactant will be present in the alveoli, the other half presumably being incorporated into intracellular pools within the lungs. Unfortunately, this behavior of surfactant together with probable differences in clearance rates depending on severity of disease and developmental maturity make attempts to measure endogenous pools and secretory rates of surfactant by dilution techniques very unreliable in humans [24]. For example, if the administered dose of surfactant stayed within the alveoli, mixed well with endogenous alveolar pools, and had known clearance rates, the treatment dose could be isotope labeled and the change in isotope concentration with time could be used to measure the endogenous alveolar surfactant pool and surfactant secretory rate. Hallman et al. [15] have used this approach to measure the rate of clearance of phosphatidylglycerol following treatment of preterm infants with RDS with surfactant isolated from amniotic fluid. No phosphatidylglycerol was present in airway samples of the infants before treatment. They found an "apparent biological half-life" of about 30 hours for the phosphatidylglycerol in airway samples. However, the measurements remain rough estimates because of potential mixing, lung association, and sampling problems.

As with surfactant treatments given to healthy adult and newborn animals, there have been no metabolic or catabolic effects of the large doses of surfactant that suggest that the exogenous surfactants would be harmful to the preterm lung [40]. Preliminary results from our recent experiments studying surfactant metabolism in preterm ventilated lambs indicate active recycling with maintenance of surfactant function [35]. The exogenously administered surfactants probably enter the recycling pathways, but different types of surfactant may yield different results.

Conclusions

Surfactant metabolic pathways are very complex and a thorough understanding of where exogenously administered surfactants go requires integration of catabolic and clearance information into the overall metabolic pathways of surfactant. The treatment dose of surfactant is not equivalent to the alveolar pool of surfactant because of the large and variable lung association that occurs. This lung association probably is desirable if it represents entrance of the exogenous surfactant into endogenous recycling pathways which presumably help maintain the functional properties of the surfactant. Catabolic pathways remain largely undefined but probably are a normal function of the alveolar epithelium. Catabolism and clearance rates change with development, and the duration of clinical responses may vary as a result of these developmental changes. As a generality, surfactant treatments do not cause metabolic effects on endogenous surfactant metabolism that are likely to be clinically significant.

References

1. Baritussio A, Carraro R, Bellina L, Rossi A, Bruni R, Pettenazzo A, Enzi G (1985) Turnover of phospholipids isolated from fractions of lung lavage fluid. J Appl Physiol 59:1055–1060
2. Baritussio A, Magoon MW, Goerke J, Clements JA (1981) Precursor-product relationship between rabbit type II cell lamellar bodies and alveolar surface active material. Surfactant turnover time. Biochim Biophys Acta 666:382–393
3. Berggren P, Lachmann B, Curstedt T, Grossmann G, Robertson B (1986) Gas exchange and lung morphology after surfactant replacement in experimental adult respiratory distress syndrome induced by repeated lung lavage. Acta Anaesthesiol Scand 30:321–328
4. Berry D, Ikegami M, Jobe A (1986) Respiratory distress and surfactant inhibition following vagotomy in rabbits. J Appl Physiol 61:1741–1748
5. Brown LAS, Bliss AS, Longmore WJ (1984) Effect of nutritional status on the lung surfactant system: Food deprivation and caloric restriction. Exp Lung Res 6:133–147
6. Chander A, Reicherter J, Fisher AB (1987) Degradation of dipalmitoyl phosphatidylcholine by isolated rat granular pneumocytes and reutilization for surfactant synthesis. J Clin Invest 79:1133–1138
7. Chevalier G, Collet AJ (1972) In vivo incorporation of choline-^{3}H, leucine-^{3}H and galactose-^{3}H in alveolar type II pneumocytes in relation to surfactant synthesis. A quantitative radioautographic study in mouse by electron microscopy. Anat Rec 174:289–310
8. Davis PA, Gunther RA, Cross CE (1987) Clearance of instilled surfactant lipid from the lungs of unanesthetized sheep: lipids are differentially transported by nonlymphatic pathways. J Lab Clin Med 109:191–200
9. Desai R, Tetley TD, Curtis CG, Powell GM, Richards RJ (1978) Studies on the fate of pulmonary surfactant in the lung. Biochem J 176:455–462
10. Egan EA, Notter RH, Kwong MS, Shapiro DL (1983) Natural and artificial lung surfactant replacement therapy in premature lambs. J Appl Physiol 55:875–883
11. Ennema JJ, Reijngoud D-J, Wildevuur CHRH, Egberts J (1984) Effects of artificial ventilation on surfactant phospholipid metabolism in rabbits. Respir Physiol 58:15–28
12. Fisher AB, Dodia C, Chander A (1987) Degradation and reutilization of alveolar phosphatidylcholine by rat lungs. J Appl Physiol 62:2295–2299
13. Geiger K, Gallagher ML, Hedley-Whyte J (1975) Cellular distribution and clearance of aerosolized dipalmitoyl lecithin. J Appl Physiol 39:759–766
14. Glatz T, Ikegami M, Jobe A (1982) Metabolism of exogenously administered natural surfactant in the newborn lamb. Pediatr Res 16:711–715
15. Hallman M, Merritt TA, Pohjavuori M, Gluck L (1986) Effect of surfactant substitution on lung effluent phospholipids in respiratory distress syndrome: Evaluation of surfactant phospholipid turnover, pool size, and the relationship to severity of respiratory failure. Pediatr Res 20:1228–1235
16. Hallman M, Epstein BL, Gluck L (1981) Analysis of labeling and clearance of lung surfactant phospholipids in rabbit. Evidence of bi-directional surfactant flux between lamellar bodies and alveolar lavage. J Clin Invest 68:742–751
17. Heath MF, Jacobson W (1984) Developmental changes in enzyme activities in fetal and neonatal rabbit lung. Cytidylyltransferase, cholinephosphotransferase, phospholipases A_1 and A_2, β-galactosidase, and β-glucuronidase. Pediatr Res 18:395–401
18. Ikegami M, Jobe A, Glatz T (1981) Surface activity following natural surfactant treatment in premature lambs. J Appl Physiol 51:306–312

19. Ikegami M, Jobe A, Nathanielsz PW (1981) The labeling of pulmonary surfactant phosphatidylcholine in newborn and adult sheep. Exp Lung Res 2:197–206
20. Jackson SC, Palmer S, Standaert TA, Murphy J, Truog WE, Woodrum DE, Hodson WA (1984) Developmental changes of surface active material in newborn nonhuman primates. Am Rev Respir Dis 129:A204
21. Jacobs H, Jobe A, Ikegami M, Conaway D (1983) The significance of reutilization of surfactant phosphatidylcholine. J Biol Chem 258:4159–4165
22. Jacobs HC, Ikegami M, Jobe AH, Berry DD, Jones S (1985) Reutilization of surfactant phosphatidylcholine in adult rabbits. Biochim Biophys Acta 837:77–84
23. Jacobs H, Jobe A, Ikegami M, Jones S (1982) Surfactant phosphatidylcholine source, fluxes, and turnover times in 3-day-old, 10-day-old, and adult rabbits. J Biol Chem 257:1805–1810
24. Jacobs H, Jobe A, Ikegami M, Jones S (1985) Accumulation of alveolar surfactant following delivery and ventilation of premature lambs. Exp Lung Res 8:125–140
25. Jacobs H, Jobe A, Ikegami M, Miller D, Jones S (1984) Reutilization of phosphatidylcholine analogues by the pulmonary surfactant system. The lack of specificity. Biochim Biophys Acta 793:300–309
26. Jobe A (1985) Difference in surfactant between the developing and adult lung. J Jpn Med Soc Biol Interface 16:9–18
27. Jobe A, Gluck L (1979) The labeling of lung phosphatidylcholine in premature rabbits. Pediatr Res 13:635–640
28. Jobe A, Ikegami M (1984) The prematurely delivered lamb as a model for studies of neonatal adaptation. In: Nathanielsz PW (ed) Animal models in fetal medicine. Perinatology, Ithaca NY, pp 1–30
29. Jobe A, Ikegami M, Jacobs H (1981) Changes in the amount of lung and airway phosphatidylcholine in 0.5–12.5-day-old rabbits. Biochim Biophys Acta 664:182–187
30. Jobe A, Ikegami M, Jacobs H, Jones S (1983) Surfactant pool sizes and severity of RDS in prematurely delivered lambs. Am Rev Respir Dis 127:751–755
31. Jobe A, Ikegami M, Sarton-Miller I, Barajas L (1980) Surfactant metabolism of newborn lamb lungs in vivo. J Appl Physiol 49:1091–1098
32. Jobe A, Ikegami M, Sarton-Miller I, Jones S, Yu G (1981) Characterization of phospholipids and localization of some phospholipid synthetic and subcellular marker enzymes in subcellular fractions from rabbit lung. Biochim Biophys Acta 666:47–57
33. Jobe AH, Jacobs HC (1984) Catabolism of pulmonary surfactant. In: Robertson B, Van Golde LMG, Batenburg JJ (eds) Pulmonary surfactant. Elsevier Science, Amsterdam, pp 271–293
34. Jobe A, Kirkpatrick E, Gluck L (1978) Lecithin appearance and apparent biologic half-life in term newborn rabbit lung. Pediatr Res 12:669–675
35. Jobe A, Ikegami M, Seidner S, Pettenazzo A, Ruffini L (1988) Surfactant phosphatidylcholine metabolism and surfactant function in preterm ventilated lambs. Am Rev Respir Dis (in press)
36. Magoon MW, Wright JR, Baritussio A, Williams MC, Goerke J, Benson BJ, Hamilton RL, Clements JA (1983) Subfractionation of lung surfactant. Implications for metabolism and surface activity. Biochim Biophys Acta 750:18–31
37. Matalon S, Holm BA, Notter RH (1987) Mitigation of pulmonary hyperoxic injury by administration of exogenous surfactant. J Appl Physiol 62:756–761
38. Nichols BA (1976) Normal rabbit alveolar macrophages. I. The phagocytosis of tubular myelin. J Exp Med 144:906–919
39. Notter RH, Egan EA, Kwong MS, Holm BA, Shapiro DL (1985) Lung surfactant replacement in premature lambs with extracted lipids from bovine lung lavage: effects of dose, dispersion technique, and gestational age. Pediatr Res 19:569–577
40. Oguchi K, Ikegami M, Jacobs J, Jobe A (1985) Clearance of large amounts of natural surfactant and DPPC from lungs of 3-day-old rabbits following tracheal injection. Exp Lung Res 9:221–235
41. Oyarzun MJ, Clements JA, Baritussio A (1980) Ventilation enhances pulmonary alveolar clearance of radioactive dipalmitoylphosphatidylcholine in liposomes. Am Rev Respir Dis 121:709–721

42. Paul GW, Hassett RJ, Reiss OK (1977) Formation of lung surfactant films from intact lamellar bodies. Proc Natl Acad Sci USA 74:3617–3620
43. Pettenazzo A, Ikegami M, Seidner S, Jobe A (1988) Clearance of surfactant phosphatidylcholine from adult rabbit lungs. J Appl Physiol 64:120–127
44. Pettenazzo A, Jobe A, Ikegami M, Seidner S (1988) Clearance of treatment doses of surfactant. Effect of lipid extraction and aggregate sizes. Biol Neonate 53:23–31
45. Pettenazzo A, Oguchi K, Seidner S, Ikegami M, Berry D, Jobe A (1986) Clearance of natural surfactant phosphatidylcholine from 3-day-old rabbit lungs: effects of dose and species. Pediatr Res 20:1139–1142
46. Seidner SR, Ikegami M, Pettenazzo A, Ruffini L, Jobe A (1987) Clearance of surfactant phosphatidylcholine from the lungs of preterm ventilated rabbits. Pediatr Res 21:465A
47. Solimano A, Bryan C, Jobe A, Ikegami M, Jacobs H (1985) Effects of high-frequency and conventional ventilation on the premature lamb lung. J Appl Physiol 59:1571–1577
48. Tarpey MM, O'Brodovich HM, Young SL (1983) Role of lymphatics in removal of sheep lung surfactant lipid. J Appl Physiol 54:984–988
49. Wilkinson A, Jenkins PA, Jeffrey JA (1985) Two controlled trials of dry artificial surfactant: early effects and later outcome in babies with surfactant deficiency. Lancet:287–291
50. Williams MC (1984) Uptake of lectins by pulmonary alveolar type II cells: subsequent deposition into lamellar bodies. Proc Natl Acad Sci USA 81:6383–6387
51. Williams MC (1987) Vesicles within vesicles: what role do multivesicular bodies play in alveolar type II cells. Am Rev Respir Dis 135:744–746
52. Wright JR, Benson BJ, Williams MC, Goerke J, Clements JA (1984) Protein composition of rabbit alveolar surfactant subfractions. Biochim Biophys Acta 791:320–332
53. Wright JR, Wager RE, Hamilton RL, Huang M, Clements JA (1986) Uptake of lung surfactant subfractions into lamellar bodies of adult rabbit lungs. J Appl Physiol 60:817–825

Distribution of Exogenous Surfactant in Premature Lambs: High-Frequency Oscillation Ventilation Versus Conventional Mechanical Ventilation

F. J. Walther

Introduction

After tracheal instillation of a surfactant preparation, both preterm animals and infants with respiratory distress syndrome (RDS) respond with a rapid improvement in pulmonary function. Surfactant treatment at birth allows for the mixing of surfactant with lung fluid and leads to an even distribution of surfactant as the lung fluid recedes, whereas the distribution of exogenous surfactant after air breathing is more uneven [1]. Not only the timing of its instillation, but also the type of ventilatory assistance may affect the distribution of exogenous surfactant.

In a recent study [2] we combined high frequency oscillation ventilation (HFOV) with conventional mechanical ventilation (CMV) in rabbits with surfactant-depleted lungs by saline lavage. The rabbits ventilated with CMV showed a significant deterioration in oxygenation over a 3 hour period, whereas the group on HFOV superimposed on CMV had a significant improvement in gas exchange at lower mean airway pressures.

In this study we compared the effects of HFOV and CMV on the distribution of exogenous surfactant in preterm lambs treated with radioactive labeled surfactant before the onset of breathing or after a period of CMV because of RDS [3].

Methods

Twenty-six lambs, 13 pairs of twins, were delivered by cesarean section at 126–132 days gestational age. Before birth a tracheotomy was done, a carotid catheter was placed for sampling arterial blood gases and pH and blood pressure monitoring, and a jugular vein catheter was inserted for glucose infusion. After birth the lambs were weighed and connected to the ventilator. Arterial blood gases and pH were measured every 15 min, arterial blood pressure and mean airway pressure (MAP) were recorded continuously.

Each lamb received 60 mg of natural sheep surfactant lipid [4] and 0.33 μCi of [1–^{14}C]palmitate dipalmitoyl-phosphatidylcholine per kg body weight. Thirteen lambs received radiolabeled surfactant by tracheal instillation at birth. Six of them (group CMV-B) were subsequently ventilated with CMV and 7 (group HFOV-B) with HFOV. Thirteen lambs were supported on CMV and did not receive radiolabeled surfactant until 30 min after birth. At treatment these lambs were all in respiratory failure as defined by elevated $PaCO_2$ levels and low pH values on at least two blood gas samples. Seven of these lambs (group CMV-R) were continued on CMV and 6 (group HFOV-B) were switched over to HFOV after instillation of surfactant. Three hours after the tracheal instillation of radiolabeled surfactant, all lambs were sacrificed. Before sacrifice inadvertent PEEP was determined in the lambs on HFOV by clamping the endotracheal tube at end-expiration for 5 seconds.

CMV was administered with constant-flow, time-cycled, pressure-limited ventilators delivering humidified and warmed oxygen. Ventilator settings were: an initial peak inspiratory pressure of 25 cm H_2O, a PEEP of 3 cm H_2O, a rate of 40 breaths/min, an inspiratory time of 0.75 s, and an FiO_2 of 1.0. Only peak inspiratory pressures were changed in an attempt to normalize $PaCO_2$. HFOV was generated by an electric piston pump and delivered at a frequency of 15 Hz and a tidal volume of 2 ml/kg. The oscillations were superimposed on the CMV waveforms by oscillating the gas flowing through the CMV ventilator circuit [2, 5].

After sacrifice, the lungs were removed intact and weighed while attached to the endotracheal tube at 15 cm H_2O distending pressure. Standardized superficial and central biopsies from all lobes were used for estimation of the amount of radiolabeled surfactant. The number of radioactive counts per biopsy was corrected for quenching and weight (dpm/g wet tissue) and divided by the mean value for the lungs of that lamb. Biopsies from both upper and lower lobes and one of the right middle lobes were prepared for morphometrics. The expansion of the alveoli in a standard surface area was quantified.

Comparisons between groups were analyzed with Student's two-tailed t test or Chi-square test.

Results

The lambs had a mean (± SD) weight of 3.0 (± 0.8) kg and a gestational age of 130.5 (± 1.3) days. Figure 1 presents the mean (± SD) PaO_2, $PaCO_2$ and MAP values of the lambs treated with radioactive labeled surfactant at birth (groups CMV-B and HFOV-B) and at 30 min of age (groups CMV-R and HFOV-R). Three hours after the administration of surfactant, the mean (± SD) oxygenation index ($OI = \Lambda AP \times FiO_2 \times 100/PaO_2$, cm H_2O/mm Hg) was 11.6 (± 4.2) in the CMV-R group, 4.7 (± 4.7) in the CMV-B group, 6.9 (± 3.5) in the HFOV-B group and 6.3 (± 2.6) in the HFOV-B group. The differences between the CMV-R group and the 3 other groups were significant ($p < 0.05$ or < 0.02), whereas the differences between the CMV-B, HFOV-B and HFOV-R groups were not significant. Background ventilation was started at a rate of 40/min, but could be reduced to 5–10 breaths/min after normocapnia was established.

Figure 2 shows the distribution of radioactive labeled surfactant in the 4 groups: 23.8% of the lung biopsies in the HFOV-R group, 28% of the biopsies in the CMV-R group, 45.9% of biopsies in the HFOV-B group and 41.3% of the biopsies in the CMV-B group received surfactant quantities within ± 25%

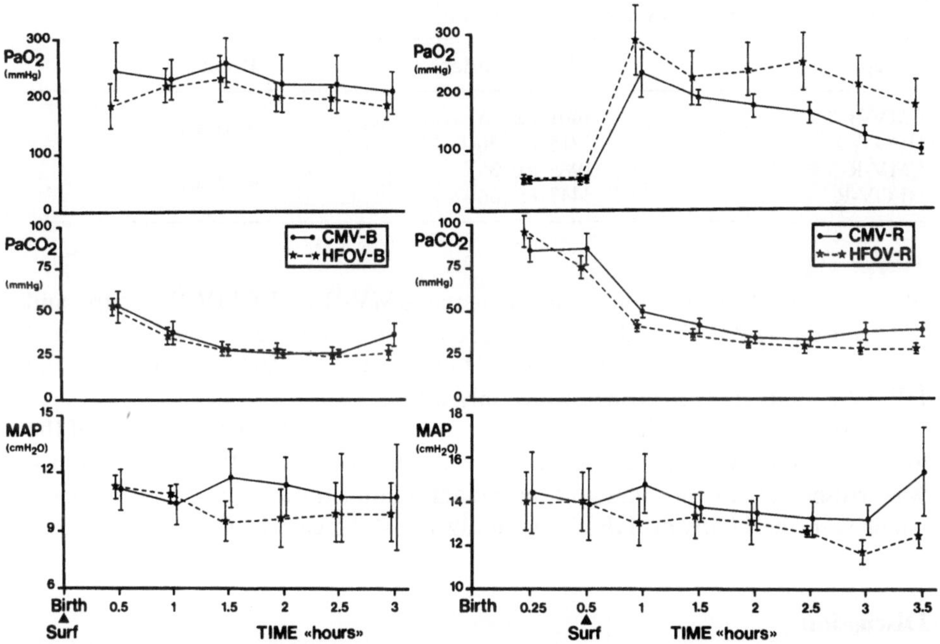

Fig. 1. Mean (± SD) values of the PaO_2, $PaCO_2$, and MAP of the lambs treated with radiolabeled surfactant at birth *(left)* and of the lambs treated 30 min after birth *(right)*

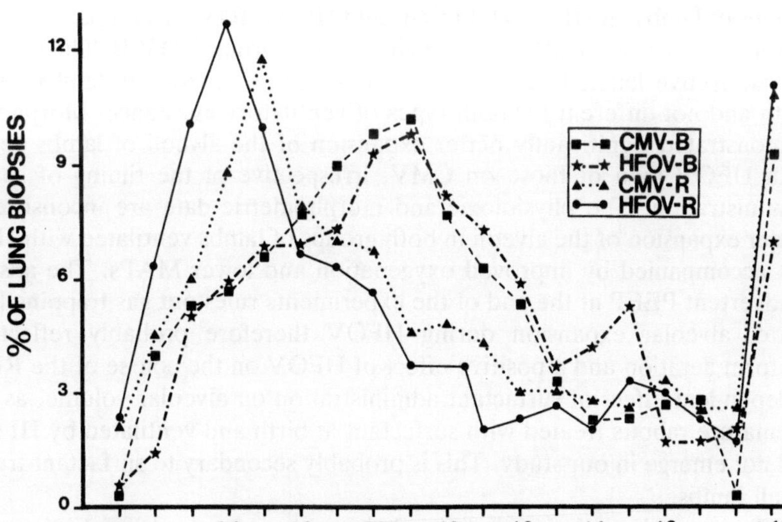

Fig. 2. Distribution of radioactive labeled surfactant in the four groups of lambs

Table 1. Mean (± SEM) alveolar surfaces in μm^2

Group	Alveolar surface	P value
CMV-B	1401 (± 25)	< 0.001
HFOV-B	3215 (± 46)	
CMV-R	1985 (± 35)	< 0.001
HFOV-R	3443 (± 56)	

of the mean. The differences between the CMV-B and CMV-R groups and between the HFOV-B and HFOV-R groups were statistically significant ($p < 0.001$). The differences between the HFOV-R and CMV-R and between the HFOV-B and CMV-B groups were only minor.

Morphometrics (Table 1) demonstrated significantly better expansion of the alveoli of lambs ventilated with HFOV than of those on CMV. The difference was irrespective of the timing of surfactant administration. Inadvertent PEEP changes in the lambs on HFOV were always < 0.5 cm H_2O.

Discussion

At comparable oxygenation, MAPs were significantly lower in the lambs treated with surfactant at birth (groups CMV-B and HFOV-B) than in lambs on CMV and treated with surfactant during RDS (group CMV-R). MAPs in both groups of lambs on HFOV (HFOV-B and HFOV-R) were comparable at values lower than in group CMV-R and higher than in group CMV-B. The distribution of radioactive labeled surfactant was more homogeneous in lambs treated at birth and not different for both types of ventilatory assistance. Morphometrics demonstrated significantly better expansion of the alveoli of lambs ventilated with HFOV than of those on CMV, irrespective of the timing of surfactant administration. The physiologic and morphometric data are inconsistent: the better expansion of the alveoli in both groups of lambs ventilated with HFOV is not accompanied by improved oxygenation and lower MAPs. The absence of inadvertent PEEP at the end of the experiments rules out gas trapping [6]. The better alveolar expansion during HFOV therefore probably reflects more uniform aeration and a positive effect of HFOV on the course of the RDS. An independent effect of surfactant administration on alveolar volume, as seen in premature rabbits treated with surfactant at birth and ventilated by HFOV [7], did not emerge in our study. This is probably secondary to surfactant treatment of all lambs.

These results indicate that prophylactic surfactant administration at birth leads to a better distribution of surfactant than rescue treatment with surfactant after the establishment of RDS and is not affected by subsequent type of ventilatory assistance. Rescue treatment with surfactant and subsequent HFOV leads to better oxygenation and alveolar expansion at comparable MAPs than rescue treatment followed by CMV.

References

1. Jobe A, Ikegami M, Jacobs H, Jones S (1984) Surfactant and pulmonary blood flow distributions following treatment of premature lambs with natural surfactant. J Clin Invest 73:848–856
2. Blanco CE, Maertzdorf WJ, Walther FJ (1987) Use of combined high-frequency oscillation and intermittent mandatory ventilation in rabbits with saline-lavaged lungs. J Intensive Care Med 2:214–217
3. Walther FJ, Kuipers IM, Gidding CEM, Willebrand D, Buchholtz RTF, Bevers EM (1987) A comparison of high-frequency oscillation superimposed onto backup mechanical ventilation and conventional mechanical ventilation on the distribution of exogenous surfactant in premature lambs. Pediatr Res 22:725–729
4. Walther FJ, Blanco CE, Houdjik M, Bevers EM (1985) Single versus repetitive doses of natural surfactant as treatment of respiratory distress syndrome in premature lambs. Pediatr Res 19:224–227
5. Boynton BR, Mannino FL, Davis RF, Kopotic RJ, Friederichsen G (1984) Combined high-frequency oscillatory ventilation and intermittent mandatory ventilation in critically ill neonates. J Pediatr 105:297–302
6. Bancalari A, Gerhardt T, Bancalari E, Suguihara C, Hehre D, Reifenberg L, Goldberg RN (1987) Gas trapping with high-frequency ventilation: jet versus oscillatory ventilation. J Pediatr 110:617–622
7. Solimano A, Bryan C, Jobe A, Ikegami M, Jacobs H (1985) Effects of high frequency and conventional ventilation on the premature lamb lung. J Appl Physiol 59:1571–1577

Immunohistochemical Study of the Distribution of Exogenous Surfactant in the Lungs of Surfactant-Deficient Rabbits Using a Monoclonal Antibody

S. Bambang Oetomo, L. de Ley, C. Schoots, B. Lachmann, and A. Okken

Introduction

Surfactant replacement therapy in surfactant deficiency has been shown to be an effective method to improve lung function. The immediate beneficial effects of surfactant therapy, however, are often followed by a relapse within several hours after treatment. In premature newborn infants factors responsible for the relapse include patency of the ductus arteriosus and inhibition of surfactant activity by proteins leaking into the alveolar space. Other factors which are possibly of significant importance are the initial distribution of the administered surfactant throughout and its further distribution in time. So far little is known about the distribution of instilled natural surfactant in surfactant deficiency.

We therefore set up a study to determine the distribution of instilled natural surfactant throughout the airways and into the alveoli of surfactant deficient animals.

Methods

Lung lavaged 3-month-old surfactant deficient rabbits (n = 12) and 27 gestational days preterm surfactant deficient rabbits (n = 12) were treated with natural porcine surfactant. The surfactant preparation used in this study was manufactured by Curstedt [1]. In both the premature and the adult lung lavaged rabbits, half of the animals received porcine surfactant endotracheally, the other half served as controls. All animals were artificially ventilated for respiratory insufficiency. The premature rabbits were sacrificed after 30 min and the adult lung lavaged rabbits were sacrificed after 4.5 hours of artificial ventilation. We were able to prepare a monoclonal antibody, using a modification of the method of Köler and Milstein [2], which reacts specifically with the instilled porcine surfactant and not with the rabbits' own surfactant.

Results

In all control animals (n = 6) none of the porcine surfactant monoclonal antibody reacted with the rabbits' own surfactant (if present).

In Fig. 1 a microscopic section is shown of the lung of a lung lavaged surfactant deficient rabbit, porcine surfactant is seen along the bronchiolar wall

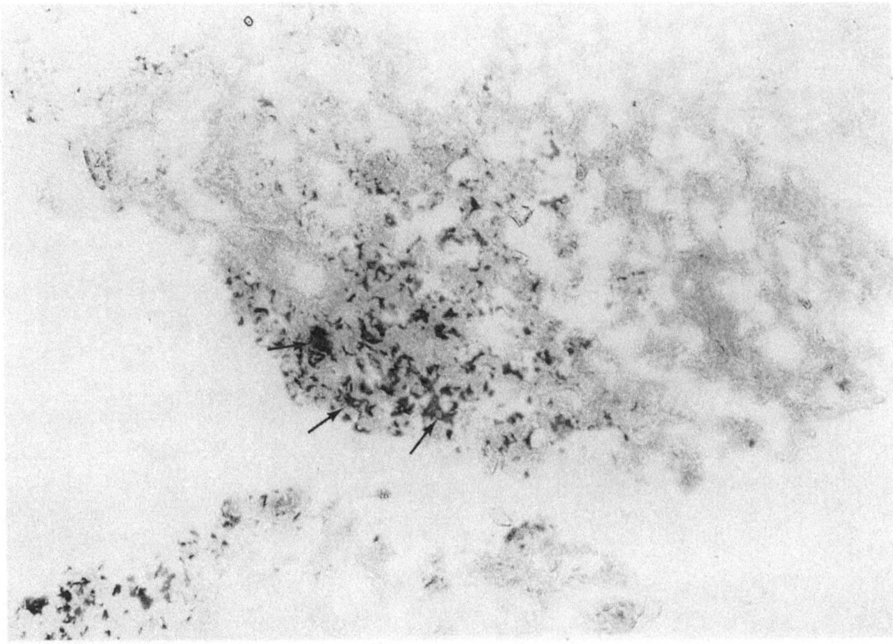

Fig. 1. Microscopic section of pulmonary tissue of a lung-lavaged surfactant-deficient rabbit that was treated with natural porcine surfactant. Surfactant *(arrows)* is visible along the bronchiolar walls and in the collapsed lung parts.

and in the subpleural parts of the lung parenchyma. Most instilled surfactant, however, is found in the collapsed lung areas.

The surfactant treated preterm rabbits that were artificially ventilated for 30 minutes showed a different pattern of surfactant distribution. Figure 2 shows that in contrast to the lung of the lung lavaged surfactant deficient rabbit it is possible to clearly distinguish the surfactant in the alveolar space. Lung collapse did not occur in these animals.

From this study we conclude:

1. With a specific monoclonal antibody it is possible to more precisely localise the endotracheal instilled surfactant in lung tissue.
2. In adult surfactant deficient rabbits, 4.5 hours after instillation, surfactant seems to be trapped in atelectic lung parts.
3. In preterm rabbits surfactant can be found in the alveolar spaces of the lung.

We speculate that endotracheal administered surfactant spreads rapidly over the lung but is only partially distributed into the alveoli and is trapped into collapsed lung areas, depending on the level of surfactant deficiency and the duration of the artificial ventilation.

This mechanism might provide an explanation for the relapse observed in surfactant deficient subjects following an initial dramatic improvement in their lung function.

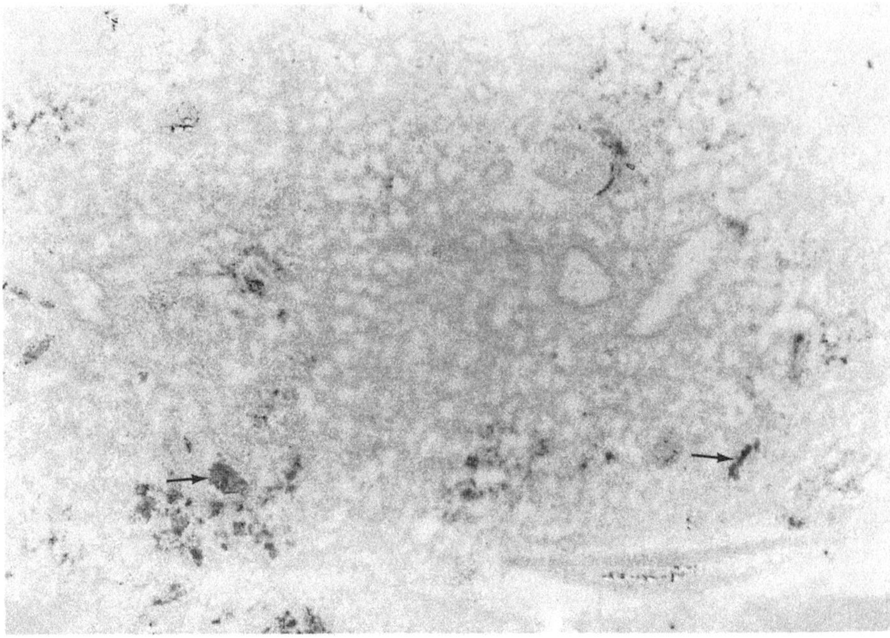

Fig. 2. Microscopic section of a preterm surfactant-deficient rabbit that was treated with natural porcine surfactant. Surfactant *(arrows)* is located in the bronchi and alveolar spaces.

References

1. Noack G, Berggren P, Curstedt T, Grossmann G, Herin P, Mortensson W, Nilson R, Robertson B (1987) Severe neonatal respiratory distress syndrome treated with the isolated fraction of natural surfactant. Acta Paediatr Scand 76:697–705
2. Köler G, Milstein C (1975) Continuous cultures of fused cells secreting antibodies of predefined specificity. Nature 256:495–497

Immunogenicity of Surfactant and Its Implications for Replacement Therapy*

D. S. Strayer, T. A. Merritt, R. Spragg, and M. Hallman

Introduction

The etiology and pathogenesis of neonatal and adult respiratory distress syndromes (RDS) differ widely but share the characteristic inability of the lungs to adequately oxygenate the blood. Neonatal RDS occurs in premature infants deficient in surfactant (SRF) [1]. SRF deficiency causes alveoli to collapse on expiration, resulting in progressive respiratory failure. Because neonatal RDS directly reflects deficiency of surfactant, various investigators have tried therapeutic administration of heterologous, artificial, and homologous surfactants [2–4]. These preparations generally lower surface tension *in vitro* and in animal models, and are reported to be efficacious in infants with RDS.

Adult RDS is more complex. It is a final common pathway for pulmonary insults of many different origins, ranging from traumatic to infectious. Although ARDS occurs in the presence of adequate SRF production, treatment with SRF is being tried therapeutically in an attempt to improve pulmonary function.

Our therapeutic trials of neonatal RDS were designed using human surfactant derived from amniotic fluid. As lipid-protein mixtures are generally highly immunogenic, we decided to study potential immunologically mediated complications of surfactant administration.

Subsequently, trials of Curosurf (porcine surfactant) as treatment for ARDS were initiated. For the same reasons, we felt it was important to determine whether patients so treated showed any manifestations of immunologic reactivity to surfactant or of immunogically mediated tissue damage.

One of the most potentially damaging aspects of immune reactivity to surfactant is production of circulating immune complexes between surfactant and anti-surfactant antibodies. Such immune complexes might mediate tissue damage such is seen in other diseases (e.g., post-streptococcal glomerulonephritis). We therefore developed an assay to detect immune complexes between surfactant and anti-surfactant antibodies [5].

* Supported in part by USPHS grants HD 10622 and HD 16292 (T. A. Merritt), FDA grant FDA-R-000112 (T. A. Merritt and D. S. Strayer), HL 23584 (R. Spragg), a grant from the March of Dimes (T. A. Merritt), and a grant from the Finnish Academy of Sciences (M. Hallman).

Another important potential for immunologically mediated damage in these circumstances would be the ability of antibodies to surfactant to inactivate surfactant functionally. Accordingly, we examined whether anti-surfactant antibodies inhibit surfactant function in vitro.

Methods

Antibodies to Surfactant. Rabbits were immunized with the same human surfactant preparations used to treat infants with RDS. Prebleed and immune sera were collected from these animals, and absorbed exhaustively in solid phase with serum proteins from normal human blood.

Antibodies to porcine surfactant are currently being prepared. The data discussed here for patients with ARDS treated with porcine surfactant (Curosurf, the kind gift of Dr. Bengt Robertson) reflect the use of anti-human surfactant antiserum for immune complex assays.

The specificity of these antisera were determined by several methods. Using Western blotting procedures, we found that anti-SRF antisera do not bind to normal human serum proteins. These sera recognize SRF-containing lamellar bodies in human type II alveolar pneumocyte carcinoma line, A549 [5]. By immunoprecipitation and autoradiography of ^{125}I-labeled surfactant preparations, we found that these antisera bound surfactant species of 35 kd, 18 kd, and 5–7 kd (D. Strayer, et al., submitted for publication).

ELISA for Surfactant-Anti-Surfactant Immune Complexes. This assay is illustrated in Fig. 1. Immune complexes containing surfactant in complex with antibody to surfactant bind to plastic plates coated with rabbit anti-surfactant antibody via immune recognition of the surfactant portion of this immune complex. The human anti-SRF portion of the immune complex is then detected using antibody to human immunoglobulin [5].

Plasma samples from infants in surfactant treated and control groups are incubated with the anti-surfactant coated plates. Plates are then washed. Enzyme-conjugated heavy chain-specific anti-human IgG was added. Originally, horseradish peroxidase-conjugated secondary antibody was used. More recently, we have used alkaline phosphatase conjugated antibody. This secondary antibody recognizes that human IgG present bound by virtue of the human IgG being part of an immune complex with surfactant which is in turn bound to the immobilized rabbit anti-SRF antibody. Binding of the enzyme-linked anti-IgG antibody is visualized by adding substrate for the enzyme in question and reading the result on an automatic ELISA reader (Dynatech). Data are recorded as absorbance at 405 nm (alkaline phosphatase) or 490 nm (horseradish peroxidase, HPO) [6].

Data from assays using plates coated with anti-SRF are compared to identical assays in which plates are coated with albumin alone. Absorbance values from the latter are substracted from the former to yield the specific absorbance readings for the sample in question.

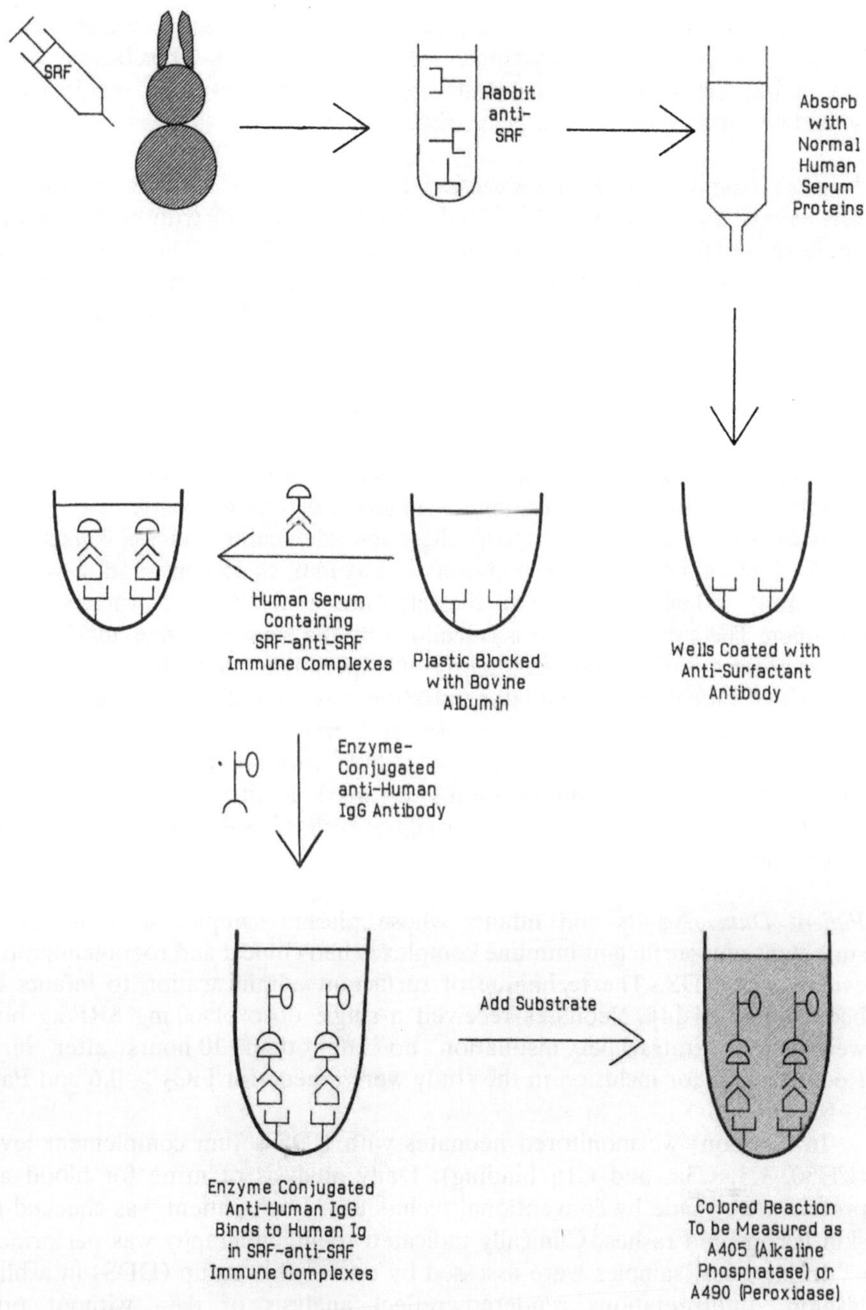

Fig. 1. The assay for surfactant-anti-surfactant immune complexes in the plasma of babies with neonatal RDS. Plastic plates are first coated with rabbit anti-SRF antibody and then bovine albumin to block nonspecific plastic binding sites for protein. The plasma sample to be tested is then added. A secondary antibody, enzyme-conjugated (either alkaline phosphatase or horseradish peroxidase) antibody directed to human IgG, follows. Finally, the substrate for the enzyme is added and the reaction is visualized at 405 nm (alkaline phosphatase) or 490 nm (horseradish peroxidase)

The only human IgG present on the plate would be that bound by virtue of being complexed with surfactant. Thus, this assay specifically detects circulating surfactant-anti-surfactant immune complexes. The specificity and sensitivity of this assay have been discussed [5].

In Vitro Assays of Surfactant Function. Surfactant isolated from amniotic fluid was slowly thawed from –90° C and mixed 1 : 1 (v : v) with control sera or antibody preparations using a Vortex mixer for 30 sec. These combinations were thereafter incubated for 30 min at 37° C with shaking. The final concentration of surfactant was 4 mg phospholipid/ml. The total protein concentration was 60–130 μg/ml.

Surface activity is measured using a pulsating bubble surfactometer, as described by Enhorning [7]. This surfactometer (Surfactometer International, Toronto, Ontario, Canada) measures the pressure gradient across air-liquid interphase of a bubble formed in a 37° C water-enclosed sample chamber. The bubble size was viewed through a microscopic optic, while the pressure changes were recorded on a strip chart recorder calibrated for 0 and –2 cm H_2O. The bubble pulsates between a maximal (0.55 mm) and a minimal (0.45 mm) radius at a rate of 20 cycles/min at 37° C in a water-saturated chamber. The surface tension is calculated by the laws of Young and La Place, and expressed as dynes/cm. In the present experiments, a bubble was formed to a maximal radius and 15 sec later pulsation was started. A continuous tracing of the pressure differences was made for 5 min. Calculations of the surface tensions were made 15 sec after the bubble formation. We calculated the surface tension at minimum radius (γ min) additionally after 5 min of pulsation. In addition, the time required to reach minimum surface tension was determined.

Patient Data. Adults and infants whose plasma samples were tested for surfactant-anti-surfactant immune complexes had clinical and roentgenographic evidence of RDS. The technique of surfactant administration to infants has been described [4]. Neonates received a single dose of 60 mg SRF/kg body weight by intratracheal instillation no later than 10 hours after birth. Requirements for inclusion in the study were a need for $FiO_2 > 0.6$ and $PaO_2 \leq 60$ mm Hg.

In addition, we monitored neonates with RDS serum complement levels (CH50, C3, C3a, and C1q binding). Daily analysis of urine for blood and protein were made by conventional techniques. Each patient was checked for skin lesions and rashes. Clinically indicated roentgenography was performed.

In all cases, samples were assessed by one of this group (DDS) in a blind fashion. Interpretations rendered reflect analysis of data without prior knowledge of clinical status, course, or treatment modalities. The relevant clinical data were supplied by others of this group (TAM, MH, RS) following such analysis.

Results

Immune Complexes in the Blood of Neonates with RDS. The results of our assays for circulating SRF-anti-SRF immune complexes in neonatal RDS have been compiled [5]. For illustrative purposes, Fig. 2 shows typical results for patients with RDS treated conventionally and with SRF replacement. Comparable analyses from a premature infant with no RDS are included for comparison. With only occasional exceptions, all infants with RDS showed evidence of circulating SRF-anti-SRF immune complexes. In these illustrations absorbance (ordinate) is shown as a measure of concentration of immune complexes, as a function of time after birth. Levels of immune complexes varied between children. In some, absorbance values not greatly higher than those observed at birth were noted. In others, considerable increases in absorbance values were found [5].

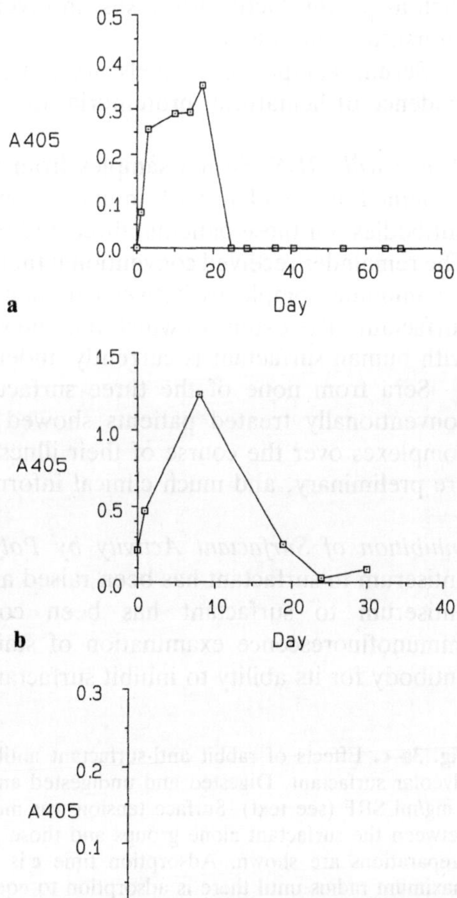

Fig. 2a–c. Appearance of SRF-anti-SRF immune complexes in plasma as a function of time for three representative patients. Absorbance is reported as relative absorbance, which is equal to the $A_{405 \text{ or } 490}$ at the time in question/$A_{405 \text{ or } 490}$ at birth. Infants were followed for a variable time after birth. **a** Analysis of plasma for an infant with RDS treated by conventional means. **b** An infant with RDS treated with surfactant. **c** A preterm infant without RDS. The procedure for performing the assays for immune complexes is described in Fig. 1

Levels of circulating immune complexes peaked at different times. They reached maximum height as early as 2–3 days in some infants, while others took as long as 2–3 weeks. Analyses of plasma from nearly all infants showed early peaks in immune complex concentrations. A subsequent decline usually followed. Considerable variation was noted between infants in all parameters. Sera from some infants showed very sharp peaks followed by comparably sharp declines. For other children, decreases in immune complex concentrations were much more gradual.

Clinical Data and Courses for Infants with Neonatal RDS. Clinical data for some of these infants are summarized elsewhere [5]. Detailed clinical histories are available on request. We have to date tested over 40 infants with neonatal RDS, both with and without SRF therapy. SRF administration produced rapid improvement in pulmonary function All infants with RDS weighed less than 1470 g at birth (range 850–1460 g), and ranged in gestational age from 27–32 weeks. All survived their episodes of RDS, though many had complications such as patent ductus arteriosus, intraventricular hemorrhage, and pulmonary interstitial emphysema.

Serum complement levels were normal in all infants. There was no evidence of hematuria, proteinuria, or rash at any time during observation.

Adults with RDS. Serum samples from nine adult patients with ARDS were examined for evidence of immune complexes between SRF and anti-SRF antibodies. Of these patients, three had received porcine surfactant (Curosurf). The remainder received conventional therapy. These serum samples were tested for immune complexes between human surfactant and antibodies to human surfactant. The extent to which porcine surfactant crossreacts immunologically with human surfactant is currently under study.

Sera from none of the three surfactant treated patients, but 2 of the 6 conventionally treated patients showed evidence of SRF-anti-SRF immune complexes over the course of their illnesses. Results from these examinations are preliminary, and much clinical information needs yet to be evaluated.

Inhibition of Surfactant Activity by Polyclonal Antibody. Rabbit polyclonal antiserum to surfactant has been raised and analyzed [5]. The reactivity of this antiserum to surfactant has been confirmed by Western blotting and immunofluorescence examination of stained lung tissues [5]. We tested this antibody for its ability to inhibit surfactant activity as measured by a pulsating

Fig. 3a–c. Effects of rabbit anti-surfactant antibodies on the functional activity of human alveolar surfactant. Digested and undigested antibody preparations were added to SRF at 4 mg/ml SRF (see text). Surface tension was measured at 1 **a** and 5 **b** min, and differences between the surfactant alone groups and those groups receiving antibody or normal serum preparations are shown. Adsorption time **c** is the time from initial bubble formation at maximum radius until there is adsorption to equilibrium surface tension prior to instituting pulsations. All differences between anti-SRF-treated preparations and those containing normal serum are statistically significant at the level of $P < 0.05$. $\Delta\gamma$ values are in dynes/cm and equal γ antibody-γcontrol ▶

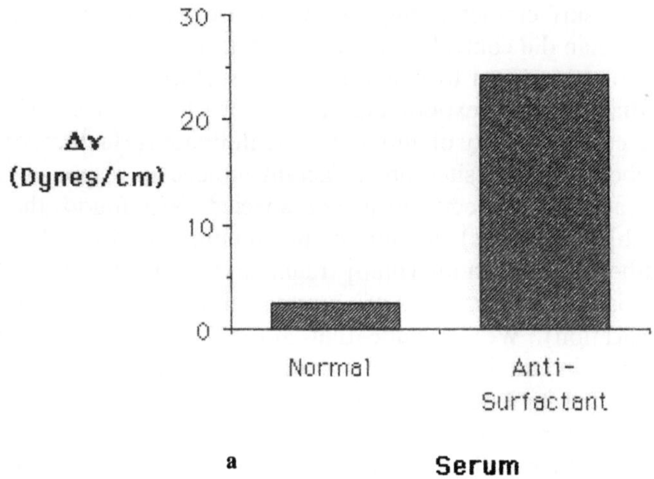
30
20
10
0
Δγ
(Dynes/cm)
Normal
Anti-
Surfactant
Serum

a

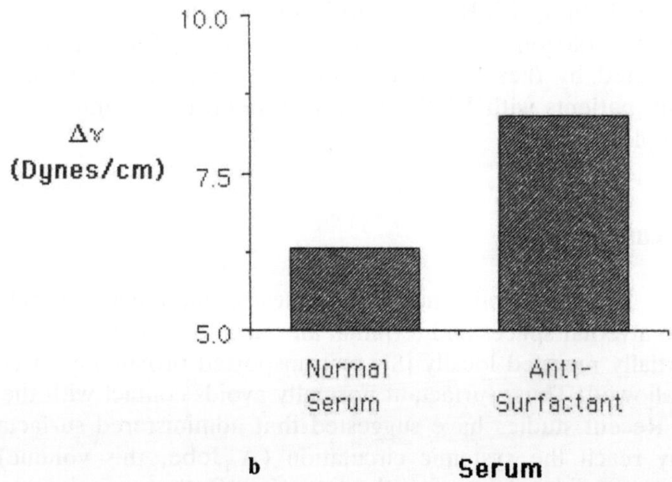
10.0
7.5
5.0
Δγ
(Dynes/cm)
Normal
Serum
Anti-
Surfactant
Serum

b

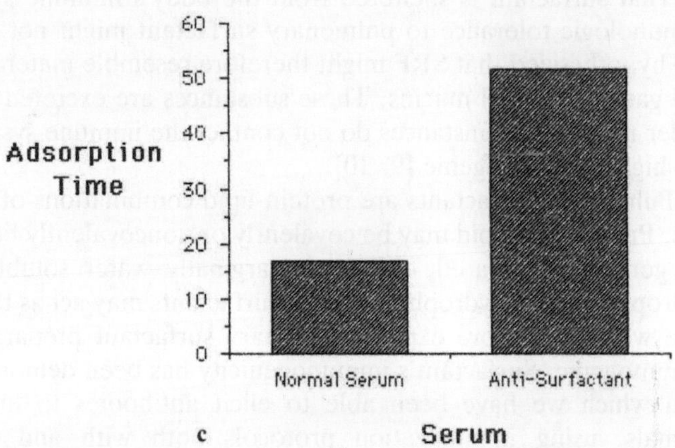
60
50
40
30
20
10
0
Adsorption
Time
Normal Serum
Anti-Surfactant
Serum

c

bubble surfactometer (Fig. 3). This antiserum clearly decreased surface activity more than did control preparations of normal rabbit serum as measured by c_{min} and time required to reach minimum surface tension.

In additional experiments (not shown), we asked whether this inhibition reflected the ability of antibody to agglutinate surfactant, or rather the ability of antibody to bind sites on surfactant molecules responsible for intermolecular interactions between surfactant species. We found that bivalent antibody ($F[ab]_2$ fragments) to surfactant inhibits biophysical activity. Monovalent antibody preparations (F[ab] fragments) inhibit surface activity no more than comparable preparations of normal rabbit serum (D. S. Strayer, submitted for publication). We conclude that most if not all of the functional inhibitory activity in these polyclonal antibody preparations reflects the ability of polyclonal antibodies to agglutinate surfactant.

Therefore, we have repeatedly detected circulating immune complexes between SRF and anti-SRF antibodies in neonates with RDS. These immune complexes develop over time, and then usually disappear. They are present only in babies with RDS, regardless of whether or not the infants were treated with homologous surfactant. We have not yet found evidence of tissue injury mediated by these immune complexes. Preliminary data suggest that some adult patients with RDS may also develop circulating SRF-anti-SRF immune complexes.

Discussion

Type II alveolar pneumocytes synthesize pulmonary surfactant and secrete it into alveolar spaces and terminal air ducts. Surfactant may be metabolized and partially recycled locally [8], or transported proximally to be expectorated or swallowed. Thus, surfactant normally avoids contact with the immune system.

Recent studies have suggested that administered surfactant phospholipids may reach the systemic circulation (A. Jobe, this volume). These studies, however, did not examine the fate of surfactant protein.

That surfactant is sheltered from the body's immune system suggests that immunologic tolerance to pulmonary surfactant might not be well developed. We hypothesized that SRF might therefore resemble materials such as keratins and gastrointestinal mucins. These substances are excreted from the body and under normal circumstances do not contact the immune system. Some of these are highly immunogenic [9, 10].

Pulmonary surfactants are protein-lipid combinations of variable molecular size. Protein and lipid may be covalently or noncovalently bound to each other. As generally prepared, they are marginally water soluble. As mixtures of hydrophobic and hydrophilic phases surfactants may act as their own adjuvants. One would therefore expect pulmonary surfactant preparations to be highly immunogenic. Surfactant's immunogenicity has been demonstrated by the ease with which we have been able to elicit antibodies to human surfactant in animals, using immunization protocols both with and without additional adjuvants (5, and D. S. Strayer, et al., submitted for publication).

When the integrity of the alveolar wall is interrupted, surfactant could then be released into the circulation. Such circumstances could accompany any pulmonary injury. Environmental toxic agents, viruses, ionizing radiation, or other disruptive insults could breach the alveolar lining enough to permit surfactant to leak into the blood. Pulmonary damage in RDS includes increases in pulmonary capillary permeability and pulmonary edema [11]. Autoimmunization by pulmonary surfactant could account for the presence of immune complexes in plasma from patients with RDS. In many infants, the time course of appearance of immune complexes suggests primary immunization. This sequence of events parallels that of endogenous antikeratin antibodies in patients with extensive burns [10], and is not unlike the appearance of antibodies to myocardial antigens reported in patients with ischemic and other myocardial diseases [12–14, however see 15 and 16].

The immune system appears to mature selectively. Responsiveness to certain antigens predates the appearance of responsiveness to others [17].

Immunoglobulins are first detected in humans at about 10 weeks of gestation [18]. Human feti produce IgM antibody to a variety of antigens. Such antibodies are used to diagnose prenatal infections [19–21]. The human cell-mediated immune system also appears to be substantially mature *in utero* [22–25]. Preterm infants are therefore sufficiently immunocompetent to recognize surfactant as an antigen and elaborate anti-surfactant antibodies.

In addition, autoimmunization may be part of the development of tolerance to autoantigens. According to the network theory of immune regulation [26], a cascade of mutually interacting antibodies follows immunization. Initial antibodies recognize the antigen in question. Subsequent antibodies, called anti-idiotypic antibodies, recognize antigen binding regions (idiotypes) of the first set of antibodies. Anti-idiotypic antibodies thereby regulate clonal expression and levels of the initial antibodies [27, 28]. Some investigators suggest that tolerance to self antigens involves first producing antibodies to these "self" antigens. Primary autoantibodies could then elicit anti-idiotypic [anti-(anti-self)] secondary antibodies, eliminating or modulating anti-self clones [29]. Anti-surfactant antibodies in immune complexes in infants with RDS may reflect transient autoresponsiveness before development of immunologic tolerance.

Anti-SRF antibodies in our infants could also be of maternal origin. Surfactant absorbed from amniotic fluid might elicit an immune response in pregnant mothers. IgG is actively transported across the human placenta. Resulting IgG anti-SRF could then cross into the fetal circulation. Its combination with surfactant which, due to RDS, is released into the neonatal circulation could account for the observed immune complexes. This type of mechanism could explain the very rapid appearance of immune complexes in some infants. Others have noted maternal anti-SRF antibody in amniotic fluid (W. Taeusch, personal communication).

None of these hypotheses necessarily excludes others. In each patient, various mechanisms may contribute differently to SRF-anti-SRF immune complex levels as detected in individuals with RDS.

Other investigators who have treated patients with xenogeneic surfactants have sought but not found circulating anti-surfactant antibodies (J. Whitsett, personal communication). We have not seen the experimental protocols used by these investigators, reviewed their assay procedures or tested their patients' samples in our assays. We also do not know at which time(s) they drew patients' sera for analysis. Thus, we cannot yet explain the differences between their findings and ours. Although, as we understand these experiments, they have been designed to detect circulating free antibody to surfactant. We have found that infants with RDS have circulating immune complexes which form at times of antigen excess or near equivalence between antigen and antibody. Immune complex formation thus precedes detectable free antibody. Our data and those of our colleagues are thus not necessarily contradictory. If most anti-SRF antibody in SRF-anti-SRF immune complexes was of maternal origin, it is possible that immune complexes account for most or all of the original maternal anti-SRF; so that free antibody could not be detected. In addition, if the pulmonary circulation remains leaky, circulating free antibody might be immediately bound by the large pool of antigen (alveolar surfactant) in the lungs.

The data shown here are illustrative of many studies on infants with neonatal RDS. We have consistently found that premature infants without pulmonary disease do not have detectable circulating SRF-anti-SRF immune complexes (5, and D. S. Strayer, et al., in preparation). We find such complexes in plasma samples from patients with RDS regardless of treatment.

We have as yet found no evidence of immune complex-mediated damage. All infants were examined for evidence of complement activation and skin and kidney damage, but none was found. Such absence of evident circulating immune complex-mediated tissue injury would be similar to the studies done in men with vasectomies. These men develop antibodies to sperm antigens and have circulating antigen-antibody complexes in their blood. Nonetheless, investigators have found no evidence of immunologically mediated tissue injury in these patients [30, 31].

We report here preliminary data suggesting that a small number of patients with ARDS may also show evidence of circulating SRF-anti-SRF immune complexes. Our data are too preliminary to permit separation of ARDS patients who develop SRF-anti-SRF immune complexes on the basis of treatment regimen(s). These studies are preliminary in nature and undertaken prospectively to evaluate the potential immunogenicity of surfactant in adult patients with pulmonary disease. Unlike neonatal RDS, adult RDS is a composite of diseases of many different etiologies that may develop similar respiratory distress syndromes. This diversity of pathogeneses and etiologies suggests that in some cases anti-surfactant immunologic reactivity and inhibition of surfactant function may play a role in the development of RDS. This possibility is strengthened by those data showing that bivalent anti-SRF antibody inactivates surfactant functionally.

The suggestion of SRF-anti-SRF immune complexes in ARDS patients is based on evaluation of a small number of patients. Additional patient accrual and data correlation are needed before these findings can be substantiated. Nevertheless, these preliminary indications, together with experimental observations on the effects of anti-SRF on SRF function, suggest that immunologically mediated mechanisms may enhance respiratory dysfunction in some cases of ARDS.

The nature of the antigens in surfactant is an important question. To treat neonatal RDS, we use a heterogeneous combination of human pulmonary surfactant species, including 34 kd apoprotein, 18 and 9 kd proteins, and 5–7 kd lipoprotein. Preparations used by others generally involve xenogeneic surfactants which may be predominantly one molecular species or another. The antigenicity of these different molecular species should not be in doubt. Our rabbit polyclonal antibody reacts with all these species as determined by radioimmunoprecipitation. In addition, we have raised murine monoclonal antibodies to human surfactant. Most of these monoclonal antibodies recognize the 5–7 kd proteolipid species (D. S. Strayer, et al., submitted for publication). The latter is of comparable size to such highly immunogenic proteins as insulin [32]. Thus, although it is smaller than the other surfactant molecules, the 5–7 kd lipoprotein is large enough to act as an independent antigen. All known molecular forms of alveolar surfactant are therefore potentially immunogenic.

We have sought to determine whether human alveolar surfactant is immunogenic as it is used in therapy of adult and neonatal respiratory distress syndromes (RDS). We have developed an enzyme linked immunosorbant assay (ELISA) to detect specific immune complexes between surfactant and antibodies directed to surfactant. Plasma samples from neonates with RDS were analyzed for such immune complexes. Premature infants with RDS were divided into two groups: one group to receive surfactant replacement and one to receive conventional therapy. Infants were also examined for clinical and serological evidence of immune complex-mediated tissue damage. Almost all infants with RDS, whether or not they had received surfactant therapy, showed evidence of circulating surfactant-anti-surfactant immune complexes. Plasma samples from premature babies without RDS showed no such immune complexes. These immune complexes generally appeared early in postnatal life. Their concentrations then peaked, and diminished afterwards. We have found no evidence of complement activation or clinically evident organ damage attributable to these immune complexes.

Sera from a small number of patients with adult RDS were similarly examined for SRF-anti-SRF immune complexes. About one-third of these patients showed such immune complexes. In addition, we found that antibodies directed against surfactant can inhibit surfactant function *in vitro*.

Thus, circulating immune complexes between surfactant and antibodies to surfactant are common in neonatal RDS and may be found in adult RDS, though less commonly. To date, they seem to cause no tissue injury in neonatal RDS. However as anti-surfactant antibodies are capable of inhibiting surfactant functionally, RDS both in neonates and adults may be potentiated by these antibodies. As both human and heterologous surfactants are now used in treating RDS, the pathogenic potential of immunologic reactivity to surfactant cannot be ignored.

In conclusion, we have shown that human surfactant is immunogenic and that circulating surfactant-anti-surfactant immune complexes are detectable in the plasma from infants and in adults with RDS. We found these immune complexes regardless of whether SRF was used in the individual treatment regimen. These immune complexes do not yet appear to cause disease in the short term. Long term effects, if any, are unknown. Indications for surfactant replacement therapy in neonatal RDS are clear. Trials of surfactant are just beginning in adult RDS. In all such situations, potential for side effects must be balanced against therapeutic efficacy and the gravity of the disease. Our data indicate that surfactants, particularly heterologous surfactants, are potent immunogens. One cannot assume that the use of homologous or heterologous surfactants in patients with RDS will always be immunologically innocuous.

Acknowledgments. The authors are grateful to Drs. Charles A. Janeway, Michael Iverson, and Stewart Sell, all of whom gave of their energy and/or resources to help with this project. The nursing and house staffs at university affiliated hospitals in San Diego, California and Helsinki, Finland provided excellent care for the infants described here. The technical assistance of Mr. Jan Dombrowski, Mrs. Kathy Holcomb, Mr. Michael Wilson, and Mrs. Sharon Packer was indispensible.

References

1. Farrell PM, Avery ME (1975) Hyaline membrane disease. Am Rev Respir Dis 111:657–669
2. Fujiwara T, Maeta J, Chida S, Morita T, Watabe Y, Abe A (1980) Artificial surfactant therapy in hyaline membrane disease. Lancet 1:55–59
3. Morley CJ, Miller N, Bangham AD, Davis JA (1983) Dry artificial lung surfactant and its effect on very premature babies. Lancet 1:64–68
4. Hallman M, Merritt TA, Schneider H, Epstein BL, Manino F, Edwards DK, Gluck L (1983) Isolation of human surfactant from amniotic fluid and pilot study of its efficacy in respiratory distress syndrome. Pediatrics 71:473–482
5. Strayer DS, Merritt TA, Lwebuga-Mukasa J, Hallman M (1986) Surfactant-anti-surfactant immune complexes in infants with respiratory distress syndrome. Am J Pathol 122:353–362
6. Engvall E, Pesce A (eds) (1981) Quantitative enzyme immunoassay. Scand J Immunol [Suppl 7]
7. Enhorning G (1977) Pulsating bubble technique for evaluating pulmonary surfactant. J Appl Physiol 43:198
8. Hallman M, Epstein B, Gluck L (1981) Analysis of labeling and clearance of lung surfactant phospholipids in the rabbit. J Clin Invest 68:742–751
9. Rott IM, Donaich D (1976) Gastric autoimmunity. In: Miescher PA, Muller-Eberhard HJ (eds) Textbook of immunopathology. Grune and Stratton, New York, pp 737–747
10. Thivolet J, Beyvin AJ (1968) Recherches par immunofluorescence d'autoanticorps serique vis-á-vis des constitutuents de l'épiderme chez les brulés. Experientia 24:945–946
11. Jeffries AL, Coates G, O'Brodovich H (1984) Pulmonary epithelial permeability in hyaline-membrane disease. N Engl J Med 311:1075–1080
12. Bauer H, Waters TJ, Talano JV (1972) Antimyocardial antibodies in patients with coronary heart disease. Am Heart J 83:612–619
13. Zabriskie JB, Hsu KC, Seegal BC (1970) Heart-reactive antibody associated with rheumatic fever: characterization and diagnostic significance. Clin Exp Immunol 7:147–159

14. Sack W, Sebvening H, Wachsmuth ED (1975) Auto-Antikörper gegen Herzmuskelsarkolemm im Serum von Patienten mit primärer Cadriomyopathie. Klin Wochenschr 53:103–110
15. Camp TF, Hess EV, Conway G, Fowler NO (1969) Immunologic findings in idiopathic cardiomyopathy. Am Heart J 77:610–617
16. Trueman T, Thompson RA, Cummins P, Littler WA (1981) Heart antibodies in cardiomyopathies. Br Heart J 46:296–301
17. Silverstein AM (1972) Fetal immune responses in congenital infection. N Engl J Med 286:1413–1414
18. Gitlin D, Biasucci A (1969) Development of γG, γA, γM, B_{1c}/B_{1A}, C'1 esterase inhibitor, ceruloplasmin, transferrin, hemopexin, haptoglobin, fibrinogen, plasminogen, α_1-antitrypsin, orosomucoid, β-lipoprotein, α2-macroglobulin, and prealbumin in the human conceptus. J Clin Invest 48:1433–1446
19. Stiehm ER, Ammann AJ, Cherry JD (1966) Elevated cord macroglobulins in the diagnosis of intrauterine infections. N Engl J Med 275:971–977
20. Alford CA, Schaefer J, Blankenship WJ et al (1966) A correlative immunologic, microbiologic and clinical approach to the diagnosis of acute and chronic infections in newborn infants. N Engl J Med 277:437–449
21. Remington JS (1969) The present status of the IgM fluorescent antibody technique in the diagnosis of congenital toxoplasmosis. J Pediatr 75:1116–1124
22. Aase JM, Noren GR, Reddy V, St. Geme JW Jr (1972) Mumps-virus infection in pregnant women and the immunologic response of their offspring. N Engl J Med 286:1377–1382
23. Field EJ, Caspary EA (1971) Is maternal lymphocyte sensitization passed to the child? Lancet ii:337–341
24. Leikin S, Oppenheim JJ (1971) Prenatal sensitization. Lancet ii:876–877
25. Cramer DV, Kunz HW, Gill TJ III (1974) Immunololgic sensitization prior to birth. Am J Obstet Gynecol 120:431–439
26. Jerne NK (1974) Towards a network theory of the immune system. Ann Immunol Inst Pasteur 125C:373–389
27. Rodkey LS (1980) Autoregulation of immune responses via idiotype network interactions. Microbiol Rev 44:631–659
28. Bona C, Paul WE (1979) Cellular basis of regulation of expression of idiotype: I. T-suppressor cells specific for MOPC-460 idiotype regulate the expression of cells secreting anti-TNP antibodies bearing 460 idiotype. J Exp Med 149:592–600
29. Strayer DS, Kohler H (1976) Immune response to phosphorylcholine II, natural "auto"-anti-receptor antibody in neonatal Balb/c mice. Cell Immunol 25:294–301
30. Shahani SK, Hattikudor NS (1981) Immunological consequences of vasectomy. Arch Androl 7:193–199
31. Wallace RB, Lee J, Gerber WL, Clarke WE, Lauer RM (1981) Vasectomy and coronary artery disease in men less than 50 years old: absence of association. J Urol 126:182–184
32. Kapp JA, Strayer DS (1978) H-2 linked Ir gene control of antibody responses to porcine insulin. I. Development of insulin-specific antibodies in some but not all nonresponder strains injected with proinsulin. J Immunol 121:978–982

Surfactant Treatment of Newborn Rabbits Impedes Intrapulmonary Phagocytic Killing of Inhaled Group B Streptococci

M. P. Sherman, E. E. Aeberhard, C. T. Barrett, J. B. D'Ambola, and L. A. Campbell

Introduction

Surfactant replacement therapy holds great promise in the treatment of the respiratory distress syndrome of the premature infant and the acquired distress syndrome of adults. However, relatively small numbers of newborns and few adults have received this therapeutic intervention, and the full range of side effects has yet to be delineated. At present, the effects of phospholipid replacement therapy on pulmonary host defenses of the newborn are poorly defined. We postulated that artificial and natural surfactant preparations might impair the *in vivo* bactericidal activity of the newborn pulmonary alveolar macrophages (PAM).

Methods

Natural surfactant was isolated from healthy adult rabbits according to the technique of Jobe et al. [1]. Two therapeutic artificial phospholipid suspensions were prepared by sonicating mixtures of dipalmitoyl L-α-phosphatidyl-choline and dipalmitoyl L-α-phosphatidyl-glycerol at 7 : 3 and 9 : 1 ratios (by weight) [2, 3]. The size and structure of the phospholipid vesicles in the three preparations were determined by electron microscopic examination [4]. Phospholipid content was determined by standard techniques [5]. Term newborn rabbits received either homologous surfactant or one of the two artificial phospholipid vesicle preparations [5 mg/kg of phospholipid in 2 ml/kg of Dulbecco's phosphate buffered saline (pH 7.4)] by direct tracheal puncture. One group of control animals had phosphate buffered saline (PBS) instilled into the surgically exposed trachea (2 ml/kg), while a second control group had no therapy or surgery. One hour after surfactant instillation treated and control animals were exposed to a concentrated aerosol of type Ia *Streptococcus agalactiae* strain H784. After a 30 min period of aerosol infection, one half of the animals were sacrificed. The remaining animals were studied after 4 hours. *In vivo* bacterial ingestion and phagocytic killing were determined in these newborn rabbits by established methods [6].

Phagocytosis of artificially prepared phospholipid vesicles by PAM were determined by differentially labeling the lipid and aqueous compartments of the liposomes [7] by adding [choline-methyl-^{14}C] dipalmitoyl phosphatidylcholine (DPPC) as a lipid phase marker and [^{3}H(G)]-inulin as an aqueous marker to the presonicated mixtures. *In vivo* intrapulmonary phagocytic uptake of phospholipid vesicles present in the natural surfactant was ascertained by phagocytosis of 0.1% acridine orange stained lipid inclusions by PAM [8]. At 15, 60 and 300 min after instillation of phospholipid vesicles at the dosage utilized to study intrapulmonary group B streptococci (GBS) clearance, animals were sacrificed, and the lungs lavaged [9]. Isolated PAM were separated from uningested, suspended radiolabelled liposomes by centrifugation of the resuspended cells through a silicone oil/0.6 M sucrose gradient [7]. Uptake of radiolabelled phospholipid vesicles by PAM was determined using the radioactive content of the cells. Ingestion of fluorescent phospholipid vesicles in the natural preparation was ascertained by epifluorescent microscopy of PAM recovered by lung lavage at 15, 60 and 300 min after instillation.

The effect of phospholipid vesicles on bacterial proliferation was studied by incubating GBS with or without addition of natural or artificial surfactants to an incubation mixture that simulated the physiologic conditions of ionic composition and pH of alveolar lining fluid. After 1 and 4 h incubation, portions of the mixture were removed, diluted and spread on blood-agar plates to determine GBS viability.

Kruskal-Wallis statistic, Mann-Whitney rank sum test adjusted for Bonferroni inequality, two-tailed unpaired t test or Chi-square statistic with continuity correction were used for data analysis depending on the data base [10]. A statistical software program was utilized for analyses (Statview 512+, Brainpower, Inc., Calabasas, CA).

Results

Newborn rabbits treated with the 7 : 3 and natural phospholipid preparations had a significant reduction in their ability to inactivate intrapulmonary GBS at 4 h after aerosol infection when they were compared to the untreated and PBS treated pups ($p < 0.02$). Animals receiving the 9 : 1 preparation also had reduced capacity to kill inspired GBS, but this difference was not statistically significant compared to the controls. The log reduction in inhaled GBS after 4 h was as follows: control = $+ 0.29 \pm 0.04$ (28), PBS = $+ 0.35 \pm 0.06$ (28), 9 : 1 = $+ 0.20 \pm 0.07$ (28), 7 : 3 = $+ 0.09 \pm 0.06$ (35), and natural = $+ 0.03 \pm 0.03$ (23) (mean ± SEM, n = no. of animals studied at 0 and at 4 h). A negative log reduction in the lungs of individual animals, indicative of intrapulmonary GBS proliferation by 4 h after aerosol infection, was noted in 7% of control, 21 of PBS, 18% of 9 : 1, 43% of 7 : 3, and 48% of natural surfactant treated animals ($p < 0.01$ for 7 : 3 and natural v. control).

The diminished inactivation of intrapulmonary GBS by the treated newborns was not explained by differences in the uptake of GBS by lung phagocytes. After 4 hours exposure, the mean number of GBS within phagocytes in histologic sections of the lungs was similar among the five treatment groups and ranged from 71 to 77%. Ingestion rates were also comparable at 0 h. The abnormalities seen in intrapulmonary bacterial inactivation following surfactant therapy could also not be explained by differences in the initial dose of inhaled GBS. Similar numbers of streptococci were cultured from the lungs of the five groups immediately after aerosol infection.

PAM recovered by lavage 1 h after surfactant therapy showed no significant alterations in either cell viability (trypan blue exclusion) or *in vitro* phagocytic function (ingestion of opsonized FITC-labelled GBS) compared to control animals (PBS or no surgery) (data not shown). At 0 h, > 98% of the cells recovered by lung lavage were PAM. Four h after GBS aerosol infection, > 88% of the lavaged cells were PAM regardless of whether the rabbit pups had received surfactant or surfactant vehicle. Thus, this study mainly examined whether surfactant therapy had an adverse effect on *in vivo* PAM bactericidal function.

When GBS were incubated in the presence of either the 7 : 3 or 9 : 1 phospholipid preparations and conditions that simulated the ionic composition and pH of alveolar lining fluid, the number of viable bacteria remained the same or declined slightly during 4 h of incubation. However, natural surfactant at phospholipid concentrations equal to and above doses used during *in vivo* studies promoted significant proliferation of GBS after 4 h of incubation (up to a 4000 fold increase).

The *in vivo* uptake and metabolism of ^{3}H-inulin and ^{14}C-DPPC labeled artificial lipid vesicles by PAM showed that phagocyte-associated ^{3}H-counts increased 7–10 fold between 15 and 300 min after instillation of either the radiolabelled 7 : 3 or 9 : 1 preparation. The rise in cell-associated ^{14}C counts was much smaller over this interval so that after 300 min the ratio of ^{3}H-inulin to ^{14}C-DPPC was 19–21 : 1. This observation was explained by the fact that the PAM ingested the radiolabelled phospholipid vesicles and were able to metabolize the DPPC, but not the inulin. Within 15 min after acridine orange-stained natural surfactant was injected into the airways, > 90% of the PAM recovered by lung lavage were filled with large and small fluorescent phospholipid vesicles. Over 70% of the PAM that had ingested phospholipid vesicles contained > 10 fluorescent bodies per cell at this time point. By 1 and 5 h after fluorescent phospholipid vesicles were instilled, the harvested PAM from individual animals had a substantial reduction in the numbers of fluorescent vesicles per phagocytic cell. This finding was consistent with the radiolabelling studies indicating that the newborn PAM readily ingested and processed phospholipid vesicles.

Electron microscopy revealed that the 7 : 3 mixture of DPPC dipalmitoyl phosphatidylglycerol (DPPG) and natural surfactant consistently had larger vesicles [0.5–2.0 μm] than those seen in the 9 : 1 suspension of DPPC/DPPG (data not shown). The 7 : 3 mixture did not have the small unilamellar liposomes [0.0002–0.2 μm] noted in the 9 : 1 suspension and natural preparations.

Discussion

When term neonatal rabbits were infected with intrapulmonary GBS after surfactant had been instilled into the airways, an *in vivo* defect in phagocytic killing by PAM was present. The diminished ability of the lung to inactivate inspired GBS appears related to the fact that PAM of neonatal rabbits avidly ingest and process phospholipid vesicles regardless of whether artificially-formulated liposomes or naturally-occurring phospholipid vesicles are instilled into the airways. Electron photomicrographs revealed that the 7 : 3 and natural preparations contained larger vesicles which were rapidly ingested by the neonatal PAM. Such large phospholipid inclusions fill the PAM recovered from alveolar proteinosis patients, and these phagocytes are known to have diminished microbicidal activity against *Candida albicans* [11]. Recent *in vitro* investigations have suggested that phagocytosis of phospholipid vesicles decreases the candidacidal activity of infantile rabbit PAM [12].

The mechanisms by which surfactant alters the microbicidal activity of the neonatal PAM are only partially clarified by the present study and this is a crucial area of future investigation. Neither PAM viability nor phagocytic behavior were decreased after phospholipid treatment. It is unknown whether ingested phospholipid vesicles either impair the respiratory burst of the PAM (O_2-dependent microbicidal activity) or deplete the PAM of lysosomal components and create abnormalities of phagolysosomal fusion (O_2-independent microbicidal killing). Furthermore, the intratracheal dose of phospholipid instilled during our studies was only 10–20% of the amount used in recent human trials of surfactant replacement [13, 14], and it is possible that larger doses of phospholipid vesicles would have elicited additional dysfunction in the neonatal PAM.

Natural surfactant preparations, but not the artificial phospholipid vesicles, significantly promoted the growth of GBS when tested at concentrations used in our studies of intrapulmonary bacterial clearance. The ability of surfactants to support or hinder bacterial growth is an important new area of evaluation because our homologous preparation has structural and biochemical characteristics that are similar to surfactants used in clinical trials and derived from bovine and human sources [15, 16].

It has been recently suggested that primary care physicians attending high risk preterm births will be able to prevent respiratory distress syndrome (RDS) by surfactant replacement in the delivery room [17]. Because 5 to 10% of preterm infants who have respiratory distress may have either RDS and bacterial pneumonitis or pneumonitis alone [18, 19], our findings indicating that surfactant administration reduces the bactericidal activity of neonatal PAM should raise some concern about delivery room surfactant replacement before a definitive pulmonary diagnosis is established. Additional investigations are needed to ascertain the effect of surfactant therapy on host defenses of the lung.

References

1. Jobe A, Ikegami M, Glatz T, Yoshida Y, Diakomanolis E, Padbury J (1981) Duration and characteristics of treatment of premature lambs with natural surfactant. J Clin Invest 67:370–375

2. Ivey HH, Kattwinkel J, Roth S (1980) Nebulization of sonicated phospholipid for treatment of respiratory distress syndrome of infancy. In: Tom S, Six H (ed) Liposomes and immunobiology. Elsevier/North Holland, New York, pp 301–314
3. Morley CJ, Miller N, Bangham AD, Davis JA (1981) Dry artificial lung surfactant and its effect on very premature babies. Lancet i:64–68
4. Obladen M, Popp D, Scholl C, Schwarz H, Jahnig F (1983) Studies on lung surfactant replacement in respiratory distress syndrome. Rapid film formation from binary mixed liposomes. Biochim Biophys Acta 735:215–224
5. Aeberhard EE, Scott ML, Barrett CT, Kaplan SA (1984) Effects of cyclic AMP analogues and phosphodiesterase inhibitors on phospholipid biosynthesis in fetal type II pneumocytes. Biochim Biophys Acta 803:29–38
6. Sherman M, Goldstein E, Lippert W, Wennberg R (1977) Neonatal lung defense mechanisms. A study of the alveolar macrophage system in neonatal rabbits. Am Rev Respir Dis 116:433–440
7. Finkelstein MC, Kuhn SH, Schierren J, Weissman G, Hoffstein S (1980) Selectivity in the uptake of liposomes by human leukocytes: a comparison of monocytes, lymphocytes and polymorphonuclear leukocytes. In: Tom S, Six H (eds) Liposomes and immunobiology. Elsevier/North Holland, New York, pp 255–270
8. van Rooijen N, van Nieuwmegen R (1978) Fluorochrome staining of multilamellar liposomes. Stain Technol 53:307–310
9. Sherman MP, Lehrer RI (1984) Superoxide generation by neonatal and adult rabbit alveolar macrophages. J Leukocyte Biol 36:39–50
10. Glantz SA (1981) Primer of biostatistics. McGraw-Hill, New York
11. Golde DW, Territo M, Finley TN, Cline MJ (1976) Defective lung macrophages in pulmonary alveolar proteinosis. Ann Intern Med 85:304–309
12. Zeligs BJ, Nerurkar LS, Bellanti JA (1984) Chemotactic and candidacidal responses of rabbit alveolar macrophages during postnatal development and the modulating roles of surfactant in these responses. Infect Immun 44:379–385
13. Kwong MS, Egan EA, Notter RH, Shapiro DL (1985) Double-blind clinical trial of calf lung surfactant for the prevention of hyaline membrane disease in extremely premature infants. Pediatrics 76:585–592
14. Merritt TA, Hallman M, Bloom BT, Berry C, Benirschke K, Sahn D, Key T, Edwards D, Jarvenpaa AL, Pohjavuori M, Kankaanpaa K, Kunnas M, Paatero H, Rapola J, Jaaskelainen J (1986) Prophylactic treatment of very premature infants with human surfactant. N Engl J Med 315:785–790
15. Hallman M, Merritt TA, Schneider H, Epstein BL, Mannino F, Edwards DK, Gluck L (1983) Isolation of human surfactant from amniotic fluid and a pilot study of its efficacy in respiratory distress syndrome. Pediatrics 71:473–482
16. Taeusch HW, Keough KMW, Williams M, Slavin R, Steele E, Lee AS, Phelps D, Kariel N, Floros J, Avery ME (1986) Characterization of bovine surfactant for infants with respiratory distress syndrome. Pediatrics 77:572–581
17. Lawson EE (1987) Exogenous surfactant therapy to prevent respiratory distress syndrome. J Pediatr 110:492–493
18. Sherman MP, Goetzman BW, Ahlfors CE, Wennberg RP (1980) Tracheal aspiration and its clinical correlates in the diagnosis of congenital pneumonia. Pediatrics 65:258–263
19. Jacob J, Edwards D, Gluck L (1980) Early-onset sepsis and pneumonia observed as respiratory distress syndrome. Am J Dis Child 134:766–768

Effect of Bovine Surfactant on Mucus Secretion from Tracheal Submucosal Glands

B. Hümmer, I. Purnama, and H.-L. Hahn

Introduction

Although the stabilizing effect of surface active material on alveolar structures is well established much less is known about possible effects of surfactant on small and, especially, on large bronchi. The question is of interest because surface active material has been identified in bronchial secretions [1, 6, 9, 11]. In patients with asthma and cystic fibrosis, lipids constituted 30–40% of the dry insoluble material obtained from lavage specimens [11] and the high lipid content of bronchial secretions has been confirmed by many investigators [2]. One would expect therefore that surface active materials have a function in conducting airways as well as in the alveoli. The closure of small (< 0.5 mm) airways may be prevented by surfactant [8]. Effects on epithelial transport functions have been considered because myelin-figured lamallae have been found in the lower zone of the mucous blanket [13] and the presence of surface active material between the sol and the gel phase of bronchial mucus seems likely from these and other studies. This may aid in maintaining continuity of the mucus layer. Stimulation of surfactant production by inhaled corticosteroids decreased the viscosity of mucus [7]. Effects on physical properties of mucus and on the interface between the ciliary layer and the mucous blanket can be expected to affect the mucociliary system of the conducting airways.

A direct effect of surfactant on gland function has never been studied, but if it exists, it could result in a very powerful regulation of mucous secretion by surfactant. Indirect effects of surfactant on mucus secretion may result from improved patency of gland ducts, from lower mucus viscosity and from an "anti-glue" function [12] of surfactant. On the other hand surfactant, being present on the mucosal surface, may never be able to penetrate deeply into gland ducts and may not affect mucus properties until mucus emerges from gland duct openings, in which case direct or indirect effects on glandular secretion would be unlikely. We decided to study the question in an epithelial tissue virtually devoid of surface secretory cells: the tracheal mucosa obtained from the ferret. In this tissue, submucosal glands are practically the only source of mucous glycoproteins, which is desirable in studies using radioactive precursors because precursors would otherwise be incorporated into secretions from submucosal glands as well as from surface secretory cells [10]. Since baseline and stimulated secretion of mucus depend on different mechanisms [3] we studied the effect of surfactant on mucus secretion under both conditions.

Methods

Fifteen ferrets were anesthetized: 9 male (mean weight 1244 ± 266 g) and 6 female (mean weight 630 ± 124 g) ($\bar{x} \pm SD$) with pentobarbital (Nembutal^R), average dose, 75.4 ± 25.8 mg kg^{-1} (males), 70.3 ± 19.9 mg kg^{-1} (females) intraperitoneally. They were exposed to an oxygen-enriched atmosphere by means of a loosely-fitting mask fabricated from a surgical glove. This served to prevent the reflex emptying of glands induced by hypoxia [4]. The trachea was exposed by midline incision and excised within seconds of a thoracotomy by median sternotomy. The trachea was transferred to an organ bath filled with medium M-199 in Earle's balanced salt solution, equilibrated with 95% oxygen and 5% carbon dioxide (pH 7.41 at 37° C). We opened the trachea along the posterior membrane and mounted segments of the anterior portion of trachea between the two halves of Ussing chambers (Fig. 1) filled with medium M-199 in Earle's balanced salt solution (Serva, Heidelberg) equilibrated with 95% oxyen and 5% carbon dioxide. 0.23% $NaHCO_3$ was added to obtain a pH of 7.41 at 37° C. 50 μCi $Na_2{}^{35}SO_4$ (Amersham) was added to the solution bathing the submucosal side and left radioactive sulfate in contact with the submucosa for the remainder of the experiment. We drained and refilled the luminal half of the chambers at 15 min intervals and collected the drained fluid (6 ml chamber content plus 1 ml rinse) in dialysis bags (Visking, pore radius 1.5–2 nm, molecular weight cutoff above 12–14 kilodaltons). We dialysed bag contents against distilled water to which sodium azide (NaN_3) was added to prevent bacterial growth and nonradioactive sulfate (Na_2SO_4) to displace radioactive sulfate bound covalently. When the radioactivity of the dialysis fluid had fallen to background levels we emptied the content of the dialysis bags into 20 ml scintillation vials (Zinsser, Frankfurt), added 10 ml scintillation fluid (Lumagel, Baker Chemicals) and counted radioactive sulfate in a beta-counter (Betamatic I. Kontron).

The postganglionic nerve endings were electrically stimulated by creating an electric field across the tracheal membrane using 2 Ag-AgCl electrodes protruding into the chamber lumen at a slightly oblique angle (not shown in Fig. 1). The tip of each electrode was centered in the chamber lumen and was

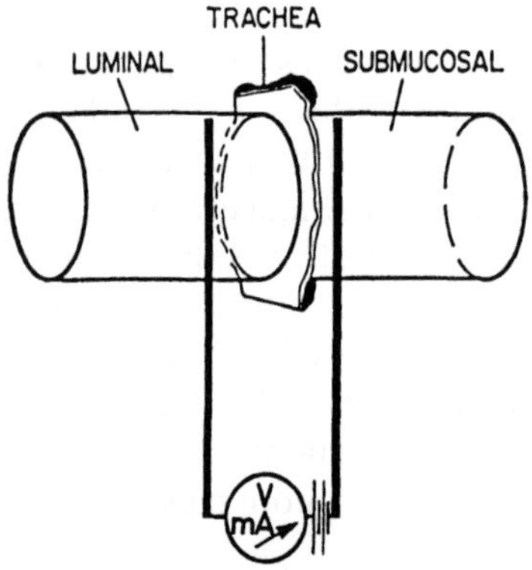

Fig. 1. The Ussing chamber with stimulating electrodes for applying electric fields. In reality stimulating electrodes were not situated parallel to the epithelial membrane but entered the chamber lumen at an oblique angle; thus, when the tips of both electrodes were centred in the middle of the membrane mounted in the chamber, they were approximately 1 mm away from each membrane surface

situated 1 mm away from the tracheal membrane. A constant stimulation frequency of 6 Hz was used. Single pulse duration was 0.2 ms, stimulation was bipolar, stimulation strength varied between ± 2 and ± 6 volts (amplitude range, 4–12 volts). Currents achieved at (±) 2, 4 and 6 volts were (±) 30 ± 5, 56 ± 7, and 76 ± 9 mA, respectively ($\bar{x} \pm SD$). Each stimulation lasted 5 min. Selection of stimulation settings was based on previous experiments in which we had established settings as those stimulating glands only through postganglionic nerves (i.e., causing effects that could be blocked by tetrodotoxin), but not directly [5].

Bovine surfactant was used (SF-RI-1, Dr. Karl Thomae GmbH, Biberach, West Germany) which contained 90% phospholipid of which 80% was phosphatidylcholine and 10% phosphatidylglycerol as the major components. The preparations also contains low molecular weight apoproteins (5 and 18 kilodaltons), but not the large 35 kilodalton apoprotein because of its antigenicity.

Surfactant was applied to the luminal side of tissues immediately after the last control collection, thus surfactant was in contact with the tissue for the full 15 min prior to sample collection. In experiments in which we assessed the effect of surfactant on secretion stimulated by an electric field, surfactant was in the chamber lumen not only during the period of stimulation, but for 2 sampling periods (30 min) prior to the period during which the electric field was applied. Thus, surfactant was in contact with the tissue for a total of 45 min. The intention was to give the material ample contact with the tissue and sufficient diffusion time to exert maximal effects. Concentration of surfactant was 10^{-2} molar throughout. Calculation of molar solutions was based on the average molecular weight (762) of all phospholipids contained in the preparation, the average being weighted according to the percentage that each phospholipid contributed to total phospholipid content.

Time course of each experiment: Labeling time after addition of radioactivity to the submucosal side was two hours. The perfusate bathing the luminal side was then drained and regular collections were started. After 5 collections to assess baseline secretion surfactant was added (or 0.9% NaCl) to the luminal side either once (in studies investigating basal secretion) or 3 times (in studies investigating stimulated secretion). Two more samples were obtained after surfactant had finally been drained from the chamber lumen. Tissue viability was subsequently assessed by studying the secretory effect of 10^{-4} M acetylcholine (ACh) applied to both sides of the tracheal membrane. In healthy glands this standard stimulus increased mucus secretion at least 8-fold and often the increase was more than ten-fold. Data obtained from tissues that did not subsequently respond to ACh were discarded. After a pause of 60 min to allow relabeling of glands a second "run" with the same time course was performed in each tissue and was again concluded with a viability test. If control solution had been applied during run 1, surfactant was studied during run 2 and vice versa, the sequence being randomized.

Statistical evaluation was for data obtained within tissues (run 1 vs. run 2) using Student's t-test for paired comparisons. Results were considered significant if the likelihood of the null hypothesis was less than 5% ($p < 0.05$).

Results

Figure 2 shows the effect of surfactant on baseline secretion. Surfactant did not affect the baseline rate of secretion of nondialyzable ^{35}S-label. Surfactant had no deleterious effect on tracheal tissues because the secretory effect of acetylcholine was identical to the effect seen under control conditions.

Figure 3 demonstrates the effect of electric field stimulation at 3 stimulation strengths in the absence and in the presence of surfactant. Again there were neither stimulatory nor inhibitory effects of surfactant on mucus secretion, nor did prolonged contact of tissues with surfactant appear to affect tissue viability. When the absolute data shown in Fig. 3 were replotted as % control (Fig. 4) there was also no significant effect of surfactant on mucus secretion at any stimulation strength ($p = 0.5$).

Discussion

The conclusion from these studies is that surfactant has no stimulatory effect on baseline mucus secretion and has neither stimulatory nor inhibitory effects on mucus secretion increased by postganglionic nerve stimulation. It should be remembered that mucociliary transport and ciliary beating were not studied in this preparation nor can we exclude an effect of surfactant on surface secretory (e.g., goblet) cells because they are lacking in ferret tracheae. We can exclude a direct effect of surfactant on the secretory rate of submucosal glands and indirect effects on mucus secretion resulting from improved patency of gland ducts, from lubricating effects at the interface between the gland or duct epithelium and the column of mucus flowing past the epithelium, from decreased viscosity of mucus or from a combination of these effects. Whether this is because such effects do not exist or because surfactant normally has no access to gland structures or because the time of contact was too short to allow adequate penetration of surfactant into gland ducts remains unknown.

The argument might be advanced that the concentration of surfactant used in our experiments was too low compared to the *in vivo* situation. This is unlikely to be the case as can be shown by the following calculation. The only

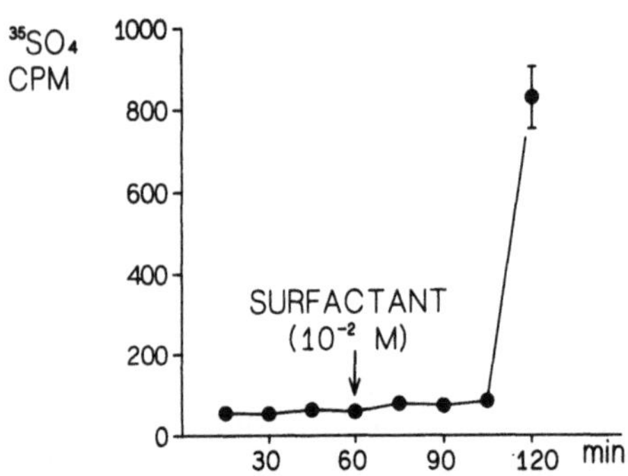

Fig. 2. Effect of surfactant (10^{-2} M) on baseline mucus secretion. Surfactant was applied at the *arrow* and was left in contact with the tissue for 15 min. Each data *point* represents the mean ± SE of seven experiments. No significant difference to baseline was found

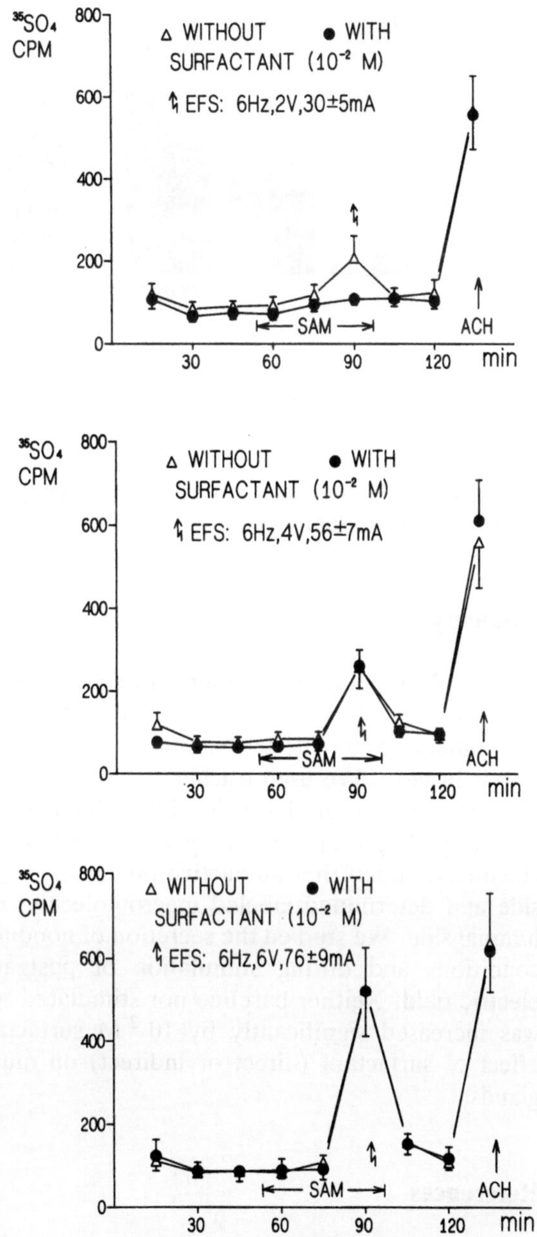

Fig. 3. Effect of surfactant (10^{-2} M) on stimulated mucus secretion. Stimulation strengths of electric field stimulation *(EFS)* were ± 2 volts *(top)*, ± 4 volts *(middle)* and ± 6 volts *(bottom)*. *Open triangles* represent control (no surfactant added), *closed circles* 10^{-2} M surfactant added to luminal side of tissue. Surfactant was present in perfusion chamber for a total of 45 min. Each data *point* represents mean ± SE from eight experiments. No significant differences between control and surfactant-treated tissues were found

assumption in this calculation is that the concentration of urea in the epithelial lining fluid is the same as in serum for which there is good evidence. If we determine the concentration of surfactant and of urea in bronchoalveolar lavage fluid and simultaneously determine serum urea we are able to calculate a dilution factor for the lavage fluid. Applying this dilution factor to the concentration of surfactant we can calculate the original concentration of

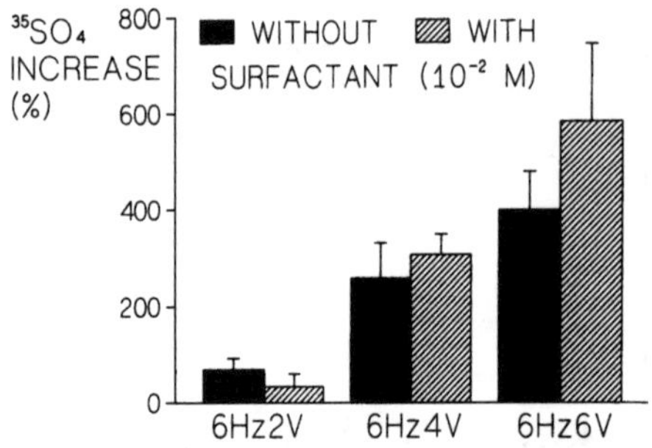

Fig. 4. Effect of surfactant (10^{-2} M) on stimulated mucus secretion. Stimulation strengths are indicated below each *column*. The same data are presented as in Fig. 3, but expressed as a percentage of pre-stimulation baseline. No significant differences between control and surfactant-treated tissues were found

surfactant in the epithelial lining layer. This turns out to be 10^{-4} M. It appears therefore that we exceeded physiological concentrations in our experiments. The fact that, using high concentrations of surfactant, we still had a negative result makes it very unlikely that lower (physiological) concentrations would stimulate mucus secretion.

Summary

Surface active material is found in secretions obtained from small and large bronchi, thus effects of surfactant may not be limited to the alveoli. We studied the influence of bovine surfactant on mucus secretion from tracheal submucosal glands in vitro. We used tracheal membranes from the ferret because they are almost devoid of goblet cells, thus submucosal glands are the only source of mucous glykoproteins in this species. We mounted tracheal segments in Ussing chambers, added the radioactive precursor $Na_2{}^{350}SO_4$ (^{35}S) to the submucosal side and determined labeled macromolecules in perfusate collected from the luminal side. We studied the secretion of nondialyzable ^{35}S-label under baseline conditions and during stimulation of postganglionic nerve endings by an electric field. Neither baseline nor stimulated secretion of ^{35}S-labeled material was increased significantly by 10^{-2} M surfactant. This excludes a stimulant effect of surfactant (direct or indirect) on mucus secretion from submucosal glands.

References

1. Bancher E, Washüttl J, Muhar F, Stachelberger H, Winker N (1976) Bestimmung von Phosphatiden im Sputum. Münch Med Wochenschr 118:1499–1500
2. Coles SJ, Bhaskar KR, O'Sullivan DD, Reid LM (1982) Macromolecular composition of secretions produced by human bronchial explants. In: Chantler EN (ed) Mucus in health and disease II. Plenum, New York, pp 357–360
3. Coles SJ, Reid L (1981) Inhibition of glycoconjugate secretion by colchicine and cytochalasin B. An in vitro study of human airway. Cell Tissue Res 214:107–118
4. Davis B, Chinn R, Gold J, Popovac D, Widdicombe JG, Nadel JA (1982) Hypoxemia reflexly increases secretion from tracheal submucosal glands in dogs. J Appl Physiol: Respirat Environ Exercise Physiol 52:1416–1419

5. Hahn HL, Lang M, Sannwald U, Purnama I (1985) Effect of electric fields on mucus secretion from ferret trachea in vitro. Fed Proc 44:640
6. Havez R (1969) Freie und gebundene Formen des Surfactant im Expektorat. Dtsch Med J 20:709–713
7. Jouanel P, Motta C, Roche G, Wahl D, Dastugue B, Molina C (1979) Les lipides et phospholipides du liquide de lavage bronchoalvéolaire: intérêt dans l'étude de la pénétration alvéolaire des corticoides par aérosol. Poumon Coeur 35:391–395
8. Macklem PT, Proctor DF, Hogg JC (1970) The stability of peripheral airways. Respir Physiol 8:191–203
9. Matthews L, Leroy S, Spector S, Lemm J, Potter J (1963) Studies on pulmonary secretions. 1. The overall chemical composition of pulmonary secretions from patients with CF, bronchiectasis and laryngectomy. Am Rev Respir Dis 88:199–204
10. Robinson NP, Venning L, Kyle H, Widdicombe JG (1986) Quantitation of the secretory cells of the ferret tracheobronchial tree. J Anat 145:153–188
11. Sahu S, Lynn WS (1977) Lipid composition of airway secretions from patients with asthma and patients with cystic fibrosis. Am Rev Respir Dis 115:233–239
12. Sanderson RJ, Paul GW, Vatter AE, Filley GF (1976) Morphological and physical basis for lung surfactant action. Respir Physiol 27:379–392
13. Yoneda K (1976) Mucous blanket of rat bronchus. An ultrastructural study. Am Rev Respir Dis 114:837–842

Part VI
Prospects for the Future

The Bronchoalveolar Surfactant System Inhibits the Cardiovascular Effects of Angiotensin II, Serotonin, and Platelet-Activating Factor

J. M. Smit, T. Hein, S. Armbruster, W. Schairer, and B. Lachmann

Introduction

A previous study has shown that bronchoalveolar lavage fluid prevented angiotensin II receptor-mediated contraction of the isolated guinea pig ileum in an organ bath (Lachmann 1985). In this former study it was speculated that an explanation for this may be that components of the surfactant system mask the receptors of smooth muscle and thus prevent contraction due to angiotensin II.

The purpose of this study was to determine whether a similar effect could also be demonstrated under in vivo conditions. In order to study this possible mechanism under in vivo conditions we gave angiotensin II via the airways before and after surfactant depletion and examined the effects on cardiovascular parameters. In addition to angiotensin II we used two other potent vasoactive substances, platelet-activating factor (PAF) and serotonin.

Materials and Methods

A total of 42 Wistar rats (T.N.O. Zeist, Netherlands; body weight, 310 ± 24 g) were anesthetized with pentobarbital (7 mg/kg) and paralyzed with pancuronium bromide (0.7 mg/kg). Following tracheotomy, catheters were placed in the carotid artery and the femoral vein. The rats were connected to a pressure-controlled ventilator (Servo 900; rate, 20/min; FiO_2, 1; peak airway pressure, 20 cm H_2O; I : E, 1 : 2). The arterial catheter was connected to a pressure transducer (Gould P231D) and blood pressure was recorded continuously (Grass Instruments, model 7D). The investigations were divided into two parts to determine whether the inhibition of the drugs was a general effect of the intact surfactant system, or whether this was a specific effect caused by components of the lung surfactant system.

Study 1. After stabilization of blood pressure, the rats (n = 21) received 0.2 ml of one of the drugs intratracheally (angiotensin II, 70 μg/kg; serotonin, 700 μg/kg; PAF, 10 μg/kg), and blood pressure was recorded for 15 min. Lung surfactant was then depleted by three lung lavages with 10 ml saline (warmed to 40° C). The interval between each lavage was 5 min. After a stabilizing period of 10 min, the drugs were instilled a second time. The lung lavage fluid was checked macroscopically to ensure that no blood was present; otherwise the animal was excluded from the study.

Study 2. The drugs angiotensin II (1.0 μg/ml), serotonin (250 μg/ml), and PAF (1.0 μg/kg) were incubated in bronchoalveolar lavage fluid from healthy rat lungs, in rat plasma, or saline in a water bath (warmed to 40° C). A quantity of 0.1 ml of the different drugs and after different incubation times was given at random to healthy rats.

Results

Study 1. The results following intratracheal instillation of the drugs before and after removal of the surfactant system, are shown in Fig. 1. The changes in blood pressure following instillation of the same amounts of angiotensin II, serotonin, or PAF as administered before lung lavage were much stronger after depletion of the surfactant system.

Study 2. As shown in Table 1, the increase in blood pressure following angiotensin II was significantly inhibited after in vitro incubation in plasma and in broncho-alveolar lavage fluid. This effect was time-related. Serotonin and PAF were not influenced by either lung lavage fluid or plasma.

Discussion and Conclusions

Blood pressure changes in the lavaged lungs were significantly stronger in comparison to unlavaged lungs despite the fact that lung lavage with saline leads to a significant decrease in their functional residual capacity and diffusion capacity (Lachmann et al. 1980).

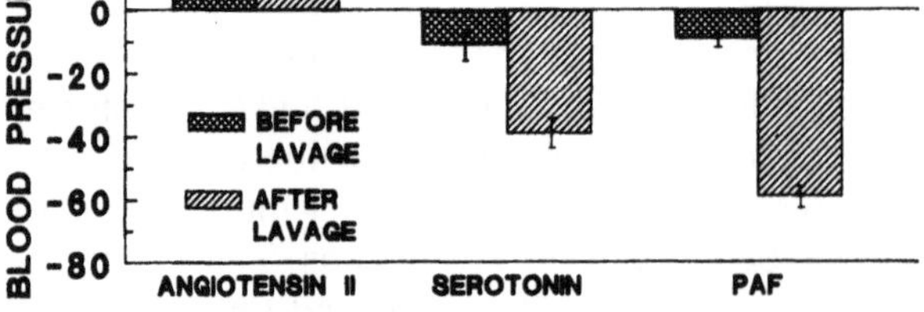

Fig. 1. Changes in blood pressure following intratracheal application before and after removing the surfactant system (mean ± SEM). Blood pressure before administration of the drugs was set to 100%. Changes in blood pressure following drug instillation before and after lung lavage were found to be significant using the Wilcoxon test

Table 1. Percentage change in blood pressure following incubation of angiotensin (ANG) II, serotonin, and platelet-activating factor (PAF) in lung lavage fluid, plasma, or saline.

Drug	Solvent	Before	Incubation time 15 min	30 min	60 min
Ang II	Lavage fluid	186%	117%	105%	102%
	Plasma	201%	122%	120%	112%
	Saline	163%	163%	164%	168%
Serotonin	Lavage fluid	65%	74%	67%	60%
	Plasma	70%	71%	85%	60%
	Saline	65%	69%	77%	69%
PAF	Lavage fluid	44%	52%	55%	66%
	Plasma	43%	55%	77%	81%
	Saline	44%	50%	56%	59%

Blood pressure just before application of each drug was set to 100%. The number of animals in each group was at least six.

This means that the surfactant system in the lung has some nonspecific properties to protect the cardiovascular system against substances which reach the lung via the airways. Futhermore, we found a potent substance in the bronchoalveolar lavage fluid which rapidly inhibits angiotensin II in its functional properties. The role of this potent angiotensin II inhibitor, which can be recovered by lung lavage, needs further investigation.

References

Lachmann B, Robertson B, Vogel J (1980) In vivo lung lavage as an experimental model of the respiratory distress syndrome. Acta Anaesthesiol Scand 24:231–236
Lachmann B (1985) Possible function of bronchial surfactant. Eur J Respir Dis 67:49–61

Artificial Surfactants Based on Different Hydrophobic Low-Molecular-Weight Proteins

T. Curstedt, H. Jörnvall, P. Berggren, and B. Robertson

Summary

Hydrophobic low-molecular weight proteins were isolated from minced pig lungs and separated into two fractions by chromatography on Sephadex LH-60. Electrophoresis of protein fraction 1 showed two major bands corresponding to *apparent* molecular weights of about 3 kDa and 14 kDa, while sequence degradation disclosed only one major structure. Electrophoretic separation of protein fraction 2 revealed one band, at an *apparent* molecular weight of about 6 kDa. The two fractions had different N-terminal structures and amino acid compositions.

Artificial surfactant preparations, obtained by recombination of protein fraction 1 or 2 with synthetic phospholipids, mixed in proportions roughly similar to those in natural surfactant, were evaluated with the pulsating bubble method and in experiments on artificially ventilated premature newborn rabbits. The addition of protein fraction 1 to the phospholipid mixture reduced minimum surface tension from more than 20 to nearly O mN/m. When this surfactant preparation was instilled into the airways of newborn rabbits, the tidal volumes at insufflation pressure 25 cm H_2O was increased about twenty-fold compared to the volumes obtained in non-treated controls. Preparations with protein fraction 1 had better in vitro and in vivo properties than those with protein fraction 2. Both these protein-containing preparations were decisively more effective than phospholipids alone.

Introduction

Pulmonary surfactant is a lipid/protein complex which reduces surface tension at the air/liquid interface of the alveoli [1]. Its major components are phosphatidylcholines, especially 1,2-dipalmitoylphosphatidylcholine, phosphatidylglycerols, hydrophilic surfactant-associated proteins (28–36 kDa, depending on the degree of glycosylation) and hydrophobic proteins (3–18 kDa) [2]. Several attempts have been made to produce a protein-free surfactant but the physiological activity of such a surfactant is consistently lower than that of natural surfactant [3, 4]. However, replacement therapy with organic solvent extracts of pulmonary surfactant has shown excellent effects, indicating that the hydrophilic 28–36 kDa protein may not be an essential component of exogenous surfactants [5–9]. These results are in agreement with the promising in vitro [10–14] and in vivo [12, 14] activities obtained by recombination of synthetic phospholipids with hydrophobic surfactant-associated proteins.

Materials and Methods

Preparation and Analysis of Surfactant and Its Protein Components

Surfactant was isolated from minced porcine lungs by a combination of washing, centrifugation, chloroform-methanol extraction and liquid-gel chromatography [14]. From the isolated phospholipid fraction, which is identical to the surfactant preparation applied in a current clinical trial [9], hydrophobic low-molecular weight proteins were isolated and separated into two fractions by Sephadex LH-60 chromatography [14]. Phospholipids and proteins were analyzed as described previously [14–16].

Recombination of Isolated Protein Fractions with Synthetic Phospholipids

Dipalmitoylphosphatidylcholine, palmitoyloleoylphosphatidylcholine and dipalmitoylphosphatidylglycerol were mixed in the relative proportions 55 : 35 : 10 (w/w/w), respectively, and used as the surfactant preparation "phospholipids". This particular combination of molecular species roughly simulates the phospholipid profile of natural surfactant preparations (see below). To aliquots of the lipid mixture, one of the two protein fractions, dissolved in chloroform/methanol 1 : 2 (v/v), was added at a protein to phospholipid ratio of 1 : 50. The solvents were evaporated to dryness and the different surfactant preparations were suspended in saline.

Evaluation of Surface Properties

The surface properties of the artificial surfactant preparations were analyzed at 37° C with a pulsating bubble [17] at a phospholipid concentration of 10 mg/ml. Surface tension at maximal and minimal bubble size was calculated from the recordings of the pressure gradient across the bubble wall during 50% cyclic surface compression at a rate of 40/min.

Animal Experiments

The in vivo activity of the artificial surfactant preparations was evaluated on premature newborn rabbits delivered by hysterotomy on day 27 of gestation (term = 31 days). Artificial surfactant was instilled via the tracheal cannula at a dose volume of 2 ml/kg (phospholipid concentration in all preparations 80 mg/ml) before the onset of ventilation; non-treated litter mates served as controls. During the period of artificial ventilation the animals were kept in a body plethysmograph system at 37° C [18]. The animals were then ventilated at 35 cm H_2O for 1 min, 25 cm H_2O for 15 min, 20 cm H_2O for 5 min, 15 cm H_2O for 5 min and finally at 25 cm H_2O for 5 min [14]. Tidal volumes were measured with a pneumotachygraph connected to each plethysmograph.

Results

Composition of Surfactant

The phospholipid composition (mean value ± S.D.) of the six batches of surfactant preparations used in our clinical trial [9] is shown in Table 1. The phosphatidylcholines account for about 75% of the total phospholipids. Analysis of molecular species revealed that 1,2-dipalmitoyl species comprise 45–50% and other disaturated species 10–15% of the phosphatidylcholines.

The surfactant preparations contained 0.9 ± 0.2% proteins (mean value ± S.D. for the six batches). The protein eluted first from the Sephadex LH-60 column (protein fraction 1) gave by SDS/polyacrylamide gel electrophoresis two major bands at *apparent* molecular weights of 3 kDa and 14 kDa. The other protein (protein fraction 2) showed one band at the *apparent* molecular weight of about 6 kDa.

Table 1. Phospholipid composition of surfactant preparations

Phospholipid	Composition (mol %)
Phosphatidylcholine	75.1 ± 3.3
Phosphatidylethanolamine	7.5 ± 2.7
Phosphatidylserine	1.2 ± 1.1
Phosphatidylinositol	7.2 ± 1.7
Phosphatidylglycerol	3.5 ± 1.3
Lysophosphatidylcholine	0.7 ± 0.5
Sphingomyelin	4.7 ± 1.7

Values are given as means ± SD for the six batches used in the clinical trial.

Table 2. N-terminal structures of the two protein fractions

Sample	Elution volume from Sephadex LH-60 (% of column volume)	N-terminal structure	Percentage of total
Fraction 1	15–35	Phe-Pro-Ile-Pro-	85–90
		Ile-Pro-	10–15
Fraction 2	40–60	Leu-Arg-Ile-Pro-	70–85
		Arg-Ile-Pro-	10–20
		Ile-Pro-	2–10

Structural Analysis

Direct sequence analysis of the mixture of proteins revealed two N-terminal sequences, Leu-Arg-Ile-Pro- and Phe-Pro-Ile-Pro-, in a ratio of about 3 : 1. Samples from the two protein fractions were also submitted to direct sequence analysis, revealing the presence of one major amino acid sequence in each fraction. Both the unseparated mixture and the two isolated fractions showed evidence of N-terminal heterogeneity where the full-length structures accounted for about 70–90% (Table 2) of the amounts in each fraction. The proteins were extremely hydrophobic and essentially resistent to digestion with proteases. These facts and the N-terminal heterogeneity complicated structural analyses. Sequence analyses of the ^{14}C-carboxymethylated proteins showed that fraction 1 contained eight half-cystine/cysteine residues, six of which were recovered before position 60, while protein fraction 2 revealed only two such residues in the 35 first positions. The positions of the half-cystine/cysteine residues are completely different in the two fractions, showing that the proteins are not derived from partly overlapping identical structures.

In Vitro Activity

Protein-containing artificial surfactants adsorbed to the air/liquid interface of the bubble and reduced surface tension to low values during surface compression (Table 3). The lowest values for both the maximal and minimal surface tension were obtained with preparations based on protein fraction 1.

In Vivo Activity

Animals treated with the phospholipids alone showed only a moderate increase in tidal volumes, clearly inferior to that obtained with each of the protein-containing preparations (Table 4). The best results were obtained with surfactant based on protein fraction 1 (p vs phospholipid + protein fraction

Table 3. Surface properties (mean values ± SD) of protein-containing artificial surfactants, compared with those of phospholipids only, after 5 min of pulsation

Surfactant	*n*	Surface tension (mN/m)	
		γ max	γ min
Phospholipids + protein fraction 1	5	35 ± 5*,**	1 ± 1*,**
Phospholipids + protein fraction 2	5	48 ± 1*	5 ± 5*
Phospholipids	5	66 ± 4	23 ± 2

The recordings were obtained with a pulsating bubble at 37° C, 50% surface compression and a rate of 40/min.
* *P* vs phospholipids = 0.01; ** *P* vs phospholipids + protein fraction 2 = 0,01.

Table 4. Tidal volumes (mean values ± SD) during artificial ventilation of immature newborn rabbits treated with surfactant preparations

Treatment group	No. of animals	Tidal volumes (ml/kg)	
		15 cm H_2O	25 cm H_2O
Phospholipids + protein fraction 1	10	4.5 ± 3.7	22.2 ± 7.7
Phospholipids + protein fraction 2	11	1.5 ± 1.2	13.0 ± 5.7
Phospholipids	10	0.7 ± 0.6	6.8 ± 4.0
Controls	10	0.5 ± 0.5	2.1 ± 1.0

The values at 15 and 25 cm H_2O were recorded after a ventilation of 25 and 30 min, respectively.

2 < 0.01 at both 25 and 30 min) but both protein-based preparations gave significantly higher volumes than treatments with phospholipids alone ($p < 0.05 - 0.001$).

Comments

Our data indicate that a physiologically active surfactant can be produced by recombination of the phospholipid mixture with either of two hydrophobic surfactant-associated proteins, especially that in protein fraction 1. Although the tidal volumes recorded in animals treated with the protein-containing preparations are high in relation to those in control animals, they are still not as large as those recorded after treatment with organic solvent extracts of natural surfactant. Further studies are required to elucidate whether an optimal surfactant substitute can be produced from other combinations of the components than those used in the present investigation, or from other phospholipid/protein mixtures, perhaps including the hydrophilic 28–36 kDa protein as well. The synergistic action of the hydrophilic and hydrophobic surfactant-associated proteins, documented in vitro [19] remains to be confirmed in animal experiments.

Acknowledgements. This study was supported by the Swedish Medical Research Council, The Swedish National Association Against Heart and Chest Diseases, The Swedish Cancer Society, The Expressen Prenatal Research Foundation, Kabi Gen AB, The General Maternity Hospital Foundation and The Research Funds of the Karolinska Institute.

References

1. Goerke J (1974) Lung surfactant. Biochim Biophys Acta 344:241–261
2. Van Golde LMG, Batenburg JJ, Robertson B (1988) The pulmonary surfactant system: biochemical aspects and functional significance. Physiol Rev 68:374-455

3. Egan EA, Notter RH, Kwong MS, Shapiro DL (1983) Natural and artificial surfactant replacement in premature lambs. J Appl Physiol 55:875–883
4. Kobayashi T, Grossmann G, Robertson B, Ueda T (1984) Effects of artificial and natural surfactant supplementation in immature newborn rabbits. J Jpn Med Soc Biol Interface 15:125–131
5. Fujiwara T, Maeta H, Chida S, Morita T, Watabe Y, Abe T (1980) Artificial surfactant therapy in hyaline-membrane disease. Lancet i:55–59
6. Enhorning G, Shennan A, Possmayer F, Dunn M, Chen CP, Milligan J (1985) Prevention of neonatal respiratory distress syndrome by tracheal instillation of surfactant: a randomized clinical trial. Pediatrics 76:145–153
7. Gitlin JD, Soll RF, Parad RB, Horbar JD, Feldman HA, Lucey JF, Taeusch HW (1987) Randomized controlled trial of exogenous surfactant for the treatment of hyaline membrane disease. Pediatrics 79:31–37
8. Raju TN, Vidyasager D, Bhat R, Sobel D, McCulloch KM, Anderson M, Maeta H, Levy P, Furner S (1987) Double-blind trial of single-dose treatment with bovine surfactant in severe hyaline membrane disease. Lancet i:651–656
9. Collaborative European Multicenter Study Group (X) An international clinical trial of surfactant replacement therapy in severe neonatal respiratory distress syndrome. Pediatrics (in press)
10. Takahashi A, Fujiwara T (1986) Proteolipid in bovine lung surfactant: its role in surfactant function. Biochem Biophys Res Commun 135:527–532
11. Yu S-H, Possmayer F (1986) Reconstitution of surfactant activity by using the 6 kDa apoprotein associated with pulmonary surfactant. Biochem J 236:85–89
12. Suzuki Y, Curstedt T, Grossmann G, Nilsson R, Nohara K, Robertson B (1986) The role of the low-molecular weight (≤ 15000 daltons) apoproteins of pulmonary surfactant. Eur J Respir Dis 69:336–345
13. Notter RH, Shapiro DL, Ohning B, Whitsett JA (1987) Biophysical activity of synthetic phospholipids combined with purified lung surfactant 6000 dalton apoprotein. Chem Phys Lipids 44:1–17
14. Curstedt T, Jörnvall J, Robertson B, Bergman T, Berggren P (1987) Two hydrophobic low-molecular-mass protein fractions of pulmonary surfactant: characterization and biophysical activity. Eur J Biochem 168:255–262
15. Berggren P, Curstedt T, Grossmann G, Nilsson R, Robertson B (1985) Physiological activity of pulmonary surfactant with low protein content: effect of enrichment with synthetic phospholipids. Exp Lung Res 8:29–51
16. Curstedt T, Johansson J, Barros-Söderling J, Robertson B, Nilsson G, Westberg M, Jörnvall H (1988) Low-molecular weight protein type 1: the primary structure of a hydrophobic 8 kDa polypeptide with 8 half-cystine residues. Eur J Biochem 172:521-525
17. Enhorning G (1977) Pulsating bubble technique for evaluating pulmonary surfactant. J Appl Physiol 43:198–203
18. Lachmann B, Grossmann G, Freyse J, Robertson B (1981) Lung-thorax compliance in the artificially ventilated premature rabbit neonate in relation to variations in inspiration: expiration ratio. Pediatr Res 15:833–838
19. Hawgood S, Benson BJ, Schilling J, Damm D, Clements JA, White RT (1987) Nucleotide and amino acid sequences of pulmonary surfactant protein SP 18 and evidence for cooperation between SP 18 and SP 28–36 in surfactant lipid adsorption. Proc Natl Acad Sci USA 84:66–70

Reconstitution of Human Surfactant Activity Using Low-Molecular-Weight Apoproteins*

S. D. Revak, T. A. Merritt, M. Hallman, and C. G. Cochrane

Introduction

We have isolated two low molecular weight (LMW) apoproteins from human amniotic fluid surfactant using organic lipid extraction methods. These proteins are distinct from a previously described 35,000 dalton apoprotein [1]. Having partially characterized these two proteins we looked at their capacities, when added to phospholipids, to lower the surface tension of a pulsating bubble and to increase lung compliance and alveolar expansion when instilled into fetal rabbits. These two distinct LMW apoproteins will be referred to in this report as SP 18 and SP 9, with "SP" signifying "surfactant protein" and the number indicating the molecular weight (in kilodaltons) of the protein as it appears in unreduced SDS-polyacrylamide gel electrophoresis.

Protein Characterization

The LMW apoproteins partitioned with lipids during organic extractions of surfactant were well separated from the 35,000 dalton surfactant apoprotein which remained in the aqueous phase. To remove the lipids, chromatography on silicic acid or lipophilic Sephadex LH-20 columns was employed [2, 3]. When analyzed by SDS gel electrophoresis under reducing conditions, two protein bands were apparent (Fig. 1). The upper band, at an apparent molecular weight of 18,000 daltons (and therefore termed SP 18) is a dimer, and with the addition of β-mercaptoethanol, reduced to 9000 daltons. The other LMW apoprotein, termed SP 9, migrates as a diffuse band between 9 and 12,000 daltons in the presence or absence of reducing agents. These two proteins could be separated by chromatography on Sephadex LH-60 (method of K. Shiffer and S. Hawgood, personal communication, and [2]) yielding the purified proteins shown in lanes B, C, E, and F in Fig. 1.

Amino acid compositions showed both proteins to be very hydrophobic with high levels of valine and leucine. Amino-terminal sequencing of SP 18 yielded the following sequence: NH_2 – phe – pro – ile – pro – leu – pro – tyr –. This sequence can be found within a large open reading frame of cDNA described by three groups [2, 4, 5]. While the precise carboxy-terminus has not

* This is publication no. 5087-IMM from the Department of Immunology, Research Institute of Scripps Clinic.

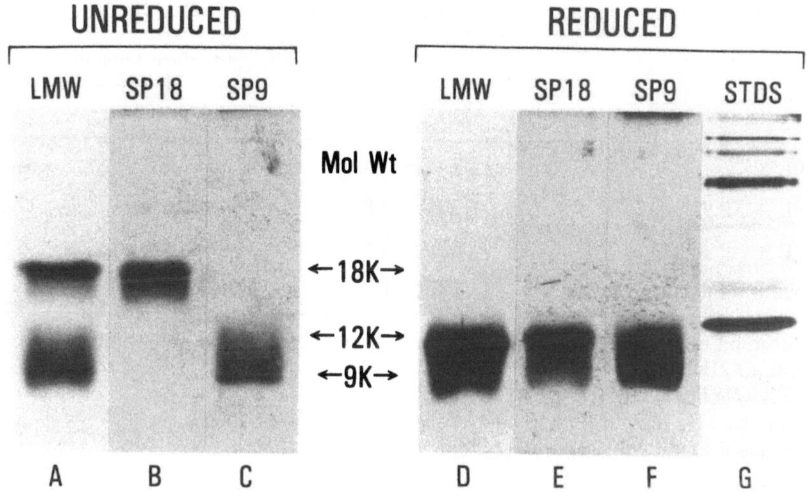

Fig. 1. Silver-stained SDS-polyacrylamide gel of low-molecular-weight (LMW) apoproteins. Lanes *A* and *D* show a sample after silicic acid or Sephadex LH-20 chromatography; both LMW proteins are present. Lanes *B, C, E,* and *F* show the resolution of SP 18 (lanes *B* and *E*) and SP 9 (lanes *C* and *F*) following chromatography on Sephadex LH-60. Mol wt standards are shown in lane *G*. Lanes *A–C* are unreduced samples, lanes *D–F* contain identical samples reduced with β-mercaptoethanol prior to electrophoresis. (From [2])

yet been determined, comparison of the amino acid composition with theoretical compositions predicted by the cDNA as well as molecular weight determinations in SDS gels, suggests a protein of approximately 81 residues.

Repeated sequencing of the purified SP 9 protein showed multiple peptides, all containing at least 6 consecutive valines preceded by a leucine-rich region. NH_2-terminal analyses consistently showed the residues phenylalanine, glycine, and isoleucine, with the relative amounts varying between preparations. The data obtained from sequencing attempts of SP 9 suggest a collection of peptides having an identical sequence but with one or two residues removed from the amino-terminus ("ragged ends").

Reconstitution of Surfactant Activity

In order to see if we could reconstitute the biophysical and biologic activities of surfactant by adding the LMW apoproteins to phospholipids (PL), samples were prepared containing 400 μg of a 3:1 (by weight) mixture of dipalmitoyl-phosphatidylcholine and phosphatidylglycerol (DPPC:PG) and 4 μg of either SP 9 or SP 18 in a final volume of 100 μl saline. Control samples of phospholipids alone, LMW apoproteins alone, and natural human surfactant diluted to a similar phospholipid concentration, were also prepared. Each sample was assayed in a pulsating bubble surfactometer [6] for its ability to lower surface tension. Table 1 shows the mean minimal surface tension at 15 sec, 1 min, and 5 min. While neither LMW apoproteins (data not shown) nor

Table 1. Minimum surface tensions in the pulsating bubble (from [2])

	Minimum surface tension		
	15 s	1 min	5 min
PL[a]	42.9 ± 1.4	41.6 ± 1.6	34.9 ± 4.9
PL + SP 9[b]	16.7 ± 0.8	14.1 ± 1.2	12.2 ± 1.0
PL + SP 18[b]	6.3 ± 0.2	5.1 ± 1.0	4.9 ± 0.6
Natural human surfactant[c]	2.0 ± 1.2	2.4 ± 1.4	0.4 ± 0.4

Pulsation of 20 cycles/min started 10 s after bubble formation. All values are in dynes.cm^{-1} and are the average of at least three determinations.
[a] Phospholipids DPPC:PG, 3:1, 4 mg/ml.
[b] 1% by weight compared to phospholipids.
[c] Diluted to 4 mg/ml.

phospholipids alone had significant surface-tension lowering capacities, a mixture of phospholipids with either SP 9 or SP 18 showed significant activity. The addition of 1% by weight of SP 18 lowered the surface tension measured to levels comparable to those obtained with an equal amount of natural human surfactant (5.1 ± 1.0 dynes/cm for PL + SP 18 at 1 min, 2.4 ± 1.4 dynes/cm for natural surfactant). SP 9 lowered surface tension less effectively (14.1 ± 1.2 dynes/cm) on an equal weight basis.

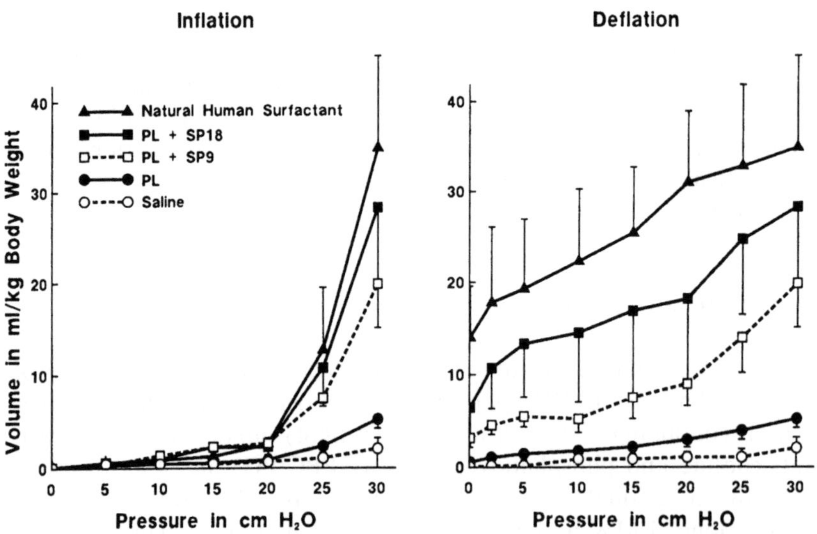

Fig. 2. Inflation and deflation pressure/volume curves of fetal rabbit lungs 30 min after intratracheal instillation of 100 µl of saline (○----○); 2 mg phospholipids (*PL*) DPPC:PG, 3:1 (●—●); PL + 10 µg SP 9 □----□); PL + 10 ;µg SP 18 (■—■); or 2 mg natural human surfactant (▲—▲). Data are expressed as the mean of four animals ± 1 SD. (From [2])

In vivo assays of reconstituted surfactant activity were performed by instilling into the airways of immature (27 day gestation) fetal rabbits saline solutions containing calcium alone or with the addition of phospholipids, phospholipids plus LMW apoproteins, or natural human surfactant. Separate studies were performed to observe either a mixture of SP 18 and SP 9 or each protein separately. For all studies, the phospholipid (DPPC:PG, 3:1) concentration was 2 mg/100 µl/animal, with the addition of 1% (by weight) mixed SP 18 + SP 9 or 0.5% SP 18 or SP 9 alone. Following instillation of the surfactant, the animals were ventilated for 30 min and then degassed by placement in a bell jar under vacuum. The lungs were then inflated to given pressures up to 30 cm H_2O and the volume of air required for each pressure was recorded. The volumes required for given pressures during deflation were similarly determined. The resulting pressure/volume curves are shown in Fig. 2 for animals which received SP 18 or SP 9 and appropriate control animals.

Greatly improved lung compliance, approaching that seen following treatment with natural surfactant was observed in those animals which received the PL + SP 18 reconstitutant. Surfactant made with SP 9 was again less effective. Results very similar to the PL + SP 18 curves seen in Fig. 2 were obtained when a mixture of SP 9 and SP 18 was used (data not shown). Following compliance measurements, lungs were inflated again to 30 cm H_2O, deflated back to 10 cm H_2O, clamped, excised, and fixed overnight in formalin. Thin sections stained with hematoxylin and eosin were examined microscopically. Figure 3 shows representative fields. Lungs treated with saline

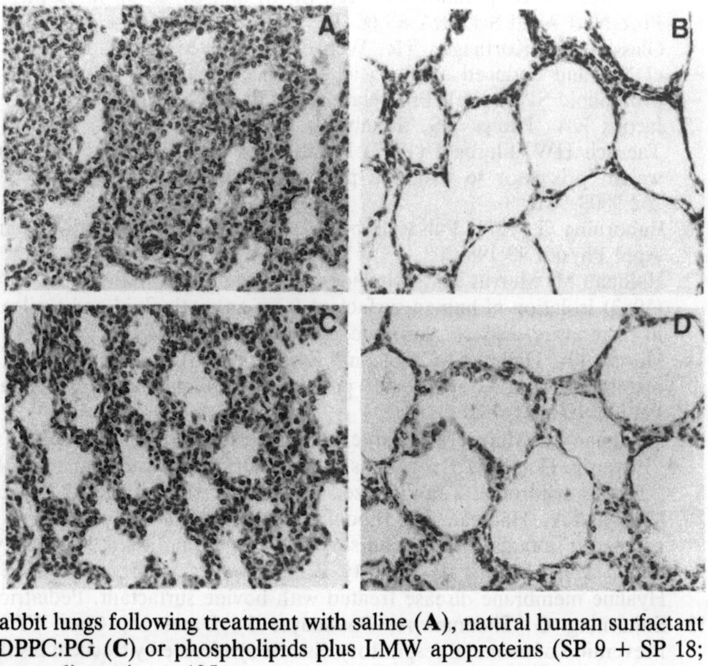

Fig. 3. A–D. Fetal rabbit lungs following treatment with saline (**A**), natural human surfactant (**B**), phospholipids DPPC:PG (**C**) or phospholipids plus LMW apoproteins (SP 9 + SP 18; **D**). (From [2]). Hematoxylin-eosin, × 125

(A) or phospholipids alone (C) appear atelectatic while those from animals which received natural (B) or reconstituted (D) surfactant show normal alveolar expansion. Morphometric analyses of the thin sections showed an interstitium to air space ratio of 0.498 for natural surfactant and 0.538 for reconstituted surfactant as compared with 4.70 for saline and 3.29 for treatment with phospholipids alone.

Improvements in lung function, as measured by a decrease in mean airway pressure and oxygen requirement, is an effect also seen in preterm human infants treated intratracheally with natural human surfactant [7–10] or lipid extracts (containing LMW apoproteins) of bovine surfactant [11–16]. The ability to reproduce the essential components of these surfactants by synthetic means (i.e., genetically-engineered protein(s) combined with synthetic phospholipids) would permit their use in the treatment of not only infant respiratory distress syndrome, but also in other pathologic conditions where a shortage or abnormality of pulmonary surfactant plays a crucial role.

References

1. Revak SD, Merritt TA, Hallman M, Cochrane CG (1986) Reconstitution of surfactant activity using purified human apoprotein and phospholipids measured *in vitro* and *in vivo*. Am Rev Respir Dis 134:1258–1265
2. Revak SD, Merritt TA, Degryse E, Stefani L, Courtney M, Hallman M, Cochrane CG (1988) Use of human surfactant low molecular weight (LMW) apoproteins in the reconstitution of surfactant biologic activity. J Clin Invest (in press)
3. Hawgood S, Benson BJ, Schilling J, Damm D, Clements JA, White RT (1987) Nucleotide and amino acid sequences of pulmonary surfactant protein SP 18 and evidence for cooperation between SP 18 and SP 28–36 in surfactant lipid adsorption. Proc Natl Acad Sci USA 85:66–70
4. Glasser SW, Korfhagen TR, Weaver T, Pilot-Matias T, Fox JL, Whitsett JA (1987) cDNA and deduced amino acid sequence of human pulmonary surfactant-associated proteolipid SPL (phe). Proc Natl Acad Sci USA 84:4007–4011
5. Jacobs KA, Phelps DS, Steinbrink R, Fisch J, Kriz R, Mitsock L, Dougherty JP, Taeusch HW, Floros J (1987) Isolation of a cDNA clone encoding a high molecular weight precursor to a 6-kDa pulmonary surfactant-associated protein. J Biol Chem 262:9808–9811
6. Enhorning G (1977) Pulsating bubble technique for evaluating pulmonary surfactant. J Appl Physiol 43:198–203
7. Hallman M, Merritt TA, Schneider H, Epstein BL, Mannino F, Edwards DK, Gluck L (1983) Isolation of human surfactant from amniotic fluid and a pilot study of its efficacy in respiratory distress syndrome. Pediatrics 71:473–482
8. Merritt TA, Hallman M, Holcomb K et al. (1986) Human surfactant treatment of severe respiratory distress syndrome: pulmonary effluent indicators of lung inflammation. J Pediatr 108:741–745
9. Hallman M, Merritt TA, Jarvenpaa AL, Boynton B, Mannino F, Gluck L, Moore T, Edwards D (1985) Exogenous human surfactant for treatment of severe respiratory distress syndrome: a randomized prospective clinical trial. J Pediatr 106:963–969
10. Merritt TA, Hallman M, Bloom BT et al. (1986) Prophylactic treatment of very premature infants with human surfactant. N Engl J Med 315:785–790
11. Smyth JA, Metcalfe IL, Duffty P, Possmayer F, Bryan MH, Enhorning G (1983) Hyaline membrane disease treated with bovine surfactant. Pediatrics 71:913–917
12. Enhorning G, Shennan A, Possmayer F, Dunn M, Chee CP, Milligan J (1985) Prevention of neonatal respiratory distress syndrome by tracheal instillation of surfactant: a randomized clinical trial. Pediatrics 76:145–153

13. Fujiwara T, Maeta H, Chida S, Morita T, Watabe Y, Abe T (1980) Artificial surfactant therapy in hyaline-membrane disease. Lancet i:55–59
14. Kwong MS, Egan EA, Notter RH, Shapiro DL (1985) A double blind clinical trial of calf lung lipid for the prevention of hyaline membrane disease in extremely premature infants. Pediatrics 76:585–592
15. Shapiro DL, Notter RH, Morin FC, Deluga K, Golub LM, Sinkin R, Weiss K, Cox C (1985) A double blind, randomized trial of a calf lung surfactant extract administered at birth to very premature infants for the prevention of respiratory distress syndrome. Pediatrics 76:593–599
16. Fujiwara T (1984) Surfactant replacement in neonatal RDS. In: Robertson B, Van Golde LMG, Batenburg J (eds) Pulmonary surfactant. Elsevier Science, Amsterdam, pp 479–503

The Macromolecular Organization of Canine Lung Surfactant Protein SP 28–36: Structural Homology with the Complement Factor C1q

T. Voss, H. Eistetter, K. P. Schäfer, and J. Engel

The complete amino acid sequence of the major pulmonary surfactant associated glycoprotein SP 28–36 has been derived from the cDNA and genomic sequences (White et al. 1985; Floros et al. 1986; Sano et al. 1987). The sequences of the human, canine and rat protein are highly homologous. The protein consists of a polypeptide chain with 228 amino acids. The M_r of this protein, when determined by SDS-Page under reducing conditions, is 32–36 kDa. It runs as a fuzzy band because of its glycolsylation. The N-terminal part of the molecule is characterized by a collagen-like sequence. The C-terminal part of the polypeptide chain contains a putative site for the N-linked glycosylation. The first subcomponent C1q of the human complement C1 (Reid 1979) has similar structural features.

Studies using human lung lavage SP 28–36 obtained from proteinosis patients indicated the existence of 70 kDa dimers which cannot be dissociated into the 36 kDa polypeptide by reductive cleavage (Bhattacharyya et al. 1976; Crawford et al. 1986). No further information about the aggregation of the polypeptides into an eventual higher order structure is available. The objective of this work was to investigate the macromolecular structure of the SP 28–36 molecule.

Circular dichroism spectroscopy (CD) has been used as a tool for the examination of the collagenous triple helical structures. For these experiments human recombinant SP 28–36 was used which was available in sufficient amounts. The obtained spectrum (Fig. 1) exhibits a shoulder at 220 nm and a strong negative extremum at 205 nm (Figure 1a). Part of the SP 28–36 structure melts in a very narrow temperature range $\Delta T = 4$ °C and a midpoint temperature of $T_m = 44$ °C (insert of Figure 1a).

The sharp transition is reminiscent of a collagen triple helix to random coil transition (Traub and Piez 1971). Comparable values for collagen type I are $T_m = 38$ °C, $\Delta T = 3$ °C (Brodsky-Doyle et al. 1976). Furthermore the difference spectrum between native and denatured SP 28–36 (Figure 1b) resembles circular dichroism spectra of collagens in its general features except that the amplitudes of the extrema at 205 nm and 230 nm are 3 to 4 times smaller than those of collagen type I. Smaller amplitudes were, however, also observed for the collagenous region in C1q (Brodsky-Doyle et al. 1976) and are probably reflections of its small length and the irregularity in its collagenous sequence.

The triple helix-to-coil transition of SP 28–36 is reversible. Other collagens (collagen type III; fragment Col 1–3) in which complete dissociation of the chains is prevented by disulfide linkage also exhibit reversible transitions

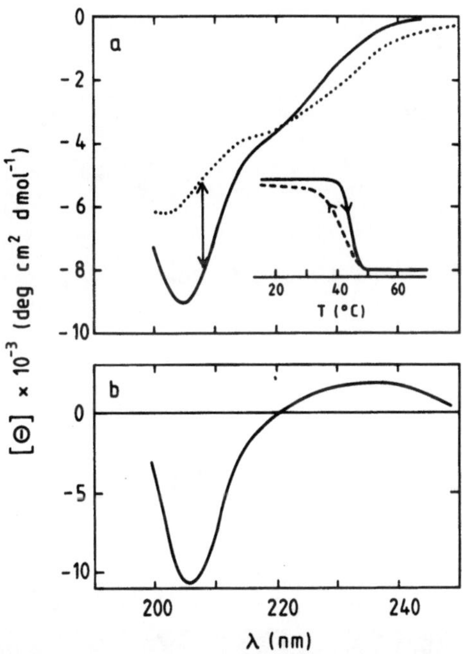

Fig. 1 a, b. Circular dichroism spectra and thermal transition of the recombinant SP 28–36 in 50 mM Tris-HCl, pH 7.5. **a** A spectrum was recorded for the native protein at 18 °C (—). Then the temperature was raised at a rate of 18°C/h and the change of ellipticity [Θ] *(arrow)* was monitored at a constant wavelength of 208 nm. A conformational transition was observed in a narrow temperature range (*insert* —>–). The midpoint "melting" temperature is T_m = 44°C, and the temperature interval between one-quarter and three-quarter "melting" is ΔT = 4°C. The spectrum of the heat-denatured protein was recorded at 70°C (...). Subsequent cooling at a rate of 18°C/h led to an almost complete recovery of the CD spectrum reminiscent of the native state (not shown) at 18 °C. The transition profile recorded during the increase of temperature (*insert* —<–) however is less steep (ΔT = 7°C) and the T_m is 40°C. **b** Difference in CD spectrum between native and heat-denatured SP 28–36. In order to facilitate the comparison with spectra of collagen type I and the collagenous pepsin fragment of C1q (Brodsky-Doyle et al. 1976), the differences between spectra of native and denatured SP 28–36 were multiplied by a factor of 3, which accounts for a weight fraction of 1/3 for the collagenous domain in SP 28–36

(Engel et al. 1980). A small hysteresis between the transition profiles for unfolding and refolding (see insert of Figure 1a) was also observed for these collagens (Engel et al. 1980).

The spectra and the sharp midpoint of temperature transition obtained for the recombinant SP 28–36 protein shows features characteristic for an intact triple helix (Traub and Piez 1971). This structural element with similar CD characteristics has been demonstrated for the first subcomponent C1q of the human complement C1 (Brodsky-Doyle et al. 1976). Hence, the triple helix is an important structural component of SP 28–36. As demonstrated by the renaturation experiment the three polypeptide chains of which the monomer structure is composed are probably disulfide linked.

Electron microscopy using the rotary shadowing technique allowed to

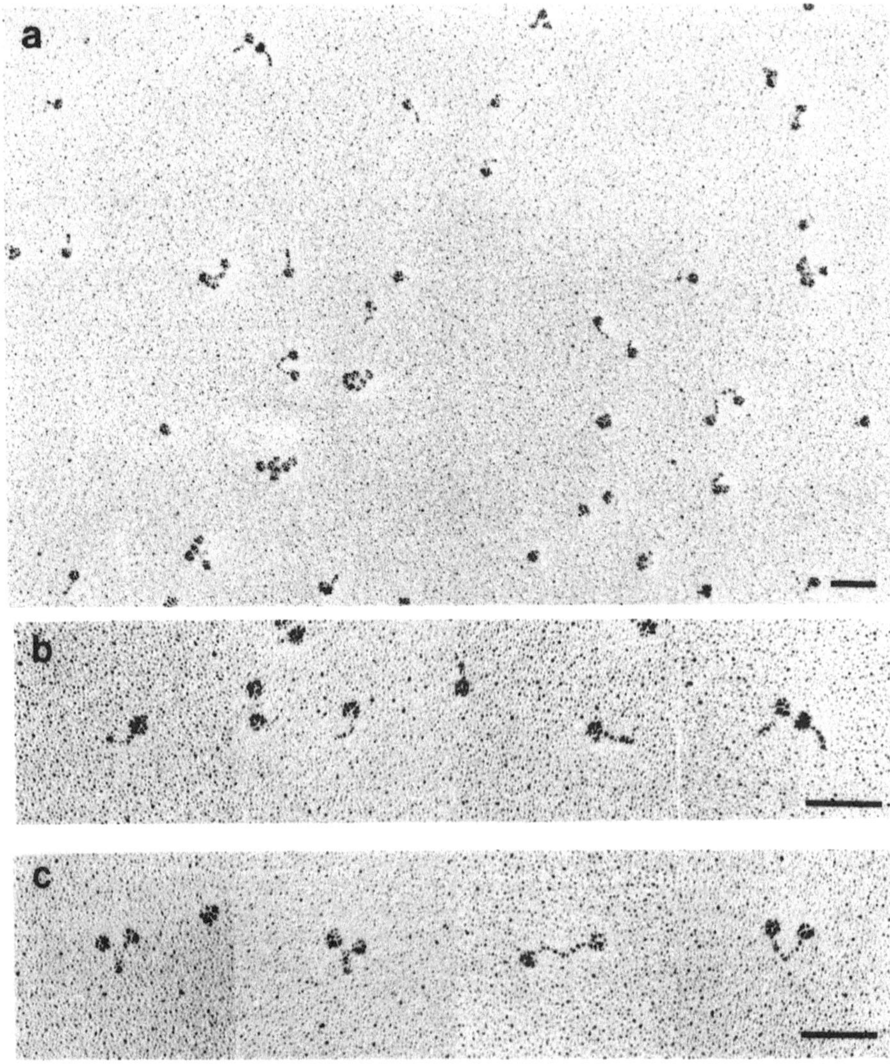

Fig. 2 a–c. Electron micrographs after rotary shadowing of human recombinant SP 28–36 after partial reductive cleavage. **a** Field demonstrating the presence of monomers and dimers. **b** Selected monomeric particles. **c** Selected dimers formed by the connection of monomers by the terminal regions of their tails. *Bars* represent a length of 50 nm

obtain structural information of the SP 28–36 molecule. By mild reduction of the human recombinant SP 28–36 the monomers composed of three polypeptides were identified as lollipop-shaped particles with a collagenous tail and a globular domain (Fig. 2). The measured contour length of 19.4 nm corresponds well with the value of 20 nm calculated from sequence data (69 residues with a translation per residue of 0.286 nm in the triple helix (Traub and Piez 1971). The diameter of the globular domains (8 nm) is overestimated by 2.5 to 5 nm (Engel et al. 1981) due to metal decoration. Therefore the diameter of 5.3 nm which is calculated from the molar mass of the globule assuming a spherical shape represents a more realistic estimate (Engel et al. 1981).

The location of the kink observed by electron microscopy corresponds well with the location of a sequence irregularity in the middle of the triple helical sequence. Kinks have been shown to correlate with similar sequence patterns in different collagen types (Reese et al. 1982; Hofmann et al. 1984). A kink caused by such a sequence irregularity is also present in the center portion of the stalks of subcomponent C1q of the first component of complement C1 (Reid 1979; Reid et al. 1982). Due to the different sequences of the three C1q chains the triple helix forms a kink at this position with a fixed angle of 60 ° C (Kilchherr et al. 1985). This angle has been observed in electron microscopical studies (Schumaker et al. 1981). It maintains the open structure of C1q which is of crucial importance for its function as a subcomponent of the first component of complement C1 (Reid and Porter 1981).

Since the three polypeptides in the triple helix of SP 28–36 all have the same sequence pattern, the kink at this position is most likely to be more flexible than the kink in C1q. A comparison of the dimensions of the two molecules supports this interpretation.

The micrographs obtained from the slightly reduced recombinant material allow a lateral view of the structure. The dimers as well as higher aggregates are usually connected at the tip of the stem. This probably occurs by the N-terminal cysteine residues. The parts of the triple helix between the tip and the kink subsequently aggregate to a stem. In this microfibrillar end piece the individual triple helices are stabilized by lateral aggregation.

Material obtained from canine lung lavage was used to study the structural organization of the native SP 28–36. As shown in Fig. 3 the monomers are associated to hexameric structures. Similar structures were observed in material isolated from human amniotic fluid (data not shown). The particles are seen from the top. They are adsorbed to the support with their globular heads. The stem which contains the aggregated N-terminal part of the six collagen triple helices is pointing upwards.

The similar hexameric structure of SP 28–36 and C1q in the electron micrographs obtained with the same technique suggest the flower bouquet like structure for the SP 28–36 molecule as shown in Fig. 4. This structure has been demonstrated for C1q molecules visualized by the negative staining technique (Brodsky-Doyle et al. 1976; Schumaker et al. 1981).

The function of C1q is well established. Its recognition of multivalent immunoglobulin clusters on the surface of a target cell mediates the binding and activation of the zymogen complex $C1r_2$-$C1s_2$. This process initializes the complement cascade which finally leads to the destruction of the target cell (Reid and Porter 1981).

There is limited knowledge about the function of the SP 28–36 protein. Supposedly it plays a role in the uptake and recycling of the surfactant lipid-protein complexes by type II alveolar pneumonocytes (Wright et al. 1987) and alveolar macrophages (Stern et al. 1986). Recently, sequence homologies with mannose binding lectins in the liver have been described (Drickamer et al. 1986; Patthy 1987). The mannose binding site of SP 28–36 is located in the

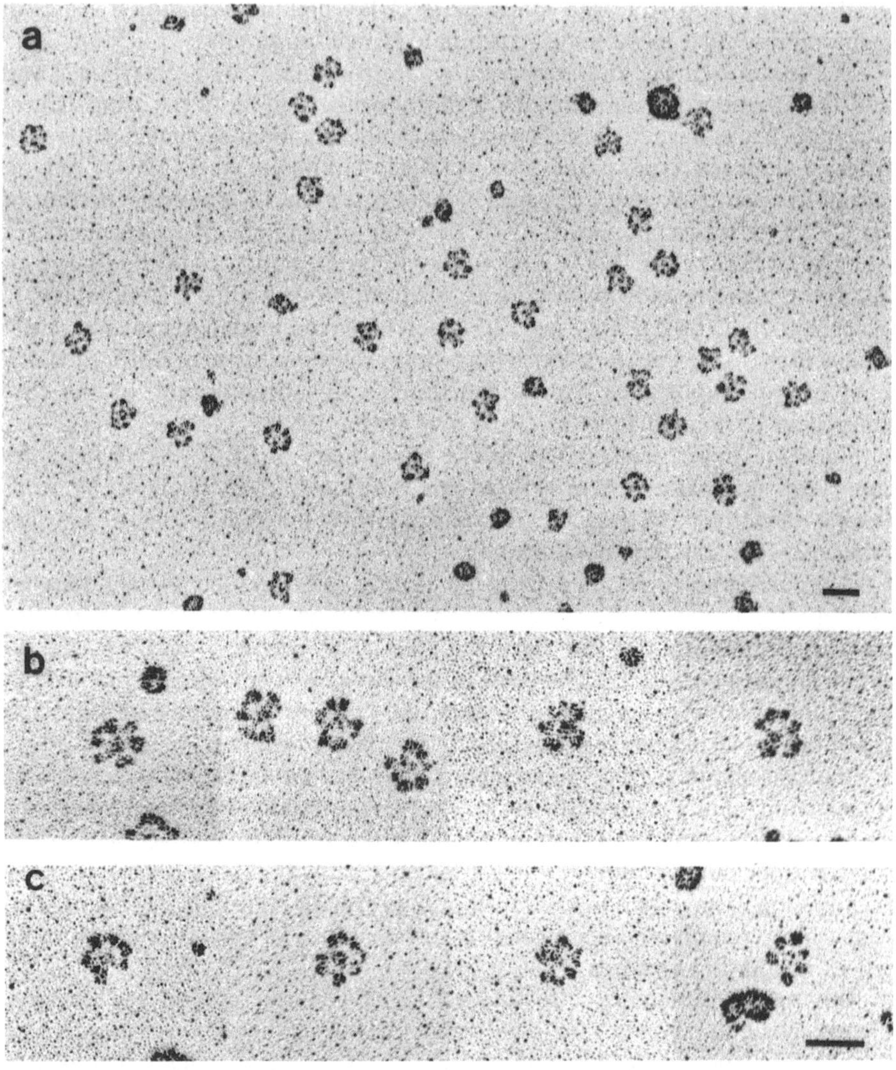

Fig. 3 a–c. Electron micrographs of hexameric SP 28–36 from saline lung lavage fluid (dog) after rotary shadowing. **a** Field demonstrating the homogeneity of the sample. **b** Isolated particles at higher magnification. *Bar* indicates a length of 50 nm. **c** Subcomponent C1q of the first component C1 of human complement

globular domain as indicated by the sequence homologies. This potential lectin-like property in connection with the structural element of the collagen stems and the bouquet like over-all structure might allow the SP 28–36 molecule to act as a anchor between the cells aligning the alveoli and the surfactant film forming the air-liquid interface in the alveoli (Meban 1987).

The structure of SP 28–36 reported here will help to get a better understanding of its function in the lung surfactant system.

Methods will be published elsewhere (Voss et al., manuscript submitted).

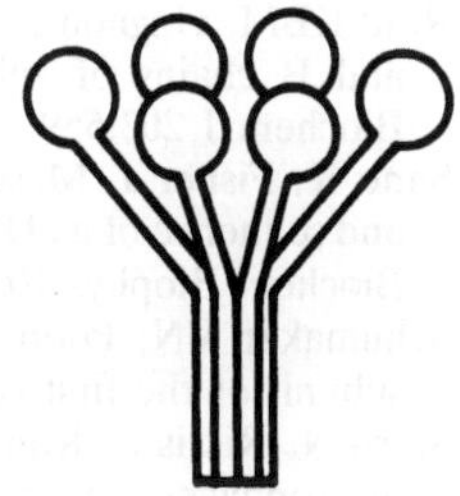

Fig. 4. Three-dimensional model of the intact hexamers of the native SP 28–36 (by homology with C1q)

Acknowledgements. We thank Ms. S. Löffler, M. Oehri, C. Rahming and C. Fauser for expert technical assistance. We are very grateful to Prof. M. Obladen (Berlin) for his generous gift of human lung surfactant isolated from amnion fluid. This work was supported in part by grant 3.254–0.81 of the Swiss National Science Foundation to J.E.

References

Bhattacharyya SN, Sahu S, Lynn WS (1976) Structural studies on a glycoprotein isolated from alveoli of patients with alveolar proteinosis. Biochim Biophys Acta 427:91

Brodsky-Doyle B, Leonard KR, Reid KBM (1976) Circular-dichroism and electron-microscopy studies of human subcomponent C1q before and after limited proteolysis by pepsin. Biochem J 159:279

Crawford SW, Mecham RP, Sage H (1986) Structural characteristics and intermolecular organisation of human pulmonary-surfactant-associated proteins. Biochem J 240:107

Drickamer E, Dordal MS, Reynolds L (1986) Mannose-binding proteins isolated from rat-liver contains carbohydrate-recognition domains linked to collagenous tails - complete primary structures and homology with pulmonary surfactant apoproteins. J Biol Chem 261:6878

Engel J, Bächinger HP, Bruckner P, Timpl R (1980) The role of disulfide bridges and cis-trans isomeration of peptide bonds in collagen folding. In: Jaenicke R (ed) Protein folding. Elsevier/North Holland, Amsterdam, p 345

Engel J, Odermatt E, Engel A, Madri J, Furthmayer H, Rohde H, Timpl R (1981) Shapes, domain organisations and flexibility of laminin and fibrocitin in two multifunctional proteins of the extra cellular matrix. J Mol Biol 150:97–120

Floros J, Steinbrink R, Jacobs K. Phelps D, Kriz R, Recny M, Salzman L, Jones S, Taeusch HW, Frank HA, Fritsch EF (1986) Isolation and characterization of cDNA clones for the 35 kD pulmonary surfactant associated protein. J Biol Chem 261:9029

Hofmann H, Voss T, Kühn K, Engel J (1984) Localization of flexible sites in thread-like molecules from electron micrographs. Comparison of interstitial, basement membrane and intima collagens. J Mol Biol 172:325

Kilchherr E, Hofmann H, Steigemann W, Engel J (1985) Structural model of the collagen-like region of C1q comprising the kink region and the fibre-like packing of the six triple helices. J Mol Biol 186:403

Meban C (1987) Lung surfactant anchor. Nature 326:252

Patthy L (1987) Is lung surfactant protein a lectin-collagen hybrid. Nature 325:490

Reese CA, Wiedemann H, Kühn K, Mayne R (1982) Characterization of a highly soluble collagenous molecule isolated from chicken hyline cartilage. Biochemistry 21:826

Reid KBM (1979) Complete amino acid sequence of the three collagen-like regions present in subcomponent C1q of the first component of human complement. Biochem J 179:367

Reid KBM, Porter RR (1981) The proteolytic activation systems of complement. Ann Rev Biochem 50:433

Reid KBM, Gagnon J, Frampton J (1982) Completion of the amino acid sequence of the A and B chains of subcomplement C1q of the first component of human complement. Biochem J 203:559

Sano K, Fisher J, Mason R, Kuroki Y, Schilling J, Benson B, Voelker D (1987) Isolation and sequence of a cDNA clone for the rat pulmonary surfactant-associated protein (PSP-A) Biochem Biophys Res Commun 144:367

Schumaker VN, Poon PH, Seegan GW, Smith CA (1981) Semi-flexible joint in the C1q subunit of the first component of human complement. J Mol Biol 148:191

Stern N, Riklis S, Kalina M, Tietz A (1986) The catabolism of lung surfactant by alveolar macrophages. Biochim Biophys Acta 877:323

Traub W, Piez KA (1971) The chemistry and structure of collagen Adv Protein Chem 25:243

White RT, Damm D, Miller J, Spratt K, Schilling J, Hawgood S, Benson B, Cordell B (1985) Isolation and characterization of the human pulmonary surfactant apoprotein gene. Nature 317:361

Wright JR, Wager RE, Hawgood S, Dobbs L, Clements JA (1987) Surfactant apoprotein Mr = 26,000–36,000 enhances uptake of liposomes by type II cells. J Biol Chem 262:2888

Two-Step Substitution of Surfactant in Fetal Minipigs

R. Rüfer, J. Wuthe, and B. Widjaja

In surfactant replacement experiments [2, 3, 4] it should be realized that surface active substances serve two functions in lung mechanics: first, reduction of the alveolar surface tension and secondly, stabilization of the alveolar system of varying diameters. Although overall reduction of the surface tension to the same value in all alveoli indeed facilitates inflation of the lungs and increases specific lung compliance, according to Laplace's law pressure differences still exist between small and large alveoli. However, using dynamic techniques in the registration of pressure-volume characteristics these differences are often undetectable. Therefore, the static pressure-volume technique was used in these experiments.

In our experiments [1, 6] the isolated lungs of fetal minipigs (day 95 = 85% of the total gestation period) were found to be almost uninflatable with air at a pressure of 25 cm H_2O (Fig. 1). The biochemical immaturity of these lungs was characterized by a lack of phospholipids, e.g. the lung tissue of 95-day fetuses compared to 110-day fetuses (values in parentheses) contained: lipid-P = 2.99 ± 0.42 mg/g dry weight (5.30 ± 0.42); phosphatidylcholine = 29.45 ± 7.11

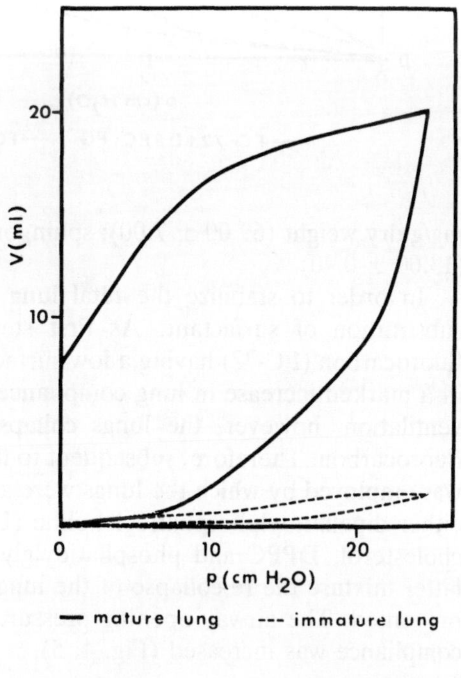

Fig. 1. Static pressure-volume curves of isolated mature and immature fetal minipig lungs. Each curve represents the mean values of six single experiments. *Continuous curve* represents 110-day fetuses, *dotted curve* 95-day fetuses

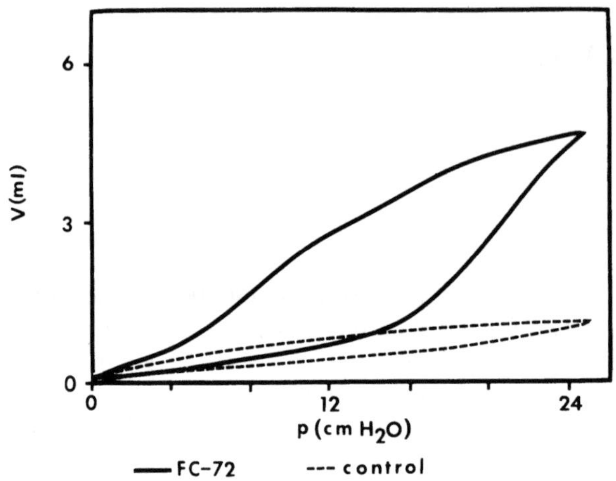

Fig. 2. Static pressure-volume curves of isolated fetal minipig lungs at day 95 of gestation. Each curve represents the mean values of six experiments

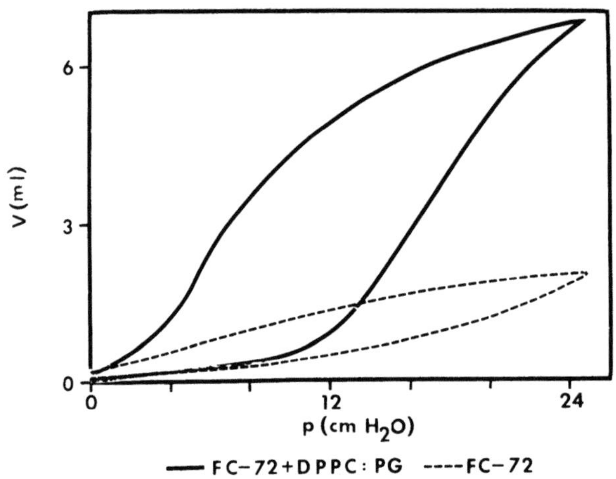

Fig. 3. Static pressure-volume curves of isolated fetal minipig lungs at day 95 of gestation. Each curve represents the mean values of six single experiments. *Continuous curve* represents lungs after lavage with fluorocarbon 72 plus 20 min aerosol ventilation with DPPC + PG (9:1); *dotted curve* indicates the same treatment without lipid substitution

mg/g dry weight (69.00 ± 7.00); sphingomyelin = 6.70 ± 1.25 mg/g dry weight (13.00 ± 0.40).

In order to stabilize the fetal lung mechanics we performed a two-step substitution of surfactant. As first step [5] the lungs were rinsed with a fluorocarbon (FC-72) having a low surface tension (12.0 mN/m) which resulted in a marked increase in lung compliance (Fig. 2). During subsequent artificial ventilation, however, the lungs collapsed again due to evaporation of the fluorocarbon. Therefore, subsequent to the fluorocarbon inflation a second step was employed by which the lungs were artificially ventilated with an aerosol of either dipalmitoylphosphatidylcholine (DPPC) or a 9:1 mixture of DPPC and cholesterol, DPPC and phosphatidylglycerol respectively (Fig. 3). Using the latter mixture the re-collapse of the lungs during artificial ventilation could be prevented. The airway opening pressure was reduced and the weight specific compliance was increased (Fig. 4, 5).

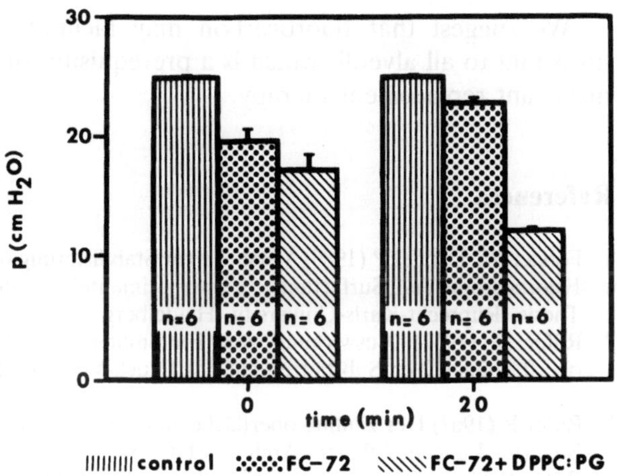

Fig. 4. Airway opening pressure of isolated fetal minipig lungs (day 95) without and with substitution of fluorocarbons alone and (according to Rüfer [3]) with a combination of surfactant and fluorocarbon 72 immediately after substitution and 20 min later

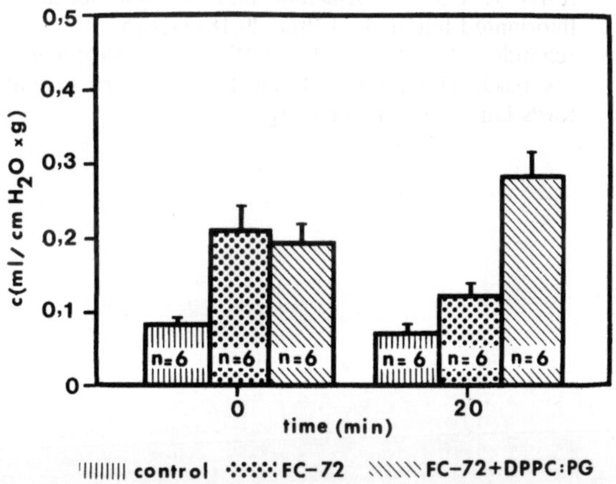

Fig. 5. Compliance per gram dry weight of isolated fetal minipig lungs (day 95) with and without substitution of surfactant and fluorocarbon 72. Compliance was measured at 25 cm H_2O

In order to compare the method of aerosol application with intratracheal instillation of surfactant a second series of replacement experiments in isolated lungs of fetal minipigs was performed. Four groups were formed in which 1 ml of a 9:1 DPPC:PG mixture (group 1) and 1 ml of 9:1 DPPC:cholesterol mixture (group II) were intratracheally administered. The same mixtures were applied intratracheally to groups III (DPPC:PG) and IV (DPPC:cholesterol) as a two-step-substitution following the fluorocarbon step described above. In all groups surfactant concentration was 10 mg/ml in 0.15 M NaCl, 0.05 M $CaCl_2$. All lungs were artificially ventilated for 40 min. To characterize lung mechanics static pressure-volume curves were recorded at time 0 and 20 min after surfactant substitution. Lung opening pressure and lung compliance per g dry weight was investigated. Most of the lungs in groups I (4/6) and II (5/6) remained uninflatable. No significant difference was observed in compliance during the observation period. Using the two-step-substitution however, lung compliance increased by more than 100% and recollapse of the lungs during the artificial ventilation was prevented (Fig. 5).

We suggest that fluorocarbon may facilitate the overall distribution of surfactant to all alveoli, which is a prerequisite for stabilization of the lungs in surfactant replacement therapy.

References

1. Förster W, Dangel P (1986) Versuche zur Stabilisierung der fetalen Alveolarmechanik mit Fluorocarbon und Surfactant – Tierexperimentelle Untersuchungen am Zwergschwein. Thesis, Ruprecht-Karls-University, Heidelberg
2. Rüfer R (1967) Auswirkung von Entfilmung und Befilmung isolierter Lungen mit oberflächenaktiven Substanzen auf das Druck-Volumen-Diagramm. Pflügers Arch 297:R 43
3. Rüfer R (1967) Der Einfluß oberflächenaktiver Substanzen auf Entfaltung und Retraktion isolierter Lungen. Pflügers Arch 298:170–184
4. Rüfer R (1968) The influence of surface active substances on alveolar mechanics in the respiratory distress syndrome. Respiration 25:441–457
5. Rüfer R (1970) Surfactant and alveolar surface forces after breathing of an inert fluorinated liquid. Fed Proc 29:1813–1815
6. Schmidt A, Seitz B (1986) Entfaltbarkeit und Phospholipidgehalt fetaler Zwergschweinlungen nach Substitution oberflächenaktiver Substanzen. Thesis, Ruprecht-Karls-University, Heidelberg

Part VII
Lipid Metabolism and Its Influence by Pharmacological Means

Regulation of Phosphatidylcholine Synthesis in Type II Alveolar Epithelial Cells

R. Burkhardt, P. von Wichert, J. J. Batenburg, and L. M. G. van Golde

Phosphatidylcholine (PC) is the most abundant phospholipid class of pulmonary surfactant. Together with the other components of surfactant it is synthesized in the alveolar type II cells. The work of several research groups with isolated alveolar type II cells has elucidated the pathways in the formation of surfactant PC and its major species, dipalmitoyl PC. The incorporation of choline into surfactant PC proceeds predominantly, if not exclusively, via cholinephosphate and CDP-choline.

Thermodynamic considerations, pulse-chase studies with radiolabeled precursors, and the measurement of the pool sizes of the intermediates have led to the suggestion that the reaction

$$\text{cholinephosphate} + \text{CTP} > \text{CDP-choline} + \text{PP}_i$$

may be rate-limiting in this pathway. Thus CTP:cholinephosphate cytidylyltransferase would be a key enzyme. The regulation of the activity of CTP:cholinephosphate cytidylyltransferase has been investigated in several cell systems. CTP:cholinephosphate cytidylyltransferase activity is present in both cytosolic and microsomal fractions. The cytosolic activity can be enhanced by the addition of lipids, e.g. phosphatidylglycerol, to the assay system. It has been suggested that cytosolic CTP:cholinephosphate cytidylyltransferase is but an inactive storage form which can be activated by its translocation to the membranes of the endoplasmic reticulum (Vance and Pelech 1984). The subcellular distribution of CTP:cholinephosphate cytidylyltransferase and thus its activity appear to be controlled by the concentration of fatty acids and by cAMP-dependent protein phosphorylation.

The development of pharmacologic agents capable of stimulating the synthesis of the components of surfactant would be of great clinical interest considering the role of surfactant deficiency in the neonatal respiratory distress syndrome and possibly in other disease states. Rational drug development can be facilitated by a better understanding of the physiologic regulatory mechanisms involved.

We studied some aspects of the physiologic regulation of the cholinephosphate pathway in alveolar type II cells isolated from adult rat lung. In a pulse-chase study alveolar type II cells and hepatocytes in primary culture were incubated for 1 hour in the presence of ^{14}C-choline. In the ensuing chase incubation period with unlabelled choline 0.5 mmol/l chlorophenylthio-cAMP, 0.2 mmol/l isobutylmethylxanthine, or both were added and the incorporation of the radiolabel into PC was followed. Chlorophenylthio-cAMP markedly

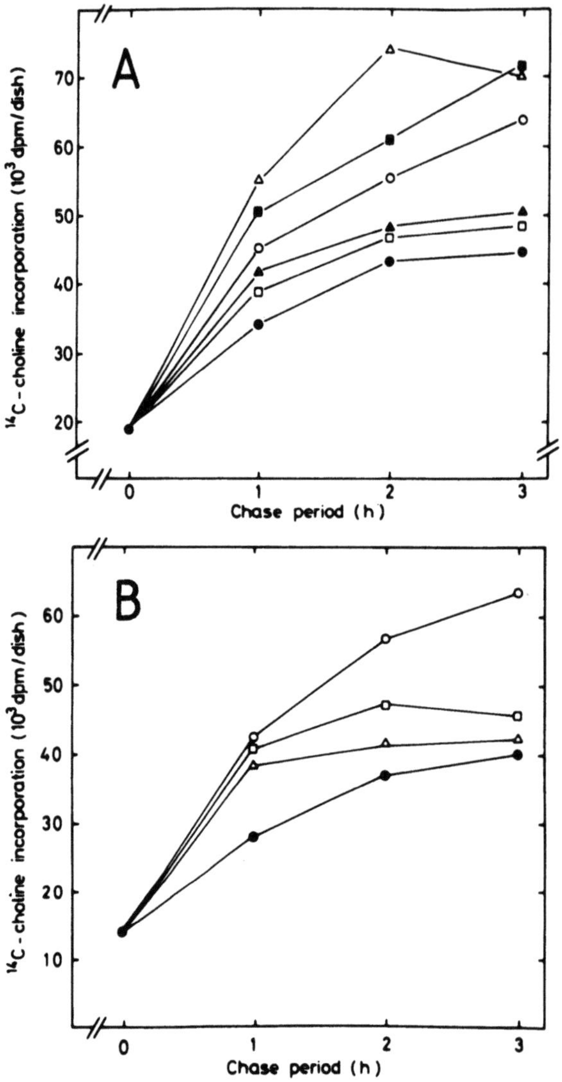

Fig. 1 A, B. Pulse-chase study of the incorporation of [*Me*-^{14}C] choline into PC by type II cells isolated from adult rat lung. Type II cells in primary culture were labeled for 1 h with [*Me*-^{14}C] choline (specific activity 59 200 dpm/nmol) and subsequently exposed to fresh medium containing unlabeled choline and fatty acids (complexed to BSA, molar ratio 5.3:1) as indicated. **A** Effects of palmitate in the following concentrations: 0.002 mmol/liter (□——□), 0.01 mmol/liter (▲——▲), 0.05 mmol/liter (○——○), 0.2 mmol/liter (■——■), and 1.0 mmol/liter (△——△), controls without fatty acid (●——●). **B** Effects of palmitate (○——○), oleate (□——□), and linoleate (△——△) in a concentration of 0.2 mmol/liter each, control without fatty acids (●——●)

inhibited the formation of PC in the hepatocytes, which confirms the results of Pelech et al. (1981). However, in alveolar type II cells neither the cAMP analogue alone nor its combination with the phosphodiesterase inhibitor had a clear effect on PC synthesis.

On the other hand the addition of fatty acids (bound to BSA) to the chase incubation medium increased the incorporation of ^{14}C-choline into PC by alveolar type II cells. The stimulation of PC synthesis was more pronounced with palmitate than with oleate or linoleate (Fig. 1). An accelerated disappearance of the radiolabel from cholinephosphate was observed in these experiments. This is consistent with the assumption that fatty acids such as palmitate exert a regulatory influence at the level of CTP:cholinephosphate cytidylyltransferase.

However, fatty acids also are substrates for PC synthesis. The fatty acyl moieties of surfactant PC can be derived both from the blood and from endogenous synthesis in type II cells from precursors such as glucose, acetate, and lactate. In the cell culture system used in vitro a deficiency of fatty acids may make the formation of diacylglyerols rate-limiting for PC synthesis. Thus the effect of the fatty acids in the culture medium could be due to compensation for a fatty acid deficiency. To test this possibility type II cells were incubated in the presence of 10 mmol/l lactate under the conditions shown by Maniscalco et al. (1983) to be optimal for endogenous fatty acid synthesis in type II cells. When type II cells in primary culture were used the presence of lactate in the incubation medium had no effect on ^{14}C-choline incorporation into PC. However, with freshly isolated type II cells incubated in suspension the inclusion of lactate in the medium resulted in enhanced PC synthesis. The

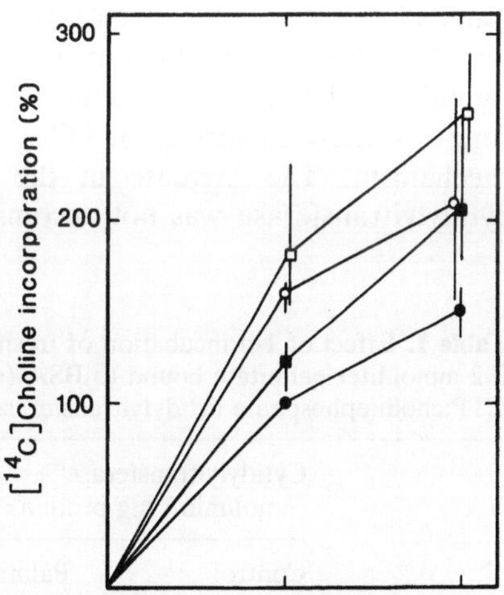

Fig. 2. Effects of lactate and palmitate on the incorporation of [*Me*-^{14}C] choline into PC by freshly isolated type II cells from adult rat lung. Type II cells in suspension were incubated in the presence of [*Me*-^{14}C] choline in Krebs-Ringer bicarbonate medium, which contained 10 mmol/liter lactate (■——■), 0.2 mmol/liter palmitate bound to BSA (o——o), or both (□——□); controls contained neither addition (●——●). Media without palmitate contained an equivalent concentration of fatty acid-free BSA (0.256%). The means and range of three experiments are given in arbitrary units derived from dpm/per dish by calculating the percentages of the control value at 90 min for each experiment

effects of lactate and palmitate appeared to be additive (Fig. 2). Thus at least part of the effect of palmitate may be due to a regulatory influence on the cholinephosphate pathway. This conclusion, however, rests on two assumptions:

1. The concentrations of palmitate and lectate used in this experiment are optimal for stimulation of PC synthesis.
2. Type II cells are not dependent on an exogenous supply of fatty acids as substrates for one of the pathways leading to the formation of PC.

To further define the site of the possible regulatory action of palmitate enzyme measurements in crude subcellular fractions of isolated type II cells were performed. Freshly isolated type II cells were incubated for 1 hour in the presence of 0.2 mmol/l palmitate (bound to BSA). The activities of choline kinase in the soluble fraction and of CTP:cholinephosphate cytidylyltransferase in the membrane fraction were significantly increased, the latter effect being more pronounced (Table 1). The soluble CTP:cholinephosphate cytidylyltransferase activity remained unchanged. The choline kinase activity in the membrane fraction was due to contamination by cytosolic protein derived from cells remaining intact after the homogenization procedure. After refinement of the fractionation scheme this contamination could be largely eliminated.

The results shown are consistent with enzymatic regulation of the cholinephosphate pathway by exogenous palmitate at the level of CTP:cholinephosphate cytidylyltransferase. A regulatory role of choline kinase seems unlikely in view of the demonstration of a large pool of cholinephosphate in alveolar type II cells (Post et al. 1984). Furthermore, our results support the hypothesis that the microsomal form of CTP:cholinephosphate cytidylyltransferase is active in surfactant PC synthesis and possibly the site of a regulatory mechanism. The increase in the membrane bound CTP:cholinephosphate cytidylyltransferase was not accompanied by a corresponding decrease in the

Table 1. Effect of 1-h incubation of freshly isolated alveolar type II cells in the presence of 0.2 mmol/liter palmitate bound to BSA (molar ratio 5.3:1) on the subcellular distribution of CTP:cholinephosphate cytidylyltransferase and choline kinase activities

	Cytidylyltransferase[a] (nmol/min · mg protein)		Choline kinase[a] (nmol/min · mg protein)	
	Control	Palmitate	Control	Palmitate
Membrane bound	2.29 (1.62–3.48)	3.74 (2.42–4.59)[b]	2.05 (1.6–3.2)	2.3 (1.75–3.05)
Soluble				
without PG	0.4 (0.24–0.57)	0.35 (0.18–0.61)	2.5 (1.15–3.55)	3.5 (2.3–4.85)[b]
with PG	1.29 (0.69–1.71)	1.18 (0.95–2.02)	–	–

PG, 1 mmol/liter phosphatidylglycerol included in the assay system as sonicated aqueous suspension.
[a] Median values of 6 experiments (25th–75th percentile).
[b] Significantly different from controls ($p < 0.05$, Wilcoxon matched pairs signed rank test).

soluble activity. Thus unequivocal evidence for the translocation hypothesis detailed above (Vance and Pelech 1984) cannot be presented.

References

Maniscalco WM, Finkelstein JN, Parkhurst AB (1983) De novo fatty acid synthesis by freshly isolated alveolar type II epithelial cells. Biochim Biophys Acta 751:462–469

Pelech SL, Pritchard PH, Vance DE (1981) cAMP analogues inhibit phosphatidylcholine biosynthesis in cultured rat hepatocytes. J Biol Chem 256:8283–8286

Post M, Batenburg JJ, Smith BT, van Golde LMG (1984) Pool sizes of precursors for phosphatidylcholine formation in adult rat lung type II cells. Biochim Biophys Acta 795:552–557

Vance DE, Pelech SL (1984) Enzyme translocation in the regulation of phosphatidylcholine synthesis. Trends Biochem Sci 9:17–20

A Controlled Trial of CPD-Choline Administration in Premature Infants with Respiratory Distress Syndrome*

A. Valls-Soler, P. Sanjurjo, and I. Lopez de Heredia

CPD-choline (Citidin-5′-diphosphate of choline) is a known precursor of DPPL (dipalmitoyl-phosphatidyl-choline), the principal component of pulmonary surfactant [1]. The CPD-choline incorporation pathway is the main mechanism of de novo synthesis of DPPL in developing mammalian lung [2]. It has been shown that the administered CPD-choline is incorporated into the DPPL synthetized by adult rat lung, and by fetal lung tissure when given to pregnant rats [3]. There is also some evidence that CPD-choline may stimulate the synthesis of phosphatidic acids in both cerebral tissue [4], and in fetal rabbit lung [5]. Katyal and Lombardi [6] have shown that fetuses born to pregnant, diet-deficient rats, died from progressive respiratory distress in the first hours of life.

Since the work of Fujiwara et al. [7], several reports have shown the effectiveness of surfactant replacement therapy in idiopathic respiratory distress syndrome (IRDS). The administration of surfactant precursors may also be beneficial. Recently Hallman et al. [8] have reported that Inositol, a precursor of phosphatil-inositol, when given to infants with IRDS resulted in amelioration of their clinical status. The administration of CPD-choline to human neonates with IRDS, has been followed by contradictory results. In an uncontrolled study, Nishimura [9] reported some benefit, but the results of other studies were inconclusive.

Thus, we undertook a preliminary prospective, randomized study to test the hypothesis that CPD-choline administration to premature infants with IRDS could improve their pulmonary status.

Material and Methods

Newborn infants were enrolled in the study if they met the following criteria: (a) birth weight below 2000 g and appropriate for gestational age, (b) clinical and X-ray findings consistent with IRDS, (c) fraction of inspired oxygen (FiO_2) requirements of 0.4 or higher to maintain a partial pressure of arterial oxygen (PaO_2) above 50 mm Hg, (d) informed parental consent.

The study was carried out in the Neonatal Intensive Care Unit, Hospital de Cruces, Bilbao, Euskadi, Spain. The study was approved by the Hospital

* Supported in part by a Grant No. 87/1067 from the Fondo de Investigaciones Sanitarias de la Seguridad Social (FIS), Spanish Ministry of Health.

Table 1. Clinical data

	Group	
	Control	CPD-choline
No. of patients	13	11
Birth weight (g)	1636 ± 138	1782 ± 140
Gestational age (weeks)	33.2 ± 0.9	32.7 ± 0.5
Apgar score 1 min	5 ± 0.9	6.3 ± 0.9
5 min	7.8 ± 0.6	8.4 ± 0.5
Arterial pH	7.15 ± 0.04	7.22 ± 0.01
$PaCO_2$ (torr)	54.9 ± 2.9	50.8 ± 2.4
PaO_2 (torr)	49 ± 3.6	50.1 ± 8.6
FiO_2	0.58 ± 0.08	0.65 ± 0.08

Values represent mean ± SE. Differences between groups were not significant.

Review Committee, and by the Spanish Department of Health (reference number 31/LS). The patients were managed by the clinical staff according to previously designed protocols, based on accepted recommendations [10].

CPD-choline was given by the i.v. route, at a dose of 100 mg/kg/d in four 6-hour aliquots. Treatment was discontinued at the 7th day of life, or when FiO_2 requirement was less than 0.3, and the patient was breathing spontaneously. Patients were randomly allocated either to treatment or placebo groups and coded accordingly. The code was unknown to the clinical staff and the investigators and was not revealed until the end of the study.

The study was designed to determine whether the administration of CPD-choline to premature infants with IRDS would ameliorate their disease process. Need and duration of intermittent positive pressure ventilation (IPPV) and time elapsed to require a FiO_2 of either 0.6, or 0.21 were chosen as principal end-points. There were no significant differences in sex, birthweight and gestational age between the groups (Table 1).

At entry in the study, arterial blood gases were determined. The gastric or tracheal lecithin to sphyngomyelin (L/S) ratio, and the palmitic acid percentage in lecithin were also measured daily, according to standard methods [11].

The statistical methods used were the Fisher exact test, and Student's t test. A type I and II errors of 50% were accepted.

Results

The clinical characteristics of the patients both in the placebo and treatment groups are given in Table 1. At entry in the study, there were no statistical differences in any of the considered parameters. The sex ratio, and the percentage of patients who required either positive respiratory pressure while on spontaneous respiration, or mechanical ventilation, were similar in both groups.

Table 2. Clinical data

	Group		
	Control	CPD-choline	P value
No. of patients	13	11	
Duration of mechanical ventilation (h)	318 ± 869	504.3 ± 12.3	NS
Maximal FiO_2	0.93 ± 0.05	0.89 ± 0.05	NS
FiO_2 above 0.21 (h)	304 ± 74.6	583.8 ± 134	< 0.05
FiO_2 above 0.6 (h)	87.4 ± 30	58.6 ± 48.7	NS
Time at death (h)	458.7 ± 122.6	511 ± 258.3	NS

Values represent mean ± SE.

Table 3. Data on morbidity and mortality

	Group			
	Control		CPD-choline	
		%		%
No. of patients	13		11	
Mechanical ventilation	12	(92)	11	(100)
Pneumothorax	5	(38)	1	(9)
BPD	3	(23)	3	(27)
PDA	5	(38)	6	(55)
IVH	6	(46)	2	(18)
Septicemia	1	(18)	2	(18)
NEC	0	—	1	(9)
Nonsurvivors	6	(46)	4	(36)

groups were not significant.
BPD, bronchopulmonary dysplasia; PDA, patent ductus arteriosus; IVH, intraventricular hemorrhage; NEC, necrotizing enterocolitis.

There were no significant differences in the total time of mechanical ventilation, duration of exposure to an FiO_2 above 0.6, maximal FiO_2 needed, and time at death (Table 2). There was a longer need for continued oxygen exposure in the patients treated with CPD-choline when compared to controls (504.3 ± 12.3, vs 318.1 ± 86.9 hours; $p < 0$ 0.05). Table 3 shows the mortality rate and the incidence of clinical complications. There were no significant differences between the groups.

As shown in Fig. 1A, the lecithin to sphyngomyelin ratio (L/S) on the first day of postnatal life was similar in both groups (3.3 ± 1.7, and 3.3 ± 1.4). In days 2, 3 and 4 the L/S ratio remained identical, but in days 5–9 the L/S ratio was significantly higher in the placebo group (9 ± 7.3, vs 5.6 ± 3.1; $p < 0.05$). In the placebo group, a significant positive correlation between postnatal age and L/S ratio was found ($y = 2.74x + 0.13$; $r = 0.36$; $p < 0.05$). This correlation was not present in the CPD-choline treated group.

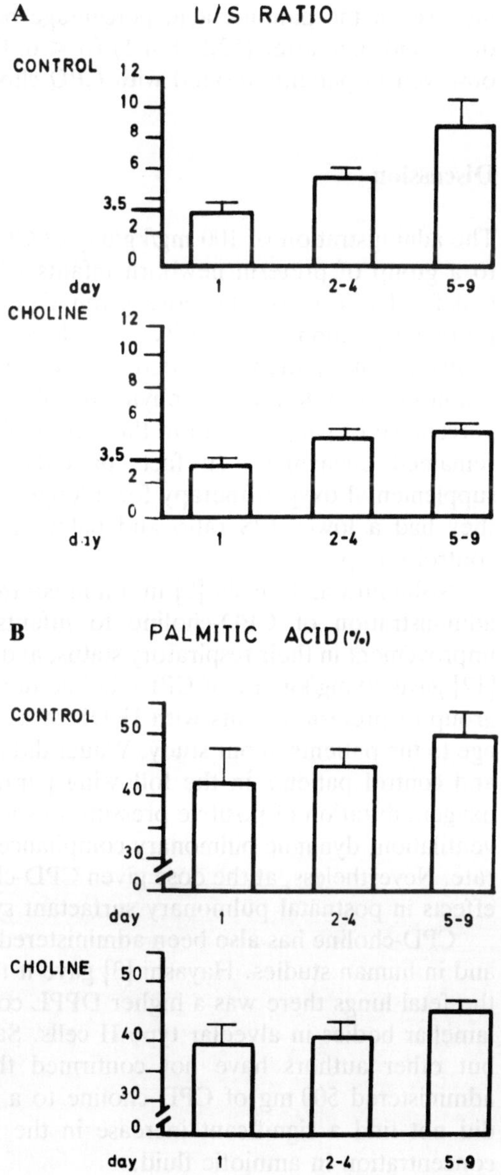

Fig. 1. A Postnatal evolution of L/S ratio in the control and CPD-treated groups. *Columns* represent mean ± SE. Values for day 1, days 2–4, and days 5–9 are indicated. **B** Postnatal evolution of palmitic acid percentage in the lecithin fraction in control, and CPD-choline-treated groups. *Columns* represent mean ± SE. Values for day 1, days 2–4, and days 5–9 are indicated

There were no significant differences between placebo and CPD-choline treated groups in palmitic acid percentage of the lecithin fraction in the first postnatal day (39.4 ± 3.7, vs 42.6 ± 5.4%). In days 2, 3 and 4 the palmitic acid percentage was also found to be similar, but in days 5–9 the percentage was higher in the placebo group (52.8 ± 4.1, vs 44.9 ± 1.9%) (Fig. 1B). A correlation between palmitic acid percentage and postnatal age was not demonstrated in either group. In the placebo group there was a significant increase in the palmitic acid percentage between days 1 and 2 (41.5 ± 3.4), day 5 and thereafter (52.8 ± 4.1) ($p < 0.05$). This positive increment was not observed in patients treated with CPD-choline.

Discussion

The administration of 100 mg/kg/day of CPD-choline by the intravenous route to a group of preterm newborn infants with IRDS was not followed by any beneficial effect. At the dose given CPD-choline did not appear to induce postnatal pulmonary maturity, and there was no modification of the clinical status of the patients treated. A relative increase in lecithin pulmonary synthesis, in relation to sphyngomyelin, could not be demonstrated. The percentage of palmitic acid in the lecithin fraction of the respiratory secretions remained unchanged. In fact, patients treated with CPD-choline needed supplemental oxygen therapy for a longer period of time and, from day 5 on, they had a lower L/S ratio and palmitic acid percentage than those in the control group.

Nishimura and Goshi [9] in a non-controlled clinical trial, claimed that the administration of CPD-choline to infants with IRDS produced a clinical improvement in their respiratory status, and a decrease in mortality rate. Wauer [12] gave 20 mg/kg/day of CPD-choline during the first seven days of life, to a group of preterm infants with IRDS of a similar birth weight and gestational age to the patients in our study. Wauer did not find differences between treated and control patients in the following parameters: total time of supplemental oxygen, duration of positive pressure in spontaneous respiration or mechanical ventilation, dynamic pulmonary compliance, chest X-ray findings, or mortality rate. Nevertheless, at the dose given CPD-choline did not show any detrimental effects in postnatal pulmonary surfactant synthesis.

CPD-choline has also been administered prenatally in experimental animals and in human studies. Hayashi [3] gave it to pregnant rats and showed that in the fetal lungs there was a higher DPPL content, and an increased number of lamellar bodies in alveolar type II cells. Sappa et al. [13] had similar results, but other authors have not confirmed these findings [14, 15]. Endo [16] administered 500 mg of CPD-choline to a group of pregnant women, but he did not find a significant increase in the L/S ratio, total lecithin, or DPPL concentration in amniotic fluid.

There is an apparent contradiction between the possible beneficial effects of prenatal CPD-choline administration in fetal lung maturity, and the lack of effect if administered postnatally. This difference could be similar to that seen with the use of corticosteroids. The prenatal administration of these drugs induce fetal pulmonary synthesis of DPPL, by activation of glycerolphosphocoline-transferase [17], key enzyme in the choline incorporation pathway [18]. Since enzyme induction is a slow process, the progression of IRDS during the immediate neonatal period could compensate any beneficial effect of CPD-choline.

Another possibility is that, at the dose given, CPD-choline inhibited the key enzyme instead of inducing it. It is known that if the cellular concentration of a substrate is below the saturation point of the corresponding enzyme, any further increase in concentration is followed by an acceleration in the velocity of the reaction [2]. Conversely, if the substrate is above the saturation point, an exogenous administration may inhibit key enzyme activity, slow down the metabolic reaction, or even reverse the direction of the pathway [3]. In any case CPD-choline administration could possibly decrease lecithin synthesis instead of enhancing it.

In summary, postnatal CPD-choline administration does not appear to be an effective aetiological approach to treat premature newborn infants with IRDS. The approach should be based on prevention of premature deliveries, prenatal corticosteroid treatment, and postnatal surfactant replacement.

References

1. Farrell PM (1973) Regulation of pulmonary lecithin synthesis. In: Villaee CA, Villae DB, Zucherman J (eds) Respiratory distress syndrome. Academic, New York, pp 311–320
2. Farrell PM, Mamosh M (1978) The biochemistry of fetal lung development. Clin Perinatol 5:197–229
3. Hayashi Y (1973) Experimental studies on the effects of CPD-choline on pulmonary surfactant. Kanasai Ika Daigaku Zasshi (Osaka) 25:227–252
4. Possmayer F (1974) CPD-choline reversal of the CMP and CTP inhibition of phosphatidic acid synthesis by rat brain preparation. Biochem Biophys Res Commun 61:1415–1421
5. Kotas RV, Clements JA (1977) Positive effects of added choline on fetal rabbit pulmonary surfactant production. Physiologist 20:52–64
6. Katyal SL, Lombardi B (1978) Effects of dietary choline and N,N-Dimethylamino-ethanol in lung phospholipid and surfactant of newborn rats. Pediatr Res 12:952–955
7. Fujiwara T, Chida S, Watabe Y, Maeta H, Morita T, Abe T (1980) Artificial surfactant therapy in hyaline-membrane disease. Lanceti:55–59
8. Hallman M, Jävenpää AL, Pohjavuori M (1986) Respiratory distress syndrome and inositol supplementation in preterm infants. Arch Dis Child 61:1076–1083
9. Nishimura K, Goshii N (1972) Treatment of idiopathic respiratory distress syndrome with nicholine. Jpn Med News (Tokyo) 9:1–4
10. Goldsmith JP, Korotkin EH (eds) (1981) Assisted ventilation of the Neonate. Saunders, Philadelphia, pp 101–127
11. Valls-Soler A, Sanjurjo P (1977) Postnatal L/S ratio and palmitic acid percentage as predictors of recovery from IRDS. Pediatr Res 11:1024
12. Wauer RR (1983) Medical treatment of neonatal hyaline-membrane disease using Bromhexine, Ambroxol, and CPD-Choline. In: Cosmi EV, Scarpelli EM (eds) Pulmonary surfactant system. Elsevier Science, Amsterdam, pp 173–188
13. Sappa M, Gramiccioni E, Manzari V, Dotta F, Matzeu M (1974) Effeto sul contenuto di fosfolipidi pel pulmone del feto di coniglio determinato dalla somministrazione de Citicolina alla madre. Mind 65:3595–3598
14. Finilä MJ, Österlind K, Lethi H (1978) Die Behandlung des idiopathischen Atemnotsyndrom mit Ambroxol. Atemwegs Lungenkr 4:62–67
15. Renovanz HD, Keuth U (1978) Respiratory Distress Syndrom und Ambroxol Therapie – Teamstudien an Kinderkliniken. Atemwegs Lungenkr 4:50–61
16. Endo H (1973) Changes in surfactant lecithin, total lecithin, L/S ratio, and total surfactant lipids in the amniotic fluid of pregnant women due to the administration of CPD-choline and Dexametasone. XI Meeting of the Japanese Academy of Neonatology (Tokyo) 73:354

17. Farrell P, Zachman RD (1972) Enhancement of lecithin synthesis and phosphorylcholine glyceride transferase activity in the fetal rabbit lung after corticosteroid administration. Pediatr Res 6:337
18. Chida N (1973) Biosynthesis of lecithin in the lung in the growing stage including behavior of choline and CPD-choline. Tohaku J Exp Med (Tokyo) 110:273–281

Effects of the β_2-Adrenoceptor-Stimulating Agent Terbutaline and Its Antagonist ICI 118.551 on Rat Lung Surfactant (Phospholipid) Metabolism

W. Bernhard, B. Müller, and P. von Wichert

The study was performed to investigate the influence of β_2-adrenergic stimulation by terbutaline and blockade by ICI 118.551 [1] on vertebrate lung phospholipid (PL) metabolism in isolated, ventilated, and perfused rat lung (IVPL). In one part of the perfusion experiments in vivo premedication with terbutaline was performed for 3 days (for dosage see Fig. 1). Total PL content, PL-subclasses, incorporation of ^{14}C-choline and ^{3}H-palmitate into phosphatidylcholine (PC), the surfactant-specific disaturated phosphatidylcholine (DSPC), and phosphatidylethanolamine (PE) were examined.

Experiments Without In Vivo Pretreatment

Terbutaline (10 µM for 3.15 h in vitro) increased total PL-content of lung tissue in IVPL for 6.3% (from 190–202 µmol/g dry weight) and reduced tissue PC, DSPC, and phosphatidylglycerol (PG) for 2.6%, 5.0% and 16.9%, respectively. The wet to dry weight ratio of IVPL preparations perfused in the presence of terbutaline was 5.1 in comparison to 6.0 in controls or perfusion in the presence of 10 µM ICI 118.551 (Fig. 1b).

The incorporation of ^{14}C-choline was increased for 12.3% into PC and for 13.9% into DSPC. ^{3}H-palmitate incorporation was increased for 18.9% into PC and for 13.9% into DSPC, but unchanged in PE. The radiotracer incorporation into intraalveolar PC was enhanced, 1.8-fold for ^{14}C-choline and 2.0-fold for ^{3}H-palmitate. Total intraalveolar PC was unchanged. Additionally, we analysed alveolar macrophages and found that total PL and PC content were increased. Incorporation of ^{3}H-palmitate into PC was enhanced 2.7-fold.

Table 1. Increase of ^{14}C-choline and ^{3}H-palmitate incorporation into tissue PC and DSPC by 3.15-h in vitro stimulation by terbutaline

	Decrease of tissue pool	Increase of incorporation		Net increase of incorporation	
		Choline	Palmitate	Choline	Palmitate
PC	2.7 ± 3.9	12.3 ± 7.3	18.9 ± 12.0	9.3	15.7
DSPC	5.5 ± 5.1	12.3 ± 7.6	13.9 ± 10.4	6.1	7.6

Data are expressed as a percentage of controls.

In tissue all the terbutaline induced changes were antagonized by adding 10 μM ICI 118.551 7 min prior to the terbutaline introduction to the perfusate besides the ^{3}H-palmitate incorporation that was also enhanced by the antagonist. In intraalveolar PC the terbutaline induced increase of ^{14}C-choline and ^{3}H-palmitate incorporation was decreased by ICI 118.551. In alveolar macrophages the terbutaline stimulated ^{3}H-palmitate incorporation into PC was blocked, but the enhancement of total PL and PC was not antagonized by ICI 118.551.

Experiments with In Vivo Pretreatment

After in vivo pretreatment with 3 × 600 μg terbutaline/kg to dry weight for 3 days the increase of tissue PL in the IVPL-preparation was 16.5% after 3.15 h perfusion in the presence of 10 μM terbutaline and 5.3% after perfusion in the presence of 10 μM ICI 118.551. Tissue PG was unchanged and DSPC was decreased only after perfusion in the presence of terbutaline but not of the antagonist. ^{3}H-palmitate incorporation into PC was enhanced but not into DSPC, and ^{14}C-choline incorporation was unchanged. Incorporation into intraalveolar PC of IVPL preparations was increased 1.4-fold for ^{14}C-choline and 1.6-fold for ^{3}H-palmitate after perfusion with 10 μM terbutaline; after perfusion with 10 μM ICI 118.551 they were unchanged. IVPL experiments with 10 μM terbutaline enhanced palmitate incorporation into macrophages PC two-fold, those with 10 μM ICI 117.551 caused no effects.

The increase of tissue PL by β_2-adrenergic stimulation in vitro was greater than the incorporation rate of ^{14}C-choline and ^{3}H-palmitate into PC and DSPC. We suggest that tissue PL, expressed as μmol/g dry weight after 3.15 h in vitro stimulation by terbutaline is increased not by synthesis but by diminished incorporation of albumin from the perfusate into the IVPL in comparison to the controls [2, 3]. After 3 d in vivo pretreatment with terbutaline total tissue PL was augmented even when the IVPL experiment was performed in the presence of 10 μM ICI 118.551. So we suggest that the enhancement was due to an increased tissue pool size after the in vivo pretreatment. Because the increase of ^{14}C-choline and ^{3}H-palmitate incorporation into tissue PC and DSPC was greater than the decrease of their pool sizes by secretion, we propose an enhanced synthesis by 3.15 h terbutaline treatment of the IVPL (Table 1). The augmented appearance of new-synthesized PC in the alveoli without an increase of total PC in this compartment suggests that the turnover, as well as secretion, was enhanced by terbutaline. Alveolar macrophages participated in the degradation, because their PC content and ^{3}H-palmitate incorporation into PC were increased.

The stimulating effect of ICI 118.551 on ^{3}H-palmitate incorporation into PC and DSPC of tissue and on PC content on alveolar macrophages may be caused by unspecific membrane effects, but not by intrinsic sympathomimetic activity [1]. This is in agreement with our finding that the β_2-blocker was incorporated into lung tissue, intraalveolar phospholipids, and alveolar macrophages in a dose-dependent manner.

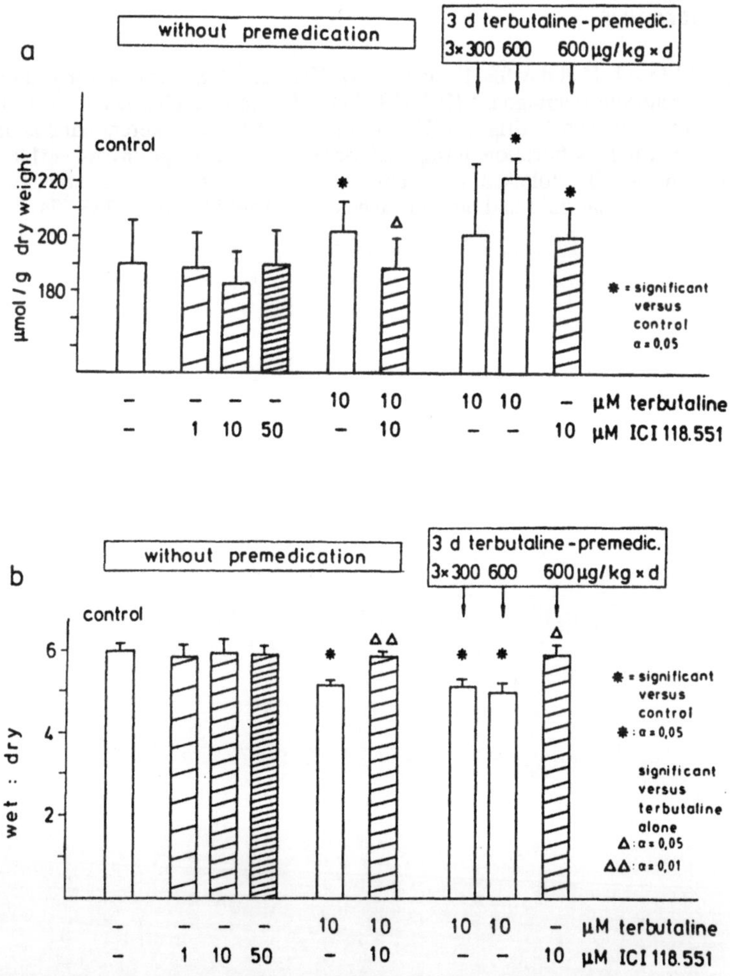

Fig. 1a, b. Influence of 3.15 h terbutaline and ICI 118.551 in IVPL experiments with or without in vivo pretreatment of animals with terbutaline for 3 days. **a** Total PL content in bronchoalveolar lavage fluid. **b** Wet-to-dry weight ratio of non-lavaged lungs

We conclude that PC and DSPC synthesis is enhanced by terbutaline in IVPL via β_2-adrenergic stimulation. The pool size of PC, DSPC, and PG is diminished by secretion in 3.15 h in vitro β_2-adrenergic stimulation of IVPL and enhanced in the pretreatment experiments. Turnover-rate of intraalveolar PC is accelerated. Furthermore we conclude that ICI 118.551 inhibits β_2-mimetic activity of terbutaline but stimulates selectively palmitate incorporation into PC and DSPC by a β_2-receptor-independent mechanism.

Acknowledgements. We thank Mrs. Chr. Skurk for her excellent technical assistance. The study was supported by the Deutsche Forschungsgemeinschaft (Wi 359/7), Astra Chemicals GmbH (Prof. Westermann), and Deutsche ICI GmbH (Ch. Blanckmeister).

References

1. Bilski AJ, Halliday SE, Fitzgerald JD, Wale JL (1983) The pharmacology of a β_2-selective adrenoceptor antagonist (ICI 118.551). J Cardiovasc Pharmacol 5:430–437
2. Foy T, Marion J, Brigham KL, Harris TL (1979) Isoproterenol and aminophylline reduce lung capillary filtration during high permeability. J Appl Physiol 46:146–151
3. Minnear FL, Johnson A, Malik AB (1986) β-Adrenergic modulation of pulmonary transvascular fluid and protein exchange. J Appl Physiol 60:266–274

Subject Index

MIX
Papier aus verantwortungsvollen Quellen
Paper from responsible sources
FSC® C105338

If you have any concerns about our products,
you can contact us on
ProductSafety@springernature.com

In case Publisher is established outside the EU,
the EU authorized representative is:
Springer Nature Customer Service Center GmbH
Europaplatz 3, 69115 Heidelberg, Germany

Printed by Libri Plureos GmbH
in Hamburg, Germany